Computational Number Theory

Computational Number Theory

Proceedings of the Colloquium on Computational Number Theory held at Kossuth Lajos University, Debrecen (Hungary), September 4–9, 1989

Editors

Attila Pethö
Michael E. Pohst
Hugh C. Williams
Horst Günter Zimmer

Walter de Gruyter · Berlin · New York 1991

Editors

Attila Pethö, Mathematical Institute, Kossuth Lajos University, H-4010 Debrecen, Hungary
Michael E. Pohst, Mathematisches Institut, Heinrich-Heine-Universität, Universitätsstr. 1, D-4000 Düsseldorf 1, FRG
Hugh C. Williams, Department of Computer Science, The University of Manitoba, Winnipeg, Mannitoba, Canada R3T 2N2
Horst Günter Zimmer, Fachbereich 9 - Mathematik, Universität des Saarlandes, D-6600 Saarbrücken, FRG

1991 Mathematics Subject Classification: 11-06, 11-04; 12-06, 12-04; 14-06, 14-04.

⊗ Printed on acid free paper which falls within the guidelines of the ANSI to ensure permanence and durability.

Library of Congress Cataloging-in-Publication Data

Colloquium on Computational Number Theory (1989 : Kossuth Lajos University)
Computational number theory : proceedings of the Colloquium on Computational Number Theory held at Kossuth Lajos University, Debrecen (Hungary), September 4–9, 1989 / editors, Attila Pethö ... [et al.].
p. cm.
Includes bibliographical references.
ISBN 3-11-012394-0 (cloth : acid-free). -- ISBN 0-89925-674-0 (cloth : acid-free)
1. Number theory--Data processing--Congresses. I. Pethö, Attila, 1950- . II. Title.
QA241.C6874 1989 91-15924
512'.7--dc20 CIP

Die Deutsche Bibliothek – Cataloging-in-Publication Data

Computational number theory : proceedings of the Colloquium on Computational Number Theory held at Kossuth Lajos University, Debrecen (Hungary), September 4–9, 1989 / ed. Attila Pethö ... – Berlin ; New York : de Gruyter, 1991
ISBN 3-11-012394-0
NE: Pethö, Attila [Hrsg.]; Colloquium on Computational Number Theory <1989, Debrecen>; Kossuth Lajos Tudományegyetem <Debrecen>

Printed in Germany.
Printing: Gerike GmbH, Berlin. Binding: Lüderitz & Bauer GmbH, Berlin.
Cover design: Thomas Bonnie, Hamburg.

Preface

This volume contains original research and survey papers presented at the Colloquium on Computational Number Theory held in the facilities of the Regional Committee of the Hungarian Academy of Sciences at Debrecen, Hungary, from September 4 to September 9, 1989.

There were 65 registered participants from 15 countries. Most of the 35 lectures which were delivered are included in these proceedings, and so are three additional papers by conference participants. The contents of these papers fall well within the areas covered by the colloquium. The contributions involve several aspects of computational number theory ranging from effective finiteness results to efficient algorithms in elementary, analytic and algebraic number theory. The topics treated, in particular, relate to finite fields, radix representations, quadratic forms, algebraic number fields, modular forms, elliptic curves and diophantine equations. In addition, two number theoretical software packages were demonstrated on various PCs. The wide range of themes of the colloquium was in complete agreement with the intentions of the organizers.

The city of Debrecen, located at the eastern edge of Hungary, provided an auspicious environment for a meeting of number theorists coming from the eastern and the western world. Indeed, it was exactly at the time of this conference that Hungary punched the first holes into the iron curtain.

This feature distinguished the colloquium from preceding conferences on computational number theory of which proceedings have been published, such as the ones held in 1969 at Oxford and in 1980 at Amsterdam. Another novelty was the above–mentioned presentation of number theoretical software packages, which allowed participants to solve some computational problems on the spot.

Many people helped in preparing the conference or in editing the proceedings. We are especially indebted to K. Györy for his various activities as a member of the organizing committee. I. Gaal was a most reliable secretary who, together with I. Nemes, took care of the scientific as well as the social program. M. Pfeifer and J. Schmitt were of great assistance in setting up the various TEX versions of these proceedings. Finally, every paper had one or two referees. We wish to thank them all.

We are grateful also to the staff of Walter de Gruyter & Co. for the cooperative manner in which the publication of this volume was managed.

Saarbrücken, February 1991

Attila Pethö
Michael Pohst
Hugh C. Williams
Horst G. Zimmer

List of contributors

Bettina Arenz, Mathematisches Institut der Heinrich-Heine-Universität Düsseldorf, Universitätsstr. 1, D-4000 Düsseldorf 1, Germany

Bryan J. Birch, Mathematical Institute, University of Oxford, 24-29 St. Giles, Oxford, OX13LB, England

Johannes Buchmann, Fachbereich 14 Informatik, Universität des Saarlandes, D-6600 Saarbrücken, Germany

Harvey Cohn, Department of Mathematics, The City College of New York, Convent Ave. at 138th St., New York, N.Y, 10031, USA

Reiner Creutzburg, Universität Karlsruhe, Institut für Algorithmen und Kognitive Systeme, PF 6980, D-7500 Karlsruhe, Germany

Stephan Düllmann, Fachbereich 14 Informatik, Universität des Saarlandes, D-6600 Saarbrücken, Germany

Veikko Ennola, Department of Mathematics, University of Turku, SF-20500 Turku 50, Finland

Jan-Hendrik Evertse, Filips van Bourgondiestraat 41A, NL-3117 SC Schiedam, The Netherlands

David Ford, Department of Computer Science, Concordia University, Montréal, Québec, Canada, H3G 1M8

István Gaal, Kossuth Lajos University, Mathematical Institute, H-4010 Debrecen Pf.12., Hungary

Aleksander Grytczuk, Department of Mathematics, Pedagogical University, Zielona Gora, 65-562, ul.Sucharskiego 18.m.14., Poland

Kálmán Győry, Kossuth Lajos University, Mathematical Institute, H-4010 Debrecen Pf.12., Hungary

Franz Halter-Koch, Institut für Mathematik, Karl-Franzens Universität Graz, Halbärthgasse 1/1, A-8010 Graz, Österreich

Christine Hollinger, Fachbereich 9 Mathematik, Universität des Saarlandes, D-6600 Saarbrücken, Germany

Béla Kovács, Kossuth Lajos University, Mathematical Institute, H-4010 Debrecen Pf.12., Hungary

Yuri V. Melnichuk, Lvovszkij Polityehnyicseszkij Insztitut, Kafedra Vicsiszlityelnoj Mat. i progrmmirovanyija, 290000 Lvov, ul. Mira 12, USSR

Richard A. Mollin, Department of Mathematics and Statistics, The University of Calgary, 2500 University Drive N.W., Calgary, Alberta, Canada T2N 1N4

Ken Nakamula, Department of Mathematics, Tokyo Metropolitan University, 2-1-1 Fukazawa, Setagaya-ku, Tokyo, 158 Japan

István Nemes, Kossuth Lajos University, Mathematical Institute, H-4010 Debrecen Pf.12., Hungary

Attila Pethö, Kossuth Lajos University, Mathematical Institute, H-4010 Debrecen Pf.12., Hungary

Michael Pohst, Mathematisches Institut der Heinrich-Heine-Universität Düsseldorf, Universitätsstr. 1, D-4000 Düsseldorf 1, Germany

Johannes Graf von Schmettow, Mathematisches Institut der Heinrich-Heine-Universität Düsseldorf, Universitätsstr. 1, D-4000 Düsseldorf 1, Germany

Ursula Schneiders, Fachbereich 9 Mathematik, Universität des Saarlandes, D-6600 Saarbrücken, Germany

Ulrich Schröter, Mathematisches Institut der Heinrich-Heine-Universität Düsseldorf, Universitätsstr. 1, D-4000 Düs-seldorf 1, Germany

Nicole Schulte, Mathematisches Institut der Heinrich-Heine-Universität Düsseldorf, Universitätsstr. 1, D-4000 Düsseldorf 1, Germany

Pascale Serf, Fachbereich 9 Mathematik, Universität des Saarlandes, D-6600 Saarbrücken, Germany

Igor E. Shparlinskiy, Institute of Radioengineering Electronics, Academy of Sciences of the USSR, K.Marx av.18, Moscow, GSP-3, 103907 USSR

Gabriele Steidl, Universität Rostock, Fachbereich Mathematik, Universitätsplatz 1, O-2500 Rostock, Germany

S.A. Stepanov, Steklov Institute of Academy Nauk USSR , ul.Vavilova 42., Moscow GSP-1, 117966 USSR

R.J. Stroeker, Econometric Institute, Erasmus University, P.O.Box 1738, 3000 DR Rotterdam, The Netherlands

Manfred Tasche, Universität Rostock, Fachbereich Mathematik, Universitätsplatz 1, O-2500 Rostock, Germany

Bogdan Tropak, Department of Mathematics, Pedagogical University, Zielona Gora, 65-562, ul.Sucharskiego 18.m.14., Poland

Nikos Tzanakis, Department of Mathematics, University of Crete, P.O.Box 1470, Iraklion, Crete, Greece

Benne M.M. de Weger, Faculty of Applied Mathematics, University of Twente, P.O.Box 217, 7500 AE Enschede, The Netherlands

Hugh C. Williams, Department of Computer Science, The University of Manitoba, Winnipeg, Manitoba, Canada R3T 2N2

Horst G. Zimmer, Fachbereich 9 Mathematik, Universität des Saarlandes, D-6600 Saarbrücken, Germany

Table of contents

On the construction of primitive elements and primitive normal bases in a finite field

S.A. Stepanov and I.E. Shparlinskiy

Abstract. The purpose of this survey is to present some recent results of the authors on finding fast algorithms for the construction of primitive elements, normal bases and primitive normal bases in a finite field. It should be noted that our results can be used for the development of other fast algorithms in finite fields.

1 Introduction

Let m, n be positive integers, $q = p^m$ a power of a prime number p, $\mathbb{F}_q$ a finite field of q elements, and $\mathbb{F}_{q^n}$ a finite extension of $\mathbb{F}_q$ of degree n. It is well known (see [1], [2]) that each field $\mathbb{F}_{q^n}$ has a *normal basis* over $\mathbb{F}_q$, i.e. a basis of the form

$$\alpha,\ \alpha^q,\ \ldots,\ \alpha^{q^{n-1}}\ .$$

Further, $\mathbb{F}_{q^n}$ has a primitive element ϑ whose powers

$$\vartheta,\ \vartheta^2,\ \ldots,\ \vartheta^{q^n-1}$$

form the multiplicative group $\mathbb{F}_{q^n}^*$ of $\mathbb{F}_{q^n}$.

A normal basis $\alpha,\ \alpha^q,\ \ldots,\ \alpha^{q^{n-1}}$ of $\mathbb{F}_{q^n}$ is called a primitive normal basis if α is a primitive element of $\mathbb{F}_{q^n}$.

The above-mentioned bases may be successfully used for the construction of very fast algorithms for multiplication and division in $\mathbb{F}_{q^n}$. Therefore these bases are widely applied in coding theory, cryptography etc. (see [1-5]).

The existence of a normal basis was first proved by K.Hensel in 1888. This result was then repeated in a series of papers (For details see comments to §3, ch.2 of [2]). However all the proofs of this fact were up to recent times completely non-constructive. For some very special fields $\mathbb{F}_{q^n}$ the first explicit construction of a normal basis over $\mathbb{F}_q$ was presented by V.M.Sidelnikov [6] in 1987. Then for a fixed q I.I.Semaev [7] found an algorithm for constructing a normal basis of $\mathbb{F}_{q^n}$ over $\mathbb{F}_q$ in time that is polynomial in n. Independently, a quite different

Computational Number Theory

way for finding the above basis was proposed by the authors of [8], [9] for all n and q satisfying the condition $(n,\ q)\ =\ 1$. The computational complexity of the algorithm obtained in this connection is evaluated by a polynomial in n and $\log q$.

The question of the existence of a primitive normal basis is considerably more difficult. This question was considered for the first time by L.Carlitz [10] and H.Davenport [11] who established the existence of primitive normal bases for all extensions $\mathbb{F}_{q^n}$ of $\mathbb{F}_q$ with q^n large enough. The above question was for all n and q solved completely by H.W.Lenstra and R.J.Schoof [12]. However all the above-mentioned results in this direction were again non-constructive and did not provide any means for the construction of a primitive normal basis short of a total enumeration of all q^n elements of the field $\mathbb{F}_{q^n}$.

In [13] an algorithm was proposed which allowed for the transition from an arbitrary primitive element $\vartheta \in \mathbb{F}_q$ to a primitive element $\alpha \in \mathbb{F}_{q^n}$, generating a primitive normal basis of $\mathbb{F}_{q^n}$ over $\mathbb{F}_q$. For a fixed n this algorithm executes in time bounded by a polynomial in $\log q$.

The question of determining a primitive element of $\mathbb{F}_{q^n}$ has been studied a great deal. In particular there are extensive tables of primitive polynomials in finite fields, the roots of which are primitive elements (see [2]). From the theoretical point of view the question of finding primitive elements has been most intensively investigated in the case of the prime field $\mathbb{F}_p$. Without loss of generality, we can identify the elements of $\mathbb{F}_p$ with numbers $0,\ 1,\ \ldots,\ p\ -\ 1$ of a complete system of residues modulo p. In this case we let ϑ represent the least primitive element of $\mathbb{F}_q$ from the set $\{1,\ 2,\ \ldots,\ p\ -\ 1\}$.

The first nontrivial result concerning the distribution of primitive elements was obtained in 1918 by I.M.Vinogradov [14] who has proved that for the least primitive element ϑ of $\mathbb{F}_p$ the following estimate

$$\vartheta \leq 2^{\nu} \frac{p-1}{\varphi(p-1)}\, p^{1/2} \log p \leq 2^{2\nu} p^{1/2} \log p$$

is true, where $\varphi(k)$ is Euler's function and ν is the number of prime divisors of $p\ -\ 1$. The above estimate has been consecutively improved by himself [15] to

$$\vartheta \leq 2^{\nu+1} \frac{p-1}{\varphi(p-1)}\, p^{1/2},$$

by Hua Loo-Keng [16] to

$$\vartheta \leq 2^{\nu+1} p^{1/2},$$

by P.Erdos [17] to

$$\vartheta \ =\ O(p^{1/2} \log^{17} p)$$

and by Wang Yuan [18] (see also [19]) to

$$\vartheta \ =\ O(p^{0.25+\varepsilon}\ ).$$

For the average value of $\vartheta\ =\ \vartheta(p)$ D.A.Burgess and P.D.T.A. Elliot [20] established that

$$\frac{1}{\pi(x)} \sum_{p \leq x} \vartheta(p) \ =\ O(\log^2 x \log^4 \log x).$$

Finally, on assuming the validity of extended Riemann hypothesis, Wang Yuan obtained the following conditional result:

$$\vartheta = O(\nu^6 \log^2 p) = O(\log^8 p).$$

The question of the distribution of primitive elements in a finite field $\mathbb{F}_{q^n}$ has been much less studied, although some natural extensions of all the above-mentioned results are likely to hold in such fields (see also comments to §1, ch.2 of [2]). In particular the following assumption appears to be true.

Conjecture. *Let $\mathbb{F}_q$ be a finite field of characteristic p of $q = p^m$ elements and $\mathbb{F}_{q^n}$ be a finite extension of $\mathbb{F}_q$ of degree n. Then there exists a primitive element $\vartheta \in \mathbb{F}_{q^n}$ such that its minimal polynomial*

$$f(x) = x^{mn} + a_1 x^{mn-1} + \ldots + a_{mn}$$

over $\mathbb{F}_p$ satisfies the condition

$$a_i = O(\log^A p), \quad i = 1, 2, \ldots, mn$$

for some positive A.

In the present paper we propose an algorithm, which in the case of a fixed p, allows us to find in time $n^{O(1)}$ a set $R \subseteq \mathbb{F}_{p^n}$ of cardinality $|R| = O(n^{10})$ that contains at least one primitive element of the field $\mathbb{F}_{p^n}$. Note that the test of primitivity for a given element $\vartheta \in \mathbb{F}_{p^n}$ requires first that we have the factorization of the number $p^n - 1$. However, at present there are no deterministic polynomial time algorithms for solving this problems (see [21]). We also propose an algorithm which, for a fixed p, allows us to determine in subexponential time a very dense sequence of natural numbers t and elements $\vartheta \in \mathbb{F}_{p^t}$ that generate primitive normal bases of the field $\mathbb{F}_{p^t}$ over $\mathbb{F}_p$.

2 On the construction of a normal basis

In this section we show that in the case $(n, q) = 1$ it is possible to pass from an arbitrary basis $\omega_1, \ldots, \omega_n$ of $\mathbb{F}_{q^n}$ to a normal basis

$$\alpha_1 = \alpha, \alpha_2 = \alpha^q, \ldots, \alpha_n = \alpha^{q^{n-1}}$$

of this field over $\mathbb{F}_q$ in time $(n \log q)^{O(1)}$.

Let us define the matrix

$$A = \| a_{i,j} \|_{1 \le i,j \le n}$$

with elements from $\mathbb{F}_q$ by the equalities

$$\omega_i^q = a_{i1}\omega_1 + \ldots + a_{in}\omega_n, \quad 1 \le i \le n. \tag{2.1}$$

Thus

$$(\omega_1^q, \ldots, \omega_n^q)^t = A(\omega_1, \ldots, \omega_n)^t.$$

Raising both parts of (2.1) to the power q^{j-1} we get

$$\omega_i^{q^j} = a_{i1}\omega_1^{q^{j-1}} + \ldots + a_{in}\omega_n^{q^{j-1}}, \quad 1 \le i \le n.$$

Then (2.1) implies that

$$\left(\omega_1^{q^j}, \ldots, \omega_n^{q^j}\right)^t = A^j(\omega_1, \ldots, \omega_n)^t. \tag{2.2}$$

Taking into account that $\omega^{q^n} = \omega$ for all $\omega \in \mathbb{F}_{q^n}$ and that $\omega_1, \ldots, \omega_n$ is a basis of $\mathbb{F}_{q^n}$ we get from the above the equality $A^n = E$, where E is the unit matrix, and that $f(\lambda) = \lambda^n - 1$ is the minimal polynomial of the matrix A.

Since $(n, q) = 1$ the polynomial $f(\lambda)$ has no multiple roots. Let

$$f(\lambda) = \prod_{\nu=1}^{m} f_\nu(\lambda) \tag{2.3}$$

be its decomposition into irreducible (over $\mathbb{F}_q$) polynomials $f_1, \ldots, f_m$ of degrees $d_1, \ldots, d_m$ respectively.

Denote by $\lambda_{\nu 1}, \ldots, \lambda_{\nu d_\nu}$ the eigenvalues of the matrix A which are the zeros of $f_\nu(\lambda)$ and find the eigenvector $z_{\nu 1}$ corresponding to $\lambda_{\nu 1}$. For this we need to find in the field $\mathbb{F}_{q^{d_\nu}}$ a nonzero solution of the system of linear equations

$$z_{\nu 1}(A - \lambda_{\nu 1}E) = 0. \tag{2.4}$$

If $z_{\nu 1} = (z_{\nu 1}(1), \ldots, z_{\nu 1}(n))$ then the vectors

$$z_{\nu r} = (z_{\nu 1}(1)^{q^r}, \ldots, z_{\nu 1}(n)^{q^r}), \quad 1 \le r \le d_\nu, \tag{2.5}$$

will be the eigenvectors corresponding to the rest of the zeros of the polynomial f_ν. Then it is clear that the vectors

$$z_\nu = z_{\nu 1} + \ldots + z_{\nu d_\nu}, \quad 1 \le \nu \le m, \tag{2.6}$$

have components in $\mathbb{F}_q$ and hence the vector

$$x = (x_1, \ldots, x_n) = z_1 + \ldots + z_m \tag{2.7}$$

also has components in $\mathbb{F}_q$.

Put

$$\alpha = x_1\omega_1 + \ldots + x_n\omega_n. \tag{2.8}$$

Lemma 2.1. *The element α, defined by the relations (2.1), (2.3)-(2.8), generates a normal basis of the field $\mathbb{F}_{q^n}$ over $\mathbb{F}_q$ and may be computed in time bounded by a polynomial in n and q.*

Proof. From (2.2) and (2.8) we deduce that

$$\alpha^{q^j} = xA^j(\omega_1, \ldots, \omega_n)^t, \quad 0 \le j \le n-1.$$

Now we show that the vectors $x, xA, \ldots, xA^{n-1}$ are linearly independent over $\mathbb{F}_q$. Indeed, in view of (2.7) the vector x has a representation $x = y_1 + \ldots + y_n$, where $y_1, \ldots, y_n$ are eigenvectors of the matrix A corresponding to pairwise distinct eigenvalues $\mu_1, \ldots, \mu_n$. Then

$$xA^j = \mu_1^j y_1 + \ldots + \mu_n^j y_n, \quad 0 \le j \le n-1,$$

and we see that the vectors $x,\ xA,\ \ldots,\ xA^{n-1}$ are linearly independent over $\mathbb{F}_q$. Hence the element α generates a normal basis of the field $\mathbb{F}_{q^n}$ over $\mathbb{F}_q$.

It is easy to verify that all computations by formulas (2.1), (2.3)-(2.8) may be realized in time which is polynomial in n and q (see [22] , [23]). The lemma is proved. □

Assume now that $q > n$. Choose in $\mathbb{F}_q$ an arbitrary subset Λ of cardinality $|\Lambda| = n + 1$ and denote by $e_i = (0, \ldots, 1, \ldots, 0)$, $1 \le i \le n$, the n-dimensional unit vectors.

Now we define a sequence of vectors $g_1, \ldots, g_n$ in the following way. Put $g_1 = e_1$. If the vector g_i is already defined then for all $\lambda \in \Lambda$ we compute the rank of the system of vectors

$$r(\lambda) = rank\{(g_i + \lambda e_{i+1}),\ (g_i + \lambda e_{i+1})A, \ldots,\ (g_i + \lambda e_{i+1})A^{n-1}\}. \quad (2.9)$$

Let its maximum be achieved for $\lambda = \lambda_0$,

$$r(\lambda_0) = \max_{\lambda \in \Lambda} r(\lambda). \quad (2.10)$$

Then we define

$$g_{i+1} = g_i + \lambda_0 e_{i+1} \quad (2.11)$$

and put

$$x = (x_1, \ldots, x_n) = g_n. \quad (2.12)$$

Lemma 2.2. *If $q > n$ then the element α, defined by the relations (2.8)-(2.12), generates a normal basis of the field $\mathbb{F}_{q^n}$ over $\mathbb{F}_q$ and may be computed in time bounded by a polynomial in n and $\log q$.*

Proof. Let $\mathcal{F}_q$ be the algebraic closure of the field $\mathbb{F}_q$ and $y_1, \ldots, y_n$ be the complete system of eigenvectors of the matrix A. Decompose over $\mathcal{F}_q$ the vectors $e_1, \ldots, e_n$ with respect to $y_1, \ldots, y_n$ and denote by I_ν, $1 \le \nu \le n$, the set of subscripts of those eigenvectors that occur in the decomposition of e_ν with nonzero coefficients. Define analogously sets J_ν, $1 \le \nu \le n$, for the vectors $g_1, \ldots, g_n$. It is clear that

$$\bigcup_{\nu=1}^{n} I_\nu = \{1, \ldots, n\}.$$

Also, taking into account the condition $|\Lambda| = n + 1$ we get, in view of (2.9)-(2.11), that

$$J_{\nu+1} = J_\nu \cup I_{\nu+1},\ 1 \le \nu \le n-1.$$

In that case we have

$$J_n = \{1, \ldots, n\}$$

and hence the vector $x = g_n$ has the form

$$x = \alpha_1 y_1 + \ldots + \alpha_n y_n,$$

where $\alpha_1, \ldots, \alpha_n$ are some nonzero elements of the field $\mathcal{F}_q$. But then the vectors $x,\ xA,\ \ldots,\ xA^{n-1}$ are linearly independent over $\mathcal{F}_q$ and hence over $\mathbb{F}_q$. Thus, element $\alpha = x_1\omega_1 + \ldots + x_n\omega_n$ generates a normal basis of the field $\mathbb{F}_{q^n}$ over $\mathbb{F}_q$. The calculation of the matrix A requires time bounded by a polynomial in n and $\log q$. In addition calculations by formulas (2.9)-(2.12) require time which is also bounded by a polynomial in n and $\log q$ (see [22]). The lemma is proved. □

Theorem 2.1. *Under the condition $(n,\ q) = 1$ a normal basis of the field $\mathbb{F}_{q^n}$ over $\mathbb{F}_q$ may be constructed in time bounded by a polynomial in n and $\log q$.*

Proof. To prove the theorem it is sufficient to apply Lemma 2.1 in the case $q < n$ and Lemma 2.2 in the case $q > n$. □

3 On the construction of a primitive normal bases

Let ϑ be a fixed primitive element of the field $\mathbb{F}_{q^n}$ defined by its decomposition $\vartheta = u_1\omega_1 + \ldots + u_n\omega_n$ with respect to some basis of $\mathbb{F}_{q^n}$ over $\mathbb{F}_q$. Our goal is to find a primitive normal basis with the generating element of the form ϑ^k.

Theorem 3.1. *There exist absolute positive constants c_1 and c_2, such that if*

$$N \geq \max\{\exp\exp(c_1 \log^2 n),\ c_2 n \log q\},$$

then among the N elements $\vartheta,\ \vartheta^2,\ \ldots,\ \vartheta^N$ there is at least one which generates a primitive normal basis of the field $\mathbb{F}_{q^n}$ over $\mathbb{F}_q$.

We remark at once that the elements

$$\vartheta^k,\ \vartheta^{kq},\ \ldots,\ \vartheta^{kq^{n-1}}$$

form a primitive normal basis if and only if $(k,\ q^n - 1) = 1$ and $\vartheta^k,\ \vartheta^{kq},\ \ldots,\ \vartheta^{kq^{n-1}}$ are linearly independent over $\mathbb{F}_q$. Successive testing of these conditions for $k = 1,\ 2,\ \ldots$ may be realized at every step in time $(n \log q)^{O(1)}$.

Now let us consider some normal basis

$$\omega_i = \omega^{q^i},\quad 0 \leq i \leq n-1, \tag{3.1}$$

of $\mathbb{F}_{q^n}$ over $\mathbb{F}_q$. We denote by $u_o(k),\ \ldots,\ u_{n-1}(k)$ the corresponding coordinates of elements ϑ^k in this basis so that

$$\vartheta^k = \sum_{i=0}^{n-1} u_i(k)\, \omega_i,\quad k = 1,\ 2,\ \ldots\ . \tag{3.2}$$

From (3.1) and (3.2) we get the relation

$$\vartheta^{kq^\nu} = \sum_{i=0}^{n-1} u_{i-\nu}(k)\, \omega_i,\quad 0 \leq \nu \leq n-1,$$

in that $u_j(k) = u_{j+n}(k)$ as soon as $j < 0$.

Denote by $A(k)$ the following circulant matrix

$$A(k) = \| u_{i-\nu}(k) \|_{0\le i,\nu\le n-1}.$$

It is evident that the elements

$$\vartheta^k, \vartheta^{kq}, \ldots, \vartheta^{kq^{n-1}}$$

form a normal basis if and only if

$$\det A(k) \neq 0. \tag{3.3}$$

Further, ϑ^k is primitive if and only if

$$(k, q^n - 1) = 1. \tag{3.4}$$

Denote by $\varphi(x) = x^n + a_{n-1}x^{n-1} + \ldots + a_0$ the minimal polynomial of the element ϑ over the field $\mathbb{F}_q$.

Lemma 3.1. *For every $i = 0, 1, \ldots, n-1$ the sequence $\{u_i(k)\}_{k=1}^{\infty}$ satisfies the following recurrence relation*

$$u_i(k+n) + a_{n-1}u_i(k+n-1) + \ldots + a_0u_i(k) = 0, \quad k = 1, 2, \ldots .$$

Proof. The above relation follows from (3.2) and the equality

$$\vartheta^{n+k} + a_{n-1}\vartheta^{n+k-1} + \ldots + a_0\vartheta^k = 0,$$

which is clear from the definition of the polynomial $\varphi(x)$. □

Define the sequence δ_l by means of reccurence relations

$$\delta_2 = 1; \quad \delta_l = \delta_{[(l+2)/2]} \,/\, [(l+2)/2], \quad l = 3, 4, \ldots,$$

and consider in an arbitrary field L the sequences

$$v(x) = \alpha_1\mu_1^x + \ldots + \alpha_l\mu_l^x, \quad x = 1, 2, \ldots,$$

where $l \ge 2$ and $\alpha_1, \ldots, \alpha_l, \mu_1, \ldots, \mu_l$ are some nonzero elements in L.

Lemma 3.2. *The number of solutions of the equation*

$$v(x) = 0, \quad 1 \le x \le N, \tag{3.5}$$

does not exceed

$$2\, lN(N^{-1} + \tau^{-1})^{\delta_l},$$

where τ is the minimal positive integer such that

$$\mu_i^\tau = \mu_j^\tau$$

for some i and j, $1 \le i < j \le l$.

Proof. It is obvious that the equation

$$\alpha_1\mu_1^x + \alpha_2\mu_2^x = 0, \quad 1 \le x \le N,$$

has at most $[N/\tau] + 1$ solutions so that for $l = 2$ the assertion is true.

Let us presuppose that it is true for $l < r$ and prove the validity of the assertion for $l = r > 2$. Put $s = [\, r/2 \,]$ and note that $1 \le r - s \le s + 1 < r$ for $r > 2$. If T denotes the number of solutions of the equation (3.5) then the number of solutions $(x_0, x_1, \ldots, x_s)$ of the system

$$\alpha_1 \mu_1^{x_i} + \ldots + \alpha_r \mu_r^{x_i} = 0, \quad 0 \le i \le s,$$
$$1 \le x_0 < x_1 < \ldots < x_s \le N$$

equals $\begin{pmatrix} T \\ s+1 \end{pmatrix}$.

Setting $x = x_0$, $y_i = x_i - x_0$, $0 \le i \le s$, we see that this number does not exceed the number of solutions of the system

$$\alpha_1 \mu_1^{x+y_i} + \ldots + \alpha_r \mu_r^{x+y_i} = 0, \quad 0 \le i \le s,$$
$$1 \le x \le N, \; y_0 = 0, \; 1 \le y_1 < \ldots < y_s \le N.$$

Denote by Y the set of all $(s+1)$-tuples of the form $(y_0, y_1, \ldots, y_s)$, $y_0 = 0$, $1 \le y_1 < \ldots < y_s \le N$. Let us fix some $(s+1)$-tuple $(y_0, y_1, \ldots, y_s) \in Y$ and denote by $c_0, c_1, \ldots, c_s$ some nonzero solution of the homogeneous linear system of equations

$$C_0 \mu_j^{y_0} + \ldots + C_s \mu_j^{y_s} = 0, \quad 0 \le j \le s.$$

If we multiply i-th equation of previous system by C_i, $0 \le i \le s$, and add, then we get an equation of the form

$$A_{s+1} \mu_{s+1}^{x} + \ldots + \alpha_r \mu_r^{x} = 0, \quad 1 \le x \le N. \tag{3.6}$$

Now let us estimate the number of $(s+1)$-tuples $(y_0, y_1, \ldots, y_s) \in Y$ for which

$$A_{s+1} = \alpha_{s+1}\left(C_0 \mu_{s+1}^{y_0} + \ldots + C_s \mu_{s+1}^{y_s}\right) = 0.$$

It is clear that for such $(s+1)$-tuples we have the relation

$$\det \| \, \mu_i^{y_j - 1} \, \|_{1 \le i,j \le s+1} = 0.$$

Let us partition all $(s+1)$-tuples in Y into classes $Y_1, \ldots, Y_s$ by setting $(y_0, y_1, \ldots, y_s) \in Y_k$ if and only if

$$\det \| \, \mu_i^{y_j - 1} \, \|_{1 \le i,j \le k} \ne 0 \tag{3.7}$$

and

$$\det \| \, \mu_i^{y_j - 1} \, \|_{1 \le i,j \le k+1} = 0. \tag{3.8}$$

Decomposing the determinant (3.8) on the last row and taking into account the inequality (3.7) we get for y_k an equation of the form (3.5) with $k + 1 \le s + 1$ terms. By the induction assumption

$$|\, Y_k \,| \le 2(s+1) N^s (N^{-1} + \tau^{-1})^{\delta_{s+1}},$$

and hence we have at most

$$\sum_{k=1}^{s} |\, Y_k \,| \le 2s(s+1) N^s (N^{-1} + \tau^{-1})^{\delta_{s+1}},$$

$(s + 1)$-tuples $(y_0, y_1, \ldots, y_s) \in Y$ for which $A_{s+1} = 0$.

For the remaining $(s + 1)$-tuples $(y_0, y_1, \ldots, y_s) \in Y$ we apply to the equation (3.6) our induction assumption and then get

$$\begin{aligned}
\binom{T}{s+1} &\leq 2s(s+1)N^s(N^{-1} + \tau^{-1})^{\delta_{s+1}} \\
&\quad + 2(s+1)N^s(N^{-1} + \tau^{-1})^{\delta_{s+1}} \\
&= 2(s+1)^2 N^s(N^{-1} + \tau^{-1})^{\delta_{s+1}}; \\
(T-s)^{s+1} &\leq 2(s+1)^2(s+1)!\, N^s(N^{-1} + \tau^{-1})^{\delta_{s+1}}; \\
T &\leq \left[2(s+1)^2(s+1)!\right]^{1/(s+1)} N(N^{-1} + \tau^{-1})^{\delta_r} + s \\
&\leq \left(\left[2(s+1)^2(s+1)!\right]^{1/(s+1)} + s\right) N(N^{-1} + \tau^{-1})^{\delta_r} \\
&\leq 2rN(N^{-1} + \tau^{-1})^{\delta_r}.
\end{aligned}$$

The lemma is proved. □

We remark that in the case $L = \mathbb{F}_p$ a similar result was obtained in [24] with $\delta_l = 2^{-l+2}$. For the sequence considered in this paper it is easy to prove by induction on l that

$$\delta_l \geq \exp(-c \log^2 l) \tag{3.9}$$

where $c > 0$ is some effective absolute constant.

Put

$$\vartheta_i = \vartheta^{q^{i-1}}, \quad 1 \leq i \leq n.$$

Lemma 3.3. *If τ is the minimal positive integer for which*

$$\vartheta_i^\tau = \vartheta_j^\tau \tag{3.10}$$

for some i and j, $1 \leq i < j \leq l$, then $\tau \geq q$.

Proof. Since ϑ is a primitive element (3.10) implies that

$$\tau(q^{j-i} - 1) \equiv 0 \pmod{q^n - 1}$$

and hence

$$\tau \geq (q^n - 1)/(q^{n-1} - 1).$$

□

Proof of Theorem 3.1. For the number $\nu(k)$ of prime divisors of an integer $k \geq 3$ we have (see [25], ch.4)

$$\nu(k) = O(\log k / \log\log k).$$

From here and from the prime number theorem we get that for some c_2 large enough and $N \geq c_2 n \log q$ the interval $[1, N]$ contains at least

$$R = N/2\log N \tag{3.11}$$

distinct prime numbers that are coprime with $q^n - 1$.

Let $\rho_1, \ldots, \rho_n$ be the various n-th roots of unity in the algebraic closure $\mathcal{F}_q$ of the field $\mathbb{F}_q$. Then for the circulant matrix $A(k)$ we have

$$\det A(k) = v_1(k) \ldots v_n(k), \quad k = 1, 2, \ldots,$$

where

$$v_i(k) = u_0(k) + \rho_i u_1(k) + \ldots + \rho_i^{n-1} u_{n-1}(k), \quad 1 \leq i \leq n.$$

From the existence of a normal basis of the field $\mathbb{F}_{q^n}$ over $\mathbb{F}_q$ it follows that each sequence $\{v_i(k)\}_{k=1}^{\infty}$, $1 \leq i \leq n$, is nonzero. By virtue of Lemma 3.1 every such sequence can be represented in the form (see [2])

$$v_i(k) = \sum_{j=1}^{n} \alpha_{ij} \vartheta_j^k, \quad k = 1, 2, \ldots,$$

where at least one of the elements $\alpha_{i1}, \ldots, \alpha_{in} \in \mathcal{F}_q$ is different from zero for all $i = 1, 2, \ldots, n$. Hence Lemma 3.2 can be applied to each of these sequences and then, taking into account the result of Lemma 3.3, we get that the number of integers $k = 1, \ldots, N$ with the condition

$$\det A(k) = 0$$

does not exceed

$$2\, n^2 N (N^{-1} + \tau^{-1})^{\delta_n}. \tag{3.12}$$

If $n \geq c_2 n \log q$ and $q \geq \exp\exp(c_1 \log^2 n)$, where c_1 is a sufficiently large effective constant we get from (3.9) that the quantity (3.12) is strictly less than the quantity (3.11). The theorem is proved. □

4 On primitive elements in finite fields of a fixed characteristic

In the sequel the constant in the symbol "O" will depend only on p.

Theorem 4.1. *Let p be a fixed prime number. Then one may find in time $n^{O(1)}$ a set $R \subseteq \mathbb{F}_{p^n}$ of cardinality $|R| = O(n^{10})$ containing at least one primitive element of the finite field $\mathbb{F}_{p^n}$.*

Proof. Put

$$r = [10 \log n / \log p] + 1$$

and denote by α some generating element of the field $\mathbb{F}_{p^{nr}}$ over $\mathbb{F}_{p^r}$ i.e.

$$\mathbb{F}_{p^{nr}} = \mathbb{F}_{p^r}(\alpha).$$

Let us consider the set

$$R = \{\beta \mid \beta = (\alpha + \lambda)^{(p^{nr}-1)/(p^n-1)}; \quad \lambda \in \mathbb{F}_{p^r}\}.$$

It is clear that

$$| R | = p^r \leq pn^{10} = O(n^{10}).$$

The fields $\mathbb{F}_{p^r}$, $\mathbb{F}_{p^{nr}}$ and the element α may be constructed in time $n^{O(1)}$ (see [7], [26]); thus, the set R may be determined in time $n^{O(1)}$.

Now we prove that for n large enough the set R contains a primitive element of the field $\mathbb{F}_{p^n}$.

For if we consider the set

$$Q = \{\gamma \mid \gamma = (\alpha + \lambda); \quad \lambda \in \mathbb{F}_{p^r}\},$$

it is sufficient to prove that Q contains a primitive element γ of the field $\mathbb{F}_{p^{nr}}$ in that

$$\vartheta = \gamma^{(p^{nr}-1)/(p^n-1)} \in R$$

is a primitive element of the field $\mathbb{F}_{p^n}$. Denote by N the number of primitive elements of the field $\mathbb{F}_{p^{nr}}$ contained in Q. Then let us fix some primitive element $\rho \in \mathbb{F}_{p^{nr}}$ and denote by indγ the index of an element $\gamma \in \mathbb{F}_{p^{nr}}$. For all divisors $d \mid p^{nr} - 1$ let us consider the sets

$$Q_d = \{\gamma \in Q \mid \text{ind}\gamma \equiv 0 \pmod{d}\}$$

and put $N_d = | Q_d |$. Further we denote by X the set of all multiplicative characters of the field $\mathbb{F}_{p^{nr}}$ and for $d \mid p^{nr} - 1$ put

$$X_d = \{\chi \mid \chi \in X, \; \chi^d = \chi_0\}$$

where χ_0 is the principal character. Then we have

$$N_d = \frac{1}{d} \sum_{\chi \in X_d} \sum_{\gamma \in Q} \chi(\gamma).$$

Now separating the terms with $\chi = \chi_0$ and using the estimate (see [27], [28])

$$\max_{\substack{\chi \in X \\ \chi \neq \chi_0}} \left| \sum_{\gamma \in Q} \chi(\gamma) \right| < np^{r/2}$$

we get

$$|N_d - p^r/d| < np^{r/2}. \tag{4.1}$$

Let $\mathbb{P}$ be the set of all prime numbers and let

$$\mathbb{P}_z = \{q \mid q \in \mathbb{P}, \; q \mid p^{nr} - 1, \; q > z\}.$$

For every $z > 1$ we define $Q(z)$ to be the set consisting of those $\gamma \in Q$ for which the number (indγ, $p^{nr} - 1$) has no prime divisors in $\mathbb{P}_z$. Put $N(z) = | Q(z) |$ and note that

$$N \geq N(z) - \sum_{q \in \mathbb{P}_z} N_q.$$

If by ω we denote the number of all prime divisors of $p^{nr} - 1$ then we get by virtue of (4.1)

$$\sum_{q \in \mathbb{P}_z} N_q \leq \sum_{q \in \mathbb{P}_z} (p^r/q + np^{r/2}) \leq \omega(p^r/z + np^{r/2}).$$

Hence taking into account the obvious inequality $2^\omega < p^{nr}$ we have

$$N \ge N(z) + O\bigl(nr(p^r z^{-1} + np^{r/2})\bigr).$$

To get a lower bound for $N(z)$ we make use of Theorem 3.1 from [29] with parameters $\lambda = \log 1.3$, $L = np^{r/2}$, $k = 1$. Taking into account the result of the exercise 9.e from ch.2 of [30] we get the inequality

$$\begin{aligned} N(z) &> cp^r \prod_{q \mid p^{nr}-1} (1 - 1/q) + O(z^{3.95} np^{r/2}) \\ &\ge cp^r / \bigl(\log z + O(1)\bigr) + O(z^{3.95} np^{r/2}) \end{aligned}$$

with some absolute constant $c > 0$. Setting $z = p^{r/9.9}$ we get, in view of the choice of the parameter r, that for some constant $c_0 > 0$ and all n large enough the estimate

$$\begin{aligned} N &\ge c_0 p^r / \log n + O\bigl(np^{r(1-1/9.9} \log n + n^2 p^{r/2} \log n\bigr) \\ &\ge c_0 n^{10} / \log n + O\bigl(np^{r(1-1/9.9)} \log n + n^7 \log n\bigr) > 0 \end{aligned}$$

is valid.

The theorem is proved. □

Theorem 4.2. *For any $\varepsilon > 0$ there exists a constant $c > 0$ such that one may find in time $p^{cn/\log\log n}$ a natural $t = n + O(n^\varepsilon)$ and an element $\vartheta \in \mathbb{F}_{p^t}$ which generates a primitive normal basis of the field $\mathbb{F}_{p^t}$ over $\mathbb{F}_p$.*

Proof. Put

$$\begin{aligned} r &= [10 \log n / \log p] + 1, \\ s &= q_1 \cdots q_\omega, \end{aligned}$$

where $q_1, \ldots, q_\omega$ are all prime numbers not exceeding $0.5\varepsilon \log n$.

We define

$$t = rs[n/rs] = n + O(n^\varepsilon)$$

and denote by $\alpha \in \mathbb{F}_{p^t}$ an element which generates a normal basis of the field $\mathbb{F}_{p^t}$ over $\mathbb{F}_{p^r}$.

Let us consider the set

$$M = \{\beta \mid \beta = (\alpha + \lambda); \quad \lambda \in \mathbb{F}_{p^r}\}.$$

As in the proof of the previous theorem we have that $\mid M \mid = O(n^{10})$ and the fields $\mathbb{F}_{p^t}$,$\mathbb{F}_{p^r}$ and the element α may be constructed at time $n^{O(1)}$ (see [7] or section 2 of this paper). So that the set M may be indicated at the time $n^{O(1)}$. We have also that for n large enough the set M contains at least two primitive elements ϑ_1, ϑ_2 of the field $\mathbb{F}_{p^t}$. Since α generates a normal basis then at least one from ϑ_1, ϑ_2 also generates a normal basis (see proof of the Theorem 3 from [31]).

It is clear that no prime divisor of $p^t - 1$ is greater than $(p + 1)^{\varphi(t)}$, where $\varphi(t)$ is Euler's function. Using the definition of s and t we find that

$$\varphi(t) \le \varphi(s)t/s = O(t/\log\log s) = O(n/\log\log n).$$

Hence we can in time $p^{cn/\log\log n}$ determine all prime divisors of $p^t - 1$ and find an element $\vartheta \in M$ which generates a primitive normal basis of the field $\mathbb{F}_{p^t}$. □

References

[1] F.J. MacWilliams and N.J.A. Sloane, *The Theory of Error-Correcting Codes*, N.H.P.C. (1977).

[2] R. Lidl and H. Niederreiter, *Introduction to Finite Fields and their Applications*, Cambridge Univ. Press, Cambridge 1986.

[3] E.R. Berlekamp, *Algebraic Coding Theory*, McGraw-Hill, New York 1968.

[4] M. Willet, *Arithmetic in a finite field*, Math. of Comp. **152** (1980), p.1353-1359.

[5] T. Itoh and S. Tsujii, *A fast algorithm for computing multiplicative inverses in $GF(2^m)$ using normal basis*, Inform. and Comp. **78** (1988), p.171-177.

[6] V.M. Sidelnikov, *On the normal bases of a finite field*, Mat.Sbornik **133** (1987), no.4, p.497-507 (Russian), Engl. Transl.: Math. USSR Sbornik **61** (1988), 485-494.

[7] I.I. Semaev, *Construction of polynomials over a finite field with linearly independent roots*, Mat.Sbornik **135** (1988), no.4, p.520-532 (Russian), Engl. Transl.: Math. USSR Sbornik **63** (1989), 507-519.

[8] S.A. Stepanov and I.E. Shparlinskiy, *On structure complexity of normal basis of a finite field*, Found. of Computation Theory, USSR, Kazan, 1987, Lect. Notes in Comput. Sci. **278** (1987), p.414-416.

[9] S.A. Stepanov and I.E. Shparlinskiy, *On construction of a normal basis of a finite field*, Acta Arithm. **49** (1987), no.2, p.189-192.

[10] L. Carlitz, *Primitive roots in a finite field*, Trans. Amer. Math. Soc. **73** (1952), p.373-382.

[11] H. Davenport, *Bases for finite fields*, J. London Math. Soc. **43** (1968), p.21-39.

[12] H.W. Lenstra and R.J. Schoof, *Primitive normal bases for finite fields*, Math. Comp. **48** (1987), no.177, p.217-231.

[13] S.A. Stepanov and I.E. Shparlinskiy, *On construction of a primitive normal basis of a finite field*, Mat.Sbornik **180** (1989), no.8, p.1067-1072 (Russian).

[14] I.M. Vinogradov, *Sur la distribution des residues et des non residues des puissances*, J. Fiz.-Mat. Ob-va Perm. Univ. **1** (1918), p.94-98 (Russian).

[15] I.M. Vinogradov, *On the least primitive root*, Dokl. Akad. Nauk. SSSR **1** (1930), p.7-11 (Russian).

[16] Hua Loo-Keng, *On the least primitive root of a prime*, Bull. Amer. Math. Soc. **48** (1942), p.726-730.

[17] P. Erdös, *On the least primitive root of a prime p*, Bull. Amer. Math. Soc. **51** (1945), p.131-132.

[18] Wang Yuan, *On the least primitive root of a prime*, Acta Math. Sinica **9** (1959), p.432-441.

[19] D.A. Burgess, *On a character sums and primitive root*, Proc. Lond. Math. Soc. **3** (1962), p.179-192.

[20] D.A. Burgess and P.D.T.A. Elliot, *The average of the least primitive root*, Mathematika **15** (1968), p.39-50.

[21] H. Riesel, *Prime Numbers and Computer Methods for Factorization*, Birkhäuser, Boston, 1985.

[22] V.I. Solodovnikov, *Upper bounds on complexity of solving systems of linear equations*, Zapiski Nauchn. Semin. LOMI Akad. Nauk. SSSR **118** (1982), p.159-187 (Russian).

[23] D.Yu. Grigoriev, *Factorising polynomials over a finite field and solving systems of algebraic equations*, Zapiski Nauchn. Semin. LOMI Akad. Nauk SSSR **137** (1984), p.20-79 (Russian).

[24] I.E. Shparlinskiy, *On arithmetical properties of recurring sequences and some applications*, Kand.dissert., M., Moscow State Pedagogical Inst., 1980.

[25] A.G. Postnikov, *An introduction to analytic number theory*, Moscow, Nauka, 1971.

[26] A.L. Chistov, *Construction of a finite field in polynomial time*, Proc. 7 All-Union Conf.on Math. Logic. Novosibirsk, 1984, p.196 (Russian).

[27] L. Carlitz, *Distribution of primitive roots in a finite field*, Quart. Journ. Math. **4** (1953), p.4-10.

[28] N.M. Katz, *An estimate for character sums*, J. Amer. Math. Soc. **2** (1989), no.2, p.197-200.

[29] H. Halberstam and H.E. Richert, *Sieve Methods*, Acad.Press, 1975.

[30] I.M. Vinogradov, *Foundations of Number Theory*, Moscow, Nauka, 1981.

[31] I.E. Shparlinskiy, *On the distribution of primitive and irreducible polynomials modulo prime*, Diskr. Mat. **1** (1989), no.1, p.117-124 (Russian).

A numerical method for the determination of the cyclotomic polynomial coefficients

A. Grytczuk and B. Tropak

Abstract. In this paper we present a method for determining the coefficients $a_k(n)$ of the cyclotomic polynomial Φ_n. The method is based on Newton's formula and Hölder's identity and provides a very simple algorithm for effecting the computations.

By using this algorithm we found the minimal k over all n values, for which $a_k(n)$ is equal to given integer $m = \pm 2, \pm 3, \ldots, \pm 9$ and 10. A table of these values k and a_k is presented at the end of the paper.

1 Introduction

The n-th cyclotomic polynomial Φ_n is defined by the formula

$$\Phi_n(z) = \prod_{\substack{(k,n)=1 \\ 1 \leq k \leq n}} \Big(z - \exp(2\pi i k/n) \Big) = \sum_{k=0}^{\phi(n)} a_k z^{\phi(n)-k} \tag{1.1}$$

where ϕ is Euler's function. It is well known that the coefficients a_k are rational integers and that $a_k = a_{\phi(n)-k}$ for all $k \leq \phi(n)$. Moreover, the set of the coefficients of Φ_n and of Φ_m, where m is the squarefree kernel of n, are the same; thus in the sequel we shall consider only squarefree n.

P. Erdös [6] showed that $a_k = 0$ or ± 1 for all k when $n < 105$ and that $a_7 = -2$ for $n = 105 = 3 \cdot 5 \cdot 7$. M. Endo [5] proved that $k = 7$ is the smallest number for which $|a_k| > 1$.

Denote by $a_k(n)$ the k-th coefficient of Φ_n.

If, for a given m such that $|m| > 2$, we want to find the minimal k with $a_k(n) = m$, then for all n we must have $a_r(n) \neq m$ if $r < k$. Hence, we need a simple method for the determination of the coefficients of Φ_n.

It is well known that one can calculate Φ_n by using the formula

$$\Phi_{mp}(z) = \begin{cases} \Phi_m(z^p) & \text{if } p|m, \\ \Phi_m(z^p)/\Phi_m(z) & \text{otherwise.} \end{cases}$$

Computational Number Theory

On the other hand we also know that

$$\Phi_n(z) = \prod_{m=1}^{\infty} (1 - z^m)^{\mu\left(\frac{n}{m}\right)}, \tag{1.2}$$

where μ denotes the Möbius function and $\mu\left(\frac{n}{m}\right) = 0$ if $\frac{n}{m}$ is not an integer.

From (1.2) it follows that, for a squarefree n, the value $a_k(n)$ depends only on the values of $\mu(n)$, $\mu\left(\frac{n}{m}\right)$ and on the primes less than $k+1$ which happen to divide n. Using (1.2) one can obtain formulae for $a_k(n)$ with fixed k.

For example we have

$$a_1(n) = -\mu(n),\ a_2(n) = \frac{1}{2}\mu(n)(\mu(n)-1) - \mu\left(\frac{n}{2}\right),$$
$$a_3(n) = \frac{1}{2}\mu^2(n) - \frac{1}{2}\mu(n) + \mu\left(\frac{n}{2}\right)\mu(n) - \mu\left(\frac{n}{3}\right), \dots .$$

This method has been used by D.H.Lehmer [8] and H.Möller [11].

The purpose of this paper is to present a method for computing of the coefficients of Φ_n. We get the following formula for a_j:

$$a_j(n) = -\frac{1}{j} \cdot \mu(n) \cdot \sum_{m=0}^{j-1} a_m(n)\mu\Big((n, j-m)\Big)\phi\Big((n, j-m)\Big) \quad \text{for } j > 0 \tag{1.3}$$

with $a_0(n) = 1$, where (k, l) denotes the gcd of k and l. We remark that this method does not use polynomial division and that it is essentially different from the methods of D.H.Lehmer and H.Möller.

By using (1.3) we found for $m = \pm 2, \pm 3, \dots \pm 9$ and 10 the minimal values k for which there exists a number n such that $a_k(n) = m$. The results are collected in a table at the end of this paper.

2 Determination of the coefficients $a_k(n)$ of Φ_n

Let $n = p_1 \cdots p_r$ be a positive squarefree integer and assume that $r \geq 2$ and $p_1, \dots, p_r$ are primes such that $p_i < p_j$ for $i < j$. From (1.1), Viete's and Newton's formulae it follows that the coefficients are given by the formulae:

$$a_0(n) = 1, \quad a_j(n) = -\frac{1}{j}\sum_{m=0}^{j-1} a_m(n) S_{j-m}(n) \qquad \text{for} \quad j > 0, \tag{2.1}$$

where

$$S_l(n) = \sum_{\substack{k=1 \\ (k,n)=1}} \exp(2\pi i k l/n). \tag{2.2}$$

From the Hölder identity we have

$$S_l(n) = \frac{\phi(n)}{\phi\left(\frac{n}{(n,l)}\right)} \cdot \mu\left(\frac{n}{(n,l)}\right). \tag{2.3}$$

Since n is a squarefree integer, by (2.3) we get

$$S_l(n) = \mu(n)\mu\Big((n,l)\Big)\phi\Big((n,l)\Big). \tag{2.4}$$

Now let k be a positive integer such that $2 \le k \le \phi(n)$; choose s to be a positive integer such that $p_s \le k < p_{s+1}$ if $p_r > k$ and $s = r$ otherwise. Let for $d = p_1 \cdots p_s$

$$T_l = \mu(n)\mu\Big((d,l)\Big)\phi\Big((d,l)\Big) \tag{2.5}$$

and

$$b_0 = 1, \quad b_j = -\frac{1}{j}\cdot\sum_{m=0}^{j-1} b_m T_{j-m} \quad \text{for } j > 0. \tag{2.6}$$

We remark that for $1 \le l \le k$ we have $(d,l) = (n,l)$ thus by (2.4) and (2.5) we get

$$T_l = \mu(n)\mu\Big((n,l)\Big)\phi\Big((n,l)\Big) = S_l(n). \tag{2.7}$$

Now it is easy to see by induction that

$$b_j = a_j(n) \qquad \text{for } j = 1,\ldots,k. \tag{2.8}$$

Thus is possible to determine the coefficients $a_j(n)$ of Φ_n for $j \le k$ for all squarefree n. We can write $n = d \cdot M$ where if $p|M$ then $p > k$.

Therefore the sequence of coefficients $a_j(n)$ of Φ_n with $j \le k$ is equal to the sequence b_j determined by d and $\mu(n)$. The greatest prime divisor of d is less than $k+1$. Hence, if r is the number of primes less than or equal to k then we need only consider 2^{r+1} choices of d and $\mu(n)$. From (2.8), (2.7) and (2.6) we get the following formulae for $a_j(n)$ of Φ_n:

$$\begin{aligned} a_0(n) &= 1, \\ a_j(n) &= -\frac{\mu(n)}{j}\sum_{m=0}^{j-1} a_m(n)\cdot\mu\Big((n,j-m)\Big)\cdot\phi\Big((n,j-m)\Big), \quad j > 0. \end{aligned} \tag{2.9}$$

For the computation by (2.5) we can rewrite (2.9) in the form

$$a_0(n) = 1, \qquad a_j(n) = -\frac{1}{j}\sum_{m=0}^{j-1} a_m(n)T_{j-m}, \quad j = 1,2,\ldots,k. \tag{2.10}$$

We can easily obtain the values T_i by using the following algorithm:

Algorithm 2.1.

(1) *put* $j = 1$ *and* $T_i = \mu(n)$ *for* $i = 1,\ldots,k$.

(2) *put* $q = p_j$

(3) *multiply* T_{qi} *by* $1-q$ *for* $i \le [k/q]$.

(4) *add* 1 *to* j.

(5) *if* $j < s+1$ *then go to* 2.

Now we can apply (2.10).

In our computations we restricted $k \leq 52$ and we present the results in the table below.

Table 2.1

a_k	k_{min}	d	μ
2	7	$2 \cdot 3 \cdot 5 \cdot 7$	1
−2	7	$3 \cdot 5 \cdot 7$	−1
3	17	$2 \cdot 7 \cdot 11 \cdot 13 \cdot 17$	1
−3	17	$7 \cdot 11 \cdot 13 \cdot 17$	−1
4	23	$2 \cdot 11 \cdot 13 \cdot 17 \cdot 19 \cdot 23$	1
−4	23	$11 \cdot 13 \cdot 17 \cdot 19 \cdot 23$	−1
5	30	$5 \cdot 7 \cdot 11 \cdot 13 \cdot 17 \cdot 19 \cdot 23$	−1
−5	35	$2 \cdot 3 \cdot 5 \cdot 7 \cdot 11 \cdot 13 \cdot 29$	1
6	36	$7 \cdot 11 \cdot 13 \cdot 17 \cdot 19 \cdot 23$	−1
−6	39	$2 \cdot 3 \cdot 5 \cdot 7 \cdot 11 \cdot 17 \cdot 23 \cdot 29$	−1
7	43	$2 \cdot 5 \cdot 7 \cdot 11 \cdot 13 \cdot 17 \cdot 19 \cdot 41 \cdot 43$	1
−7	43	$5 \cdot 7 \cdot 11 \cdot 13 \cdot 17 \cdot 19 \cdot 41 \cdot 43$	−1
8	46	$2 \cdot 3 \cdot 5 \cdot 7 \cdot 13 \cdot 17 \cdot 23 \cdot 29 \cdot 41 \cdot 43$	−1
−8	47	$5 \cdot 7 \cdot 11 \cdot 13 \cdot 17 \cdot 19 \cdot 23 \cdot 43$	−1
9	47	$2 \cdot 5 \cdot 7 \cdot 11 \cdot 13 \cdot 17 \cdot 19 \cdot 23 \cdot 43 \cdot 47$	1
−9	47	$5 \cdot 7 \cdot 11 \cdot 13 \cdot 17 \cdot 19 \cdot 23 \cdot 43 \cdot 47$	−1
10	52	$3 \cdot 5 \cdot 11 \cdot 13 \cdot 23 \cdot 29 \cdot 41 \cdot 43 \cdot 47$	−1

In this table k_{min} is a value such that for all n and for all j less than k the value of $a_j(n)$ does not take the value of a_k. The symbol d represents the value of the smallest integer n such that $a_k(n) = a_k$ from the table when $\mu(d) = \mu$. In the case of $\mu(d) = -\mu$, then we get the smallest value for n by multiplying d by the smallest $p > k$.

For further informations on the coefficients of the cyclotomic polynomials we refer the reader to [1], [2], [3], [4], [7], [9], [10], [12] and [13].

References

[1] A.S. Bang, *Om ligningen* $\Phi_n(x) = 0$, Nyt. Tidsskrift for Math. (1985), 6-12.

[2] P.T. Bateman, *Note on the coefficients of cyclotomic polynomial*, Bull. Amer. Math. Soc. **55** (1949), 1180-1181.

[3] P.T. Bateman, C. Pomerance and R.C. Vaughan, *On the coefficients of cyclotomic polynomial*, Coll. Math. Soc. J.Bolyai. 34.Topics in classical number theory, (1981), 171-202, Budapest.

[4] D.M. Bloom, *On the coefficients of the cyclotomic polynomials*, Amer. Math. Monthly **75** (1968), 372-377.

[5] M. Endo, *On the coefficients of the cyclotomic polynomials*, Comment. Math. Univ. St.Pauli **23** (1974/ 75), 121-126.

[6] P. Erdös, *On the coefficients of the cyclotomic polynomials*, Bull. Amer. Math. Soc. **52** (1946), 179-181.

[7] P. Erdös and R.C. Vaughan, *Bounds for n-th coefficients of cyclotomic polynomial*, J.London Math.Soc. (1974), 393-400.

[8] D.H. Lehmer, *Some properties of cyclotomic polynomial*, J.Math.Anal.Appl. **15** (1966), 105-117.

[9] H.W. Lenstra,Jr, *Vanishing sums of roots of unity*, "Proc.Bicentennial Congress Wiskundig Genootshapp." Part II, Math.Centre Tracts, 101, Math.Centrum, Amsterdam, 1979, pp. 249-268.

[10] A. Migotti, *Zur Theorie der Kreisteilungsgleichung*, Sitz. Akad. Wiss. Wien **87** (1883), 7-14.

[11] H. Möller, *Über die i-ten Koeffizienten der Kreisteilungspolynome*, Math.Ann. **188** (1970), 26-38.

[12] H.L. Montgomery and R.C. Vaughan, *The order of the m-th coefficients of cyclotomic polynomials*, Glasgow Math.J. **27** (1985), 143-159.

[13] W. Narkiewicz, "Elementary and analytic theory of algebraic numbers." second edition, PWN, Warszawa, 1990.

Number systems

B. Kovács *

1 Introduction

Let R be an integral domain (with unit element), $\alpha \in R$, $\mathbb{Z}$ the set of rational integers and $N = \{0, k_1, ..., k_n\} \subset \mathbb{Z}$. $\{\alpha, N\}$ is called a number system in R if every $\gamma \in R$ can be written uniquely in the form

$$\gamma = a_0 + a_1\alpha + ... + a_k\alpha^k, \ a_i \in N \ (0 \le i \le k), \ a_k \ne 0 \text{ if } k \ne 0. \tag{1.1}$$

If $N = \{0, 1, \cdots, n\}$, $\{\alpha, N\}$ is called a canonical number system in R.

This is a natural generalization of the negative base number systems in $\mathbb{Z}$, which have been discussed by several authors. In connection with the above defined notion of a canonical number system, I should point out that there are a number of special results concerning particular rings. (See, for example, [1], [2], [3], [4], [5], [6], [7].)

In this paper I will present a brief summary of the most important results on number systems which we have discovered during the last two years and which, I hope, will appear in print in the near future.

2 Results

I should first mention the following two theorems.

Theorem 2.1. *Let R be an integral domain of characteristic 0. There exists in R a canonical number system if and only if*

$$R \cong \mathbb{Z}[\alpha]$$

for some element α which is algebraic over $\mathbb{Q}$ (the field of rationals).

B. Kovács [8] proved:

*Research supported in part by Grants 273 and 400 from the Hungarian National Foundation for Scientific Research.

Theorem 2.2. *There exists a number system in an integral domain R if and only if*

(i) $R \cong \mathbb{Z}[\alpha]$ *for an* α*, algebraic over* $\mathbb{Q}$ *, if char* $R = 0$,

(ii) $R \cong \mathbb{F}_p[x]$*, where* x *is transcendental over* $\mathbb{F}_p$*, if char* $R = p$*, where* p *is a prime.*

In [9] Pethö and Kovács gave necessary and sufficient conditions for $\{\alpha, N\}$ to be a canonical number system in an order $\mathcal{O}$ and, as well, computed the representatives of all classes of bases of canonical number systems in rings of integers of some totally real cubic fields.

These results led us in a natural way to the following question:

Let $\{\alpha, N\}$ be a canonical number system in a ring $\mathbb{Z}[\beta]$, where β is an algebraic integer over $\mathbb{Q}$. Can every complex or real number γ be written in the

$$\gamma = a_k\alpha^k + \cdots + a_1\alpha + a_0 + a_{-1}\alpha^{-1} + \cdots,\ a_i \in N,\ i = k, k-1, \cdots, \tag{2.1}$$

(α is a non-real or real number, respectively).

First we study the case of α real. In order to get a complete answer to our question, we first of all need the following generalization of interval - filling sequences.

Let M be a positive integer, let Λ_M denote the set of all real sequences (λ_n) for which $|\lambda_n| > |\lambda_{n+1}| > 0$ for all $n \in \mathbb{N}$ (the set of natural numbers), and

$$\sum_{n=1}^{\infty} |\lambda_n| < \infty.$$

Define the real numbers L^-, L^+ and L by

$$L^- = \sum_{n=1}^{\infty} \frac{|\lambda_n| - \lambda_n}{2},\ L^+ = \sum_{n=1}^{\infty} \frac{|\lambda_n| + \lambda_n}{2} \text{ and } L = L^- + L^+$$

A sequence $(\lambda_n) \in \Lambda_M$ is called an interval-filling sequence of order M if for all $x \in [-ML^-, ML^+]$ there exists a sequence $(a_n) : \mathbb{N} \to \{0, 1, \cdots, M\}$ such that

$$x = \sum_{n=1}^{\infty} a_n\lambda_n$$

If $\lambda_n > 0$ for all $n \in \mathbb{N}$ and M $= 1$, we have the concept of the interval-filling sequence introduced by Z. Daróczy, I. Kátai and A. Járai ([10]).

Gy Maksa and I proved the following characterization theorem.

Theorem 2.3. *Suppose that* $(\lambda_n) \in \Lambda_M$*. Then* (λ_n) *is an interval-filling sequence of order M if and only if*

$$|\lambda_n| \le M \sum_{k=n+1}^{\infty} |\lambda_k| \text{ for all } n \in \mathbb{N}. \tag{2.2}$$

This was proved for the particular case mentioned above in [10], [11], [12]. Also, Zoltán Boros (unpublished) proved this for the special case of $M = 1$. Using this theorem we were able to prove the following theorem.

Theorem 2.4. *Let β be a real algebraic element over $\mathbb{Q}$ and let $\{\alpha, N\}$ be a canonical number system in $\mathbb{Z}[\beta]$. Then every real number γ has a representation in the form (2.1).*

(Gy. Maksa and B. Kovács, [13]).

Theorem 2.4 is a real analogue of a theorem of Daróczy and Kátai [14], furthermore, it includes the similar results of [1], [4] and [5].

Unfortunately the case of non-real α appears to be more difficult than the case of real α.

First we give a necessary and sufficient condition for a number system $\{\alpha, N\}$ such that any complex number γ can be written in the form (2.1). If $\{\alpha, N\}$ is a number system in $\mathbb{Z}[\beta]$ and

$$\gamma = a_0 + a_1\alpha + \cdots + a_k\alpha^k,\ a_i \in N\ \ (0 \le i \le k),\ a_k \neq 0 \text{ if } k \neq 0$$

then the exponent k is denoted by $L(\gamma, \alpha)$. With this notation we have:

Theorem 2.5. *Let $\{\alpha, N\}$ be a number system in $\mathbb{Z}[\beta]$, (β is an algebraic integer over $\mathbb{Q}$). A real or complex number γ can be written in the form (2.1) - according to α being real or non-real - if and only if there exist sequences with the following properties:*

(i) $\gamma \cdot \alpha^k = \gamma(k) + \delta(k)$ *for every positive integer k,*

(ii) $\gamma(k) \in \mathbb{Z}[\beta]$ *and* $L(\gamma(k), \alpha) \le k + c_1$ *where c_1 is an appropriate constant which does not depend on k,*

(iii) $\delta(k)/a^k \to 0$ *if $k \to \infty$.*

Using Theorem 2.5 we proved:

Theorem 2.6. *$\{\alpha, N\}$ be a number system in $\mathbb{Z}[\beta]$, where β is an algebraic integer of degree $n \ge 1$ over $\mathbb{Q}$ and let us suppose that $|\alpha| \le |\alpha^{(i)}|$ for every conjugate of α over $\mathbb{Q}$. Then every complex number γ has a representation in the form (2.1) if α is non-real and every real number γ has a representation in the form (2.1) if α is real.*

From this theorem one can deduce the analogous theorems in [1], [4] and [6]; moreover, in our case this theorem is stronger than the result of Daróczy and Kátai [14].

In order to prove Theorems 2.5 and 2.6, I had to apply the following theorem of Pethö and Kovács [15].

Theorem 2.7. *Let β be an algebraic integer over $\mathbb{Q}$ of degree $n \ge 1$ and let $\{\alpha, N\}$ be a number system in $\mathbb{Z}[\beta]$. Then there exist effective computable constants $C_1(\alpha, N), C_2(\alpha, N)$, depending only on α and N such that*

$$\max_{1\le i\le n} \frac{\log|\gamma^{(i)}|}{\log|\alpha^{(i)}|} + C_1(\alpha, N) \le L(\gamma, \alpha) \le \max_{1\le i\le n} \frac{\log|\gamma^{(i)}|}{\log|\alpha^{(i)}|} + C_2(\alpha, N)$$

holds for every $0 \neq \gamma \in \mathbb{Z}[\beta]$, where $\alpha^{(i)}$ and $\gamma^{(i)}$ denote the i-th conjugates of α and γ, respectively.

I remark that this theorem is a generalization of a theorem of Grossman [4] and it follows from our theorem that his upper bound is not sharp.

Finally, with the aid of Theorems 2.5 and 2.6 we can deduce:

Theorem 2.8. *Let α be a non-real algebraic integer of degree 3 (over $\mathbb{Q}$). If $\{\alpha, N\}$ is a canonical number system in a $\mathbb{Z}[\beta]$ then every complex number γ has a representation in the form (2.1).*

One can find the last four theorems in [15] and [16].

References

[1] I. Kátai and J. Szabó, *Canonical number systems for complex integers*, Acta Sci. Math. Szeged **37** (1975), 255-260.

[2] W.J. Gilbert, *Geometry of radix representation, The geometrie vein*, Springer, New York - Berlin, 1981, 129-139.

[3] B. Kovács and A. Pethö, *Canonical systems in the ring of integers*, Publ. Math. Debrecen **30** (1983), 39-45.

[4] E.H. Grossman, *Number bases in quadratic fields*, Studia Sci. Math. Hungar. **20** (1985), 55-58.

[5] I. Kátai and B. Kovács, *Kanonische Zahlensysteme in der Theorie der quadratischen Zahlen*, Acta Sci. Math. Szeged **42** (1980), 99-107.

[6] I. Kátai and B. Kovács, *Canonical number systems in imaginary quadratic fields*, Acta Math. Hungar. **37** (1981), 159-164.

[7] B. Kovács, *Canonical number systems in algebraic number fields*, Acta Math. Acad. Sci. Hungar. **97** (4) (1981), 405-407.

[8] B. Kovács, *Integral domains with canonical number systems*, to appear (Publ. Math. Debrecen).

[9] B. Kovács and A. Pethö, *Number systems in integral domains, especially in orders of algebraic number fields*, to appear (Acta Sci. Math. Szeged).

[10] Z. Daróczy, A. Járai and I. Kátai, *Intervallfüllende Folgen und volladditive Funktionen*, Acta Sci. Math. Szeged **50** (1986), 337-350.

[11] S. Kakeya, *On the partial sums of an infinite series*, Science Reports Tohoku, Imp. Univ. (1) **3** (1914), 159-163.

[12] P. Ribenboim, *Representation of real numbers by means of Fibonacci numbers*, L'Enseignement Math. **31** (1985), 249-255.

[13] B. Kovács and Gy. Maksa, *Interval-filling sequences of order N and a representation of real numbers in canonical number systems*, to appear (Publ. Math. Debrecen).

[14] Z. Daróczy and I Kátai, *Generalized number systems in the complex plane*, Acta Math. Hungar. **51** (1988), 409-416.

[15] B. Kovács and A. Pethö, *On a representation of algebraic integers*, to appear (Studia Sci. Math. Hungar.).

[16] B. Kovács, *Representation of complex numbers in number systems*, to appear (Acta Math. Hungar.).

Fast converging series representations of real numbers and their implementations in digital processing

Yu. V. Melnichuk

Abstract. The algorithms of Engel, Ostrogradsky and the new one, based on them, for representation of real numbers by the fast converging series are discussed as well as applications of those algorithms to measuring procedures and digital processing.

Let x be a real number and $x \in [0,1]$. Known series expansions due to Engel [1] and Ostrogradsky [2]

$$x = \sum_{i=0}^{\infty} ((g_0+1)(g_1+1)\cdots(g_i+1))^{-1} \tag{1}$$

$$x = \sum_{i=0}^{\infty} (-1)^i (p_0 p_1 \cdots p_i)^{-1} \tag{2}$$

are obtained with the help of the iteration of the into mappings T and S of the unit interval:

$$Tx = x([x^{-1}]+1)-1,\ x_0 = x,\ x_n = Tx_{n-1},\ g_{n-1} = \left[x_{n-1}^{-1}\right],\ g_n \geq g_{n-1} \geq 1 \tag{3}$$

$$Sx = 1-[x^{-1}], \quad y_0 = x, \quad y_n = Sy_{n-1}, \quad p_{n-1} = \left[y_{n-1}^{-1}\right], \quad p_n > p_{n-1} \tag{4}$$

where $n \geq 1$ and [a] denote the integral part of the number a. The maximal possible absolute error of the first n terms approximation of the series (1) and (2) does not exceed 2^{-n} and $((n-1)!)^{-1}$ respectively.

Let us consider the set $A = \{x \in [0,1] : Sx > Tx\}$ and let $\chi(A)$ be its characteristic function. We define the into mapping of the unit interval

$$Rx = \begin{cases} Tx, & x \in A \\ Sx, & x \in [0,1]\backslash A. \end{cases}$$

We define the so called combined series expansion of x as follows

$$z_0 = x, \quad z_n = RZ_{n-1}, \quad t_{n-1} = \left[z_{n-1}^{-1}\right], \quad t_n \geq 2t_{n-1}+1. \tag{5}$$

Computational Number Theory

Defining $k_n = \chi(z_n)$, we get

$$z_{n-1} = (t_{n-1} + k_{n-1})^{-1} + (-1)^{1+k_{n-1}} z_n (t_{n-1} + k_{n-1})^{-1}$$

and after iteration

$$x = \sum_{i=0}^{\infty} (-1)^{i+k_0+k_1+\cdots k_{i-1}} ((t_0 + k_0)(t_1 + k_1) \cdots (t_i + k_i))^{-1}. \tag{6}$$

It is quite obvious that for every rational number its combined series expansion consists of finite number of terms. The maximal possible absolute error of the approximation of x by the first n terms of the series (6) does not exceed $2^{-(n^2+n)2^{-1}+1}$. This maximal error is achieved on the small subsets of the unit interval and for the suppressing subset of numbers is efficiently smaller.

Presentation of the result of measuring procedure as the sum of the first terms of the binary or decimal power series, or of the series of Engel, Ostrogradsky, combined type has its meaningful physical realization as the way of comparing unknown analog value with the unit of measurement. Each such presentation dictates the algorithm of measuring procedure, and further, the hardware of measuring apparatus.

As series of combined type are converging faster than others mentioned above, so the same precision is reached by less number of terms, that is in less steps of iterations.

Implementation of the representations of real numbers by series (3), (4) and (6) to measuring procedures raised the new problem, which was not common in previous investigations. That means that we need the estimation of the error of representation depending not on number of terms but connected with the bound on the elements g_i, p_i or t_i.

Let Q be any fixed integer. As t_i form the growing sequence $t_i \geq 2t_{i-1} + 1$, so we are calculating the terms in (5) and (6) until $t_i \leq Q$. For different values of x the number of steps then certainly differs. The magnitude of Q has the natural interpretation: in the measuring process all values less than Q^{-1} are taken as 0.

For the needs of applications the numerical testing of the algorihms (1), (2) and (5) was made. For the arbitrary specified measuring range, included in $[0, 1]$, and arbitrary Q the program for the computer was developed for determining the maximal possible error of approximation for numbers from that range, that is for evaluating algorithmic error of analog-digital transformation. This program also makes possible to choose the range, where the error is efficiently smaller than outside.

The described algorithms may find their application in those types of digital processing, where the arithmetical operations with values x presented by the first terms of those series are practically not needed. We mean: condensing and restoring information; casting away comparatively small terms in representation; casting away the values in the sequence of data not exceeding certain threshold, or falling outside limits; data ordering; classification or dividing into clusters.

References

[1] J. Galambos, *Representations of real numbers by infinite series*, Springer-Verlag, Berlin, Heidelberg, New York, 1976.

[2] E. Remez, *About sign alternating series, which can be connected with two algorithms of Ostrogradsky for approximation of irrational numbers*, (Russian), Uspekhy Math. Nauk **6** (1954), pp 33-42.

On a polynomial transformation and its application to the construction of a public key cryptosystem

A. Pethö *

Abstract. Generalizing results of Kovács and Pethö [5] we study when a certain map is injective (or bijective). Based on this map we propose a to the Merkle-Hellman knapsack scheme related cryptosystem.

1 Introduction

Let $g(t) = t^n + g_{n-1}t^{n-1} + \ldots + g_0 \in \mathbb{Z}[t]$ and $\mathcal{N}$ a finite subset of $\mathbb{Z}$. We denote by $\mathbb{Z}$ as usual the ring of integers. Let us define the sequence $\underline{x}_i \in \mathbb{Z}^n$, $i = 0, 1, \ldots$ by

$$\underline{x}_0 = (1, 0, \ldots, 0), \qquad \underline{x}_{i+1} = \underline{x}_i G \quad \text{if} \quad i \geq 0, \tag{1.1}$$

where the nxn matrix G is given by

$$\begin{pmatrix} 0 & 1 & 0 & \cdots & 0 & 0 \\ \vdots & \vdots & \vdots & \ddots & \vdots & \vdots \\ 0 & 0 & 0 & \cdots & 0 & 1 \\ -g_0 & -g_1 & -g_2 & \cdots & \cdots & -g_{n-1} \end{pmatrix}. \tag{1.2}$$

Let $\mathcal{A}_{\mathcal{N}}$ be the set of all finite words $w = w_0 \ldots w_l$ over $\mathcal{N}$ with the convention that if $0 \in \mathcal{N}$ then either $w_l \neq 0$ or $l = 0$ and $w_0 = 0$. We call $l(w) = l + 1$ the length of w.

The pair $\{g(t), \mathcal{N}\}$ will be called *(weak) number system (WNS) NS* if the map $T : \mathcal{A}_{\mathcal{N}} \to \mathbb{Z}^n$, which is defined by

$$T(w_0, \ldots w_l) = w_0\underline{x_0} + \ldots + w_l\underline{x_l} \tag{1.3}$$

is (injective) bijective.

In the special case $g(t) = t + g_0$ with $g_0 \geq 2$ and $\mathcal{N} = \{0, \ldots, g_0 - 1\}$ (1.3) defines the g_0-adic radix representation of the integers. You find results about the

---

*Research supported in part by Hungarian National Foundation for Scientific Research Grant No. 273/86

radix representation of integers with extensive literature in Knuth [2]. Kovács and Pethö [4] proved that there exist infinitely many finite subsets $\mathcal{N}$ of $\mathbb{Z}$ such that $\{t + g_0, \mathcal{N}\}$ is a number system in $\mathbb{Z}$. They proved moreover in [5], that if $g(t)$ is irreducible then it is decidable whether the pair $\{g(t), \mathcal{N}\}$ is a number system in the ring $\mathbb{Z}[t]/g(t)\mathbb{Z}[t]$. We generalize their results for arbitrary polynomials in sections 6. and 7.

If T is injective then we can use it in a private key cryptosystem. Such a system is described in section 3.

Merkle and Hellman proposed in [6] a public key cryptographic scheme based on the knapsack problem. Starting from a superincreasing sequence $m_0, m_1, \ldots, m_n$ of integers, they chosed suitable integers w and k with $(w, k) = 1$ and took

$$\overline{m}_i \equiv w m_i \pmod{k} \quad i = 0, \ldots, n.$$

The numbers $\overline{m}_0, \ldots, \overline{m}_n$ were proposed for apply as public keys. Shamir [7] proved that knowing the public keys it is possible to compute under natural conditions in polynomial time w, k and $m_0, \ldots, m_n$. Cryptosystems based on the knapsack problem are extensively studied in Horster [1].

We propose in section 4 a to the Merkle-Hellman cryptosystem related public key cryptographic scheme based on weak number systems. For a WNS we choose a suitable integer M and an nxn invertible matrix C in $\mathbb{Z}_M$ which do not commute with G. For public keys we use $\underline{x}_n C, \ldots, \underline{x}_{n+l} C \pmod{M}$ and $\mathcal{N}$, however C^{-1}, $g(t)$ and M remain secret.

2 Preliminary results about the map T

Let $\mathcal{N}[t]$ denote the set of polynomials in t with coefficients from $\mathcal{N}$ and let $T_1 := \mathcal{A}_{\mathcal{N}} \to \mathcal{N}[t]$, defined by

$$T_1(w) = w_0 + w_1 t + \ldots + w_l t^l$$

for any $w_0 \ \ldots \ w_l = w \in \mathcal{A}_{\mathcal{N}}$. Let T_2 be the natural ring homomorphism $T_2 : \mathbb{Z}[t] \to \mathbb{Z}[t]/g(t)\mathbb{Z}[t] = R_g$. By the definition of $g(t)$ it is clear that R_g is a commutative ring which is as a $\mathbb{Z}$-module isomorphic to $\mathbb{Z}^n$. Putting $\phi = T_2(t)$ we may write the elements of R_g uniquely in the form

$$\sum_{i=0}^{n-1} y_i \phi^i \text{ , with } y_i \in \mathbb{Z} \text{ for } i = 0, \ldots, n-1.$$

Let T_3 be the $\mathbb{Z}$-module isomorphism $R_g \to \mathbb{Z}^n$.

Proposition 2.1. *We have with the above notations*

$$T = T_3 \circ T_2 \circ T_1. \tag{2.1}$$

Furthermore, T is bijective iff the restriction of T_2 to $\mathcal{N}[t]$ is bijective.

Proof. We have $g(\phi) = 0$ in R_g, hence

$$\phi^n = (-g_0, -g_1, \ldots, -g_{n-1})\Phi^T,$$

where Φ^T denotes the transposition of the vector $\Phi = (1, \phi, \ldots, \phi^{n-1})$. Using this identity and (1.1) it is easy to prove

$$\phi^i = \underline{x}_i \Phi^T \ , \qquad i = 0, 1, \ldots .$$

Therefore, if

$$T(w_0 \ldots w_l) = (y_0, \ldots, y_{n-1}),$$

then

$$\begin{aligned} T(w_0 \ldots w_l)\Phi^T &= \sum_{i=0}^{l} w_i \underline{x_i} \Phi^T = \sum_{i=0}^{l} w_i \phi^i \\ &= (y_0, \ldots, y_{n-1})\Phi^T = T_3^{-1}(y_0, \ldots, y_{n-1}). \end{aligned}$$

We know that T_3 is a $\mathbb{Z}$-module isomorphism, thus (2.1) is proved. The second assertion follows easily from (1.3). □

Proposition 2.2. *Let $|g_0| > 1$ and assume that the elements of $\mathcal{N}$ are pairwise incongruent modulo g_0. Then $\{g(t), \mathcal{N}\}$ is a WNS.*

Proof. We need to prove that T is injective. By Proposition 2.1 it is enough to show that the restriction of T_2 to $\mathcal{N}[t]$ is injective. Assume contrary that there exist polynomials $p_i(t) = p_{i0} + \ldots + p_{im_i} t^{m_i} \in \mathcal{N}[t]$, $i = 1, 2$ such that

$$T_2(p_1) = T_2(p_2), \tag{2.2}$$

and p_1 has the smallest degree among such polynomials. There exists by (2.2) a polynomial $h(t) \in \mathbb{Z}[t]$ with

$$p_1(t) = p_2(t) + h(t)g(t). \tag{2.3}$$

Comparing the constant terms in (2.3) we get

$$p_{10} \equiv p_{20} \pmod{g_0}.$$

By the assumption on $\mathcal{N}$ this is possible only if $p_{10} = p_{20}$, thus $h(t) = th_1(t)$. Hence

$$T_2(p_{11} + \ldots + p_{1m_1} t^{m_1 - 1}) = T_2(p_{21} + \ldots + p_{2m_2} t^{m_2 - 1}),$$

which contradicts the choice of $p_1(t)$. □

3 Private key cryptosystem

Let $g(t) = t^n + g_{n-1}t^{n-1} + \ldots + g_0 \in \mathbb{Z}[t]$ with $|g_0| > 1$ and assume that the elements of $\mathcal{N}$ are pairwise incongruent modulo g_0, then it is possible to define a private key cryptosystem based on the WNS $\{g(t), \mathcal{N}\}$.

Assume that the individual A will send a message $w = w_0 \dots w_l \in \mathcal{A}_{\mathcal{N}}$ to the individual B. He chooses integers $u \leq v$ and random integers $l_1, l_2, \dots$ from the interval $[u, v]$. He cuts then w into consecutive blocks $W_1, W_2, \dots$ of length $l_1, l_2, \dots$ such that the last letters of the blocks are not 0, and send the encrypted message $T(W_1), T(W_2), \dots$ to the individual B. Knowing the polynomial $g(t)$ and the set $\mathcal{N}$, B is able to decrypt the original message w from $T(W_1), T(W_2), \dots$. We describe the method formally in the following two algorithms.

Algorithm 3.1.

Input: $w_0 \dots w_l \in \mathcal{A}_{\mathcal{N}}$; $u, v \in \mathbb{Z}$ *with* $0 \leq u \leq v$.

Output: $\underline{y}_0 \dots \underline{y}_h \in \mathcal{A}_{\mathbb{Z}^n}$.

(1) *Set* $W = w_0 \dots w_l$, $h = -1$.

(2) *Repeat until* $l < 0$

Set $h = h + 1$ *and* $i = random[u, v]$. *If* $l < i$ *then set* $i = l$.

While $w_i = 0$ *and* $i > 0$ *do* $i = i - 1$.

Set $\underline{y}_h = T(w_0 \dots w_i)$; $W = w_{i+1} \dots w_l$ *and* $l = l - i - 1$.

The individual B can decrypt the message $\underline{y}_0 \dots \underline{y}_h \in \mathcal{A}_{\mathbb{Z}^n}$ using the following algorithm.

Algorithm 3.2.

Input: $\underline{y}_0 \dots \underline{y}_h \in \mathcal{A}_{\mathbb{Z}^n}$ *with* $\underline{y}_i \in T(\mathcal{A}_{\mathcal{N}})$.

Output: $W_0 \dots W_h = w_0 \dots w_l \in \mathcal{A}_{\mathcal{N}}$.

(1) *Set* $s = 0$.

(2) *Repeat until* $s > h$

(3) *Put* $l(s) = n - 1$, $p(t) = \underline{y}_s(1, t, \dots, t^{n-1})^T = z_0 + \dots + z_{l(s)} t^{l(s)}$; $j = 0$.

(4) *Repeat until* $j > l(s)$

If $z_j \notin \mathcal{N}$ *then determine* $x_j \in \mathcal{N}$ *such that* $z_j \equiv x_j \pmod{g_0}$

and set $p(t) = p(t) - g(t)t^j(z_j - x_j)/g_0$, *set* $l(s) = n + j$,

if $z_i = 0$ *for* $j + 1 \leq i \leq l(s)$ *then set* $l = j$, $j = l(s)$.

Set $j = j + 1$.

(5) *Set* $j = l$.

(6) *While* $z_j = 0$ *and* $j > 0$ *do* $j = j - 1$.

(7) *Set* $W_s = z_0 \dots z_j$, $s = s + 1$.

Theorem 3.1. *The Algorithms 3.1 and 3.2 terminate successfully after* $c_1 nl$ *operations on integers, where* c_1 *is a constant depending only on the size of* $\mathcal{N}$.

Proof. The most time consuming operation in Algorithm 3.1 is the computation of y_j in the repeat loop. We have

$$\underline{y}_j = w_0 \underline{x}_0 + \dots + w_i \underline{x}_i$$

by the definition of T, where the vectors $\underline{x}_i$ are defined by (1.1). Hence $\underline{y}_j$ can be computed by $2n(i+1)$ arithmetical operations. Summing up these numbers we get that Algorithm 3.1 terminates after $O(nl)$ operations.

In the inner repeat loop of Algorithm 3.2 the inverse of the map T_2 is computed. The vectors $\underline{y}_s$ have by the assumption preimages, and by Proposition 2.2 they are unique. Hence Algorithm 3.2 terminates also. If the length of W_s is $j+1$, then the inner repeat loop must be performed at most $j+2$ times, and the computation of $p(t)$ requires each times at most $g_0 + n + 2$ operations. This proves the upper bound for the number of operations needed in Algorithm 3.2. □

Remark 1. For a binary message, i.e. for a message consisting of 0's and 1's, it is most convenient to choose g_0 a power of 2, say $g_0 = 2^m$, and $\mathcal{N}$ a complete residue system modulo g_0. In this case we split the message into consecutive blocks of length m, change each block with that element of $\mathcal{N}$, which it is congruent modulo g_0 and then apply Algorithm 3.1 for the encryption. In order to decrypt the message one must apply naturally the inverse change.

Remark 2. Theorem 3.2 gives information about the time complexity of the encryption/decryption algorithm, but we have said nothing about their space complexity. If $g(t)$ is squarefree and its roots are of absolute value larger then 1, we examine the space complexity of Algorithms 3.1 and 3.2 in Theorem 6.3.

Remark 3. We have freedom in Algorithm 3.1 for the choice of the integers u and v. It is convenient to choose $u > n$ because otherwise we get the same words after the transformation as in the plaintext.

4 The public key cryptosystem

We shall now modify the cryptosystem described in the previous section so that the parameters required for the encryption are public, but for the efficient decryption one needs further secret parameters. The idea behind our cryptosystem is similar to that used by Merkle and Hellman [6].

Let $\{g(t),\mathcal{N}\}$ be a WNS. If for a $w = w_0 \dots w_l \in \mathcal{A}_{\mathcal{N}}$ we have $T(w) = (y_0, \dots, y_{n-1}) \in \mathbb{Z}^n$, then $m(w) = \max\{|y_0|, \dots, |y_{n-1}|, 1\}$ will be called the height of w.

Let $\mathcal{A}_{\mathcal{N}}^L$ denote the set of all words from $\mathcal{A}_{\mathcal{N}}$ with length at most $L+1$, and let M be an integer with

$$M > 2\max\left\{m(w) \ : \ w \in \mathcal{A}_{\mathcal{N}}^{L+n}\right\}. \tag{4.1}$$

Denote $\mathbb{Z}_M$ the residue class ring modulo M, and $\mathbb{Z}_M(n,m)$ the ring of nxm matrices over $\mathbb{Z}_M$.

Let $C \in \mathbb{Z}_M(n,n)$ be invertible which does not commute with G, i.e. $GC \neq CG$ in $\mathbb{Z}_M(n,n)$. Let

$$\hat{\underline{x}}_i \equiv \underline{x}_{n+i}C \pmod M \quad \text{for } i = 0, \dots, L, \tag{4.2}$$

and let the map $\hat{T} : \mathcal{A}_{\mathcal{N}}^L \to \mathbb{Z}^n$ be defined by

$$\hat{T}(w_0 \dots w_l) = w_0\hat{\underline{x}}_0 + \dots + w_l\hat{\underline{x}}_l \quad (l \leq L). \tag{4.3}$$

The individual B, who creates the public key system publishes $\mathcal{N}$, $\hat{\underline{x}}_0, \dots, \hat{\underline{x}}_L$ but not $g(t)$, C, and M. The individual A can send messages to B using Algorithm 3.1 with $u = 0$, $v = L$ and computing $\hat{T}(w_0 \dots w_i)$ instead of $T(w_0 \dots w_i)$. Knowing C and M, the individual B is able to compute C^{-1} and so decrypt the message. We have namely by (4.3) and (4.2)

$$\hat{T}(w_0 \dots w_l) = w_0\hat{\underline{x}}_0 + \dots + w_l\hat{\underline{x}}_l \equiv (w_0\underline{x}_n + \dots + w_l\underline{x}_{n+l})C \pmod M,$$

hence

$$\hat{T}(w_0 \dots w_l)C^{-1} = (\hat{y}_0, \dots, \hat{y}_{n-1})C^{-1} \equiv w_0\underline{x}_n + \dots + w_l\underline{x}_{n+l} \pmod M.$$

Therefore

$$(y_0, \dots, y_{n-1}) = T(O^n w_0 \dots w_l) \equiv \hat{T}(w_0 \dots w_l)C^{-1} \pmod M. \tag{4.4}$$

We have $|y_i| < M/2$ by (4.1), hence (4.4) determines the integers $y_0, \dots, y_{n-1}$ uniquely performing the computations in the absolute smallest residue system modulo M. Knowing $(y_0, \dots, y_{n-1})$ the individual B is able to compute $w_0 \dots w_l$ applying Algorithm 3.2 to $(y_0, \dots, y_{n-1})$ and dropping the first n invaluable letters from the result.

5 Remarks about the parameters

We have in the public key cryptosystem defined in the previous section two important parameters: the polynomial $g(t)$ and the matrix C.

It is convenient to choose $L < n^2 + n - 1$ because otherwise taking $g_1, \dots, g_{n-1}$ and the entries of C as unknowns we would get at least as many equations by (4.2) and (1.1) as we have unknowns.

There exist in $\mathbb{Z}_M(n,n)$ at least $M^{n-1}\phi(M)$ invertible elements. It is easy to prove that if M is prime then the exact number of invertible elements in $\mathbb{Z}_M(n,n)$ is $(M^n-1)*\ldots*(M^n-M^{n-1})$.

The most important assumption in our cryptosystem is

$$GC \neq CG. \tag{5.1}$$

Suppose namely that (5.1) is not satisfied, i.e. $GC = CG$. Then we get by (4.2) and (1.1)

$$\hat{\underline{x}}_i = \underline{x}_{n+i}C = \underline{x}_{n+i-1}GC = \underline{x}_{n+i-1}CG = \hat{\underline{x}}_{i-1}G,$$

from which G is easy to compute. Knowing G it is possible to compute C from equation (4.2).

Hence it is important to know how many matrices from $\mathbb{Z}_M(n,n)$ satisfy (5.1). The next proposition shows that at least half of the invertible elements of $\mathbb{Z}_M(n,n)$ satisfy (5.1).

Proposition 5.1. *Let* $\mathbb{Z}_M^*(n,n)$ *be the group of the invertible elements of* $\mathbb{Z}_M(n,n)$ *and* G *the matrix defined by (1.2). Denote* $C(G)$ *the set of elements of* $\mathbb{Z}_M^*(n,n)$ *which do not satisfy (5.1). Then*

$$|\mathbb{Z}_M^*(n,n)| \geq 2\,|C(G)|\,.$$

Proof. By the definition $C(G)$ is the centralizer of G in $\mathbb{Z}_M^*(n,n)$, which is a subgroup. We have $|C(G)| \mid |\mathbb{Z}_M^*(n,n)|$ by Lagrange's theorem, hence it is enough to prove that there exists at least one element of $\mathbb{Z}_M^*(n,n)$ which does not commute with G.

Let P be the following permutation matrix

$$P = \begin{pmatrix} 0 & 1 & 0 & \cdots & 0 \\ 1 & 0 & 0 & \cdots & 0 \\ 0 & 0 & & & \\ \vdots & \vdots & & E_{n-2} & \\ 0 & 0 & & & \end{pmatrix}$$

where E_{n-2} denotes the $(n-2)$x$(n-2)$ unit matrix. Then PG differs from G such that the first two rows are exchanged, while GP such that the first two columns are exchanged. Hence $P \notin C(G)$ except when $n=2$, $g_0 \equiv -1$ and $g_1 \equiv 0 \pmod{M}$. In the exceptional case taking $P = \left(\begin{smallmatrix} 0 & 1 \\ 1 & 1 \end{smallmatrix}\right)$ one can prove by an easy computation that $P \notin C(G)$. The proposition is proved. □

6 Characterisation of number systems

We have given in section 2 a general sufficient condition for $\{g(t), \mathcal{N}\}$ to be a WNS. In the following theorem we are given a characterization of number systems when $g(t)$ is squarefree, i.e. without multiple roots. Denote in the sequel $A = \max\{|a| : a \in \mathcal{N}\}$, D_g the discriminant of g and $|\bar{g}|$ the maximum of the absolute values of the roots of g.

Theorem 6.1. *Let $g(t) = t^n + g_{n-1}t^{n-1} + \ldots + g_0 \in \mathbb{Z}[t]$ be squarefree. Then $\{g(t), \mathcal{N}\}$ is a number system if and only if*

(i) *Any roots α_i $(i = 1, \ldots, n)$ of $g(t)$ satisfy $|\alpha_i| > 1$,*

(ii) *$\mathcal{N}$ is a complete residue system modulo $|g_0|$ containing 0,*

(iii) *Any $\underline{y} = (y_0, \ldots, y_{n-1}) \in \mathbb{Z}^n$ with $|y_i| < c_2$, $(i = 1, \ldots, n)$ has a unique preimage under the map T, where*

$$c_2 = \left|D_g^{-2}\right| (n|\bar{g}|^{n-1})^{n/2} A \max_{1 \le i \le n} \frac{|\alpha_i|}{(|\alpha_i| - 1)}$$

Remark. This is the generalization of Theorem 3 of Kovács and Pethö [5].

Proof. We shall use the notations of section 2. Let $w = w_0 \ldots w_l \in \mathcal{A}_{\mathcal{N}}$ and $T(w) = (z_0, \ldots, z_{n-1}) \in \mathbb{Z}^n$. We shall prove first the necessity of the conditions.

(i) Let α be a complex root of $g(t)$. Then we have

$$T_1(w)_\alpha = w_0 + w_1\alpha + \ldots + w_l\alpha^l = z_0 + z_1\alpha + \ldots + z_{n-1}a^{n-1}. \tag{6.1}$$

If $|\alpha| < 1$, then the absolute value of the number staying on the left hand side of (6.1) is bounded independently of w, while the supremum of the right hand side is infinite, hence $|\alpha| \ge 1$. Assume now that $|\alpha| = 1$. Then α must be a root of unity (see the proof of Theorem 3. in Kovács and Pethö [5]). Let m be the smallest positive integer with $\alpha^m = 1$ and let $g_1(t) = g(t)/\gcd(g(t), t^m - 1)$. The degree of $g_1(t)$ is less then n and

$$g_1(t) \equiv g_1(t)t^m \pmod{g(t)}. \tag{6.2}$$

Suppose that

$$(T_3 \circ T_2)^{-1} g_1(\phi) = w_0 + x_1 t + \ldots + w_l t^l$$

with $w_i \in \mathcal{N}$, $w_l \ne 0$. Then we have

$$(T_3 \circ T_2)^{-1} g_1(\phi)\phi^{hm} = w_0 t^{hm} + \ldots + w_l t^{hm+l}$$

by (6.2) for any $h = 0, 1, \ldots$, hence T can not be bijective.

(ii) Let $\underline{x}_i = (x_{i0} \ldots x_{i,n-1})$, $i = 0, 1, \ldots$, where $\underline{x}_i$ is defined by (1.1). It follows $g_0 | x_{i0}$ immediately from (1.1), hence $z_0 \equiv w_0 \pmod{g_0}$ thus $\mathcal{N}$ contains a complete residue system modulo g_0.

Assume that $0 \ne w = w_0 \ldots w_l \in T^{-1}(0)$. Then $T_2 \circ T_1(w) = 0$ and $T_2 \circ T_1(ww) = T_2 \circ T_1(w) + t^{l+1} T_2 \circ T_1(w) = 0$, hence $T(w) = T(ww)$ and T is not even injective, so $0 \in \mathcal{N}$.

Assume finally that there exist $n_1, n_2 \in \mathcal{N}$ with $n_1 \equiv n_2 \pmod{g_0}$ but $n_1 \ne n_2$. Let $n_1 = n_2 + sg_0$ with an integer s. If $T^{-1}(sg_0) = h_0 \ldots h_l$, then $h_0 \equiv sg_0 \equiv 0 \pmod{g_0}$, hence $h_0 = 0$ and so

$$n_1 = (T_2 \circ T_1)^{-1}(n_1) = (T_2 \circ T_1)^{-1}(n_2 + sg_0) = n_2 h_1 \ldots h_l$$

which contradicts the bijectivity of T.

(iii) is obviously necessary for $\{g(t),\mathcal{N}\}$ to be an NS.

We shall now prove the sufficiency of (i),(ii) and (iii). If (i) and (ii) hold, then T is injective by Proposition 2.2. For a polynomial $p(t) = p_0+p_1t+\ldots+p_{n-1}t^{n-1} \in \mathbb{Z}[t]$ let

$$\begin{aligned} R(p(t)) &= w_{p(t)} = w, \quad \text{with } w \in \mathcal{N} \text{ and } w \equiv p_0 \pmod{g_0}, \\ S(p(t)) &= (p(t) - (p_0 - w)g(t)/g_0 - w)/t. \end{aligned}$$

The polynomial $S(p(t))$ has integer coefficients, it is of degree at most $n-1$ and it satisfies

$$p(t) = w_0 + w_1 t + \ldots + w_k t^k + S^{k+1}(p(t))t^{k+1} \tag{6.3}$$

for any $k = 0, 1, \ldots$, where $w_i = R(S^i(p(t)))$.

Let $\varepsilon > 0$ and choose $k = k(\varepsilon)$ so large that

$$\left|p(\alpha_i)/\alpha_i^{k+1}\right| < \varepsilon$$

hold for $i = 1, \ldots, n$. This is possible because of (i). We have by (6.3)

$$p(\alpha_i) = \sum_{j=0}^{k} w_j \alpha_i^j + S^{k+1}(p(t))_{\alpha_i} \alpha_i^{k+1},$$

which implies

$$\left|S^{k+1}(p(t))_{\alpha_i}\right| < \frac{|\alpha_i|}{|\alpha_i| - 1} A + \varepsilon, \qquad i = 1, \ldots, n. \tag{6.4}$$

Let $S^{k+1}(p(t)) = p_{k+1,0} + \ldots + p_{k+1,n-1}t^{n-1}$, where $p_{k+1,j} \in \mathbb{Z}$. The matrix (α_i^j) $i = 1, \ldots, n$; $j = 0, \ldots, n-1$ is invertible , because the roots of $g(t)$ are distinct. Hence there exists only finitely many possibilities for the polynomials $S^{k+1}(p(t))$. We may take ε arbitrarily small, hence choosing k large enough we get

$$\left|S^{k+1}(p(t))_{\alpha_i}\right| \le \frac{|\alpha_i|}{|\alpha_i| - 1} A. \tag{6.5}$$

Considering (6.5) as a system of linear inequalities in $p_{k+1,0}, \ldots, p_{k+1,n}$ and using Hadamard's inequality we get $|p_{k+1,j}| \le c_3$. The polynomial $S^{k+1}(p(t))$ has therefore a preimage under $T_1 \circ T_2$ by (iii). Inserting the resulting word into (6.3) we get the preimage of $p(t)$ which proves the theorem. □

In the sequel let

$$|\underline{g}| = \min\{|\alpha| \ : \ \alpha \text{ a root of the polynomial } g(t)\}.$$

Theorem 6.2. *Let $g(t)$ be a squarefree polynomial such that $|\underline{g}| > 1$ and let $\{g(t),\mathcal{N}\}$ be a WNS. Then there exists a constant c_4 depending only on $n, g(t)$ and A such that*

$$m(w) \le A n^{n/2} |\bar{g}|^{l(w)+1+n(n-1)/2} (|\underline{g}| - 1)^{-1} \tag{6.6}$$

and

$$l(w) \le \frac{\log\ m(w)}{\log\ |\underline{g}|} + c_4 \tag{6.7}$$

hold for any $w \in \mathcal{N}$.

Proof. Let $\alpha_1, \dots, \alpha_n$ denote the roots of $g(t)$, let $w = w_0 \dots w_l$ and $T(w) = (y_0, \dots, y_{n-1})$. Then we have

$$y_0 + \dots + y_{n-1}\alpha_i^{n-1} = w_0 + \dots + w_l\alpha_i^l\ , \qquad i = 1, \dots, n.$$

Hence

$$\left|y_0 + \dots + y_{n-1}\alpha_i^{n-1}\right| \le \frac{|\alpha_i|^{l+1} - 1}{|\alpha_i| - 1}.$$

Using again Hadamard's inequality we get

$$m(w) = \max_{1 \le i \le n} |y_i| \le An^{n/2}|\bar{g}|^{l(w)+1+n(n-1)/2}(|\underline{g}| - 1)^{-1}.$$

For $w = 0$ the inequality (6.7) is obviously true with $c_4 = 1$, hence we may assume $w \ne 0$ in the following. Take

$$k = \left[\frac{\log\ m(w)}{\log\ |\underline{g}|} - \min_{1 \le i \le n} \frac{\log(|\alpha_i| - 1)}{\log\ |\alpha_i|}\right] + n + 1.$$

then

$$\frac{1}{|\alpha_i|^k}\left|y_0 + \dots + y_{n-1}\alpha_i^{n-1}\right| \le \frac{m(w)|\alpha_i|^n}{|\alpha_i|^k(|\alpha_i| - 1)} < 1$$

holds for any $i = 1, \dots, n$. This inequality implies (6.4) with $\varepsilon = 1$. There exists only finitely many polynomials in $\mathbb{Z}[t]/g(t)\mathbb{Z}[t]$ for which (6.4) holds. Let c_5 denotes the maximum of the length of the preimages of this polynomials. Then

$$l(w) \le k + c_5 = \frac{\log\ m(w)}{\log\ |\underline{g}|} + c_4,$$

and the theorem is proved. □

The quantity $\mu(w) = n \log\ m(w)$ is a natural measure for the size of $w \in \mathcal{A}_{\mathcal{N}}$. Theorems 3.1 and 6.2 implies immediately the following

Theorem 6.3. *Let* $\{g(t), \mathcal{N}\}$ *be as in Theorem 6.2. Then the Algorithms 3.1 and 3.2 terminate successfully after* $O(\mu(w))$ *operations on integers of size* $O(\mu(w))$ *where the O-constant depends only on the choice of the number system, namely on* $g(t)$ *and* $\mathcal{N}$.

7 A special case

Let a polynomial $g(t) = t^n + g_{n-1}t^{n-1} + \ldots + g_0 \in \mathbb{Z}[t]$ with $|g_0| \geq 2$ and a finite subset $\mathcal{N}$ of $\mathbb{Z}$ be given. To verify that T is injective, which is most important for the applications in cryptography, $O(|g_0|^2)$ operations are needed by Proposition 2.2. In Theorem 6.1 we proved necessary and sufficient conditions for $\{g(t),\mathcal{N}\}$ to be an NS. It is easy to check whether $g(t)$ is squarefree as well as the conditions (i) and (ii), but to verify (iii) generally extensive computations are needed. The example $g(t) = x^3 - 2x^2 - x + 4$, $\mathcal{N} = \{0, 1, 2, 3\}$ shows that $\{g(t),\mathcal{N}\}$ is a WNS for which (i) and (ii) hold, but it is not an NS because for example -1 does not have a preimage. Hence (iii) is necessary too.

For a wide class of pairs $\{g(t),\mathcal{N}\}$ we are able to prove however that it is an NS. The idea goes back to Kovács [3]. We assume in the sequel that $0 < g_{n-1} \leq \ldots \leq g_0$, $g_0 \geq 2$ and none of the roots of $g(t)$ are roots of unity. We take finally $\mathcal{N} = \{0, \ldots, g_0 - 1\}$. We are given now a variant of the decryption Algorithm 3.2. however, for simplicity, only for a single input block.

Algorithm 7.1.

Input: $\underline{x} = (x_0, \ldots, x_{n-1}) \in \mathbb{Z}^n$,

Output: $w = w_0 \ldots w_{l(w)} \in \mathcal{A}_{\mathcal{N}}$

(1) *Put* $l(w) = n - 1$, $w_i = x_i$ $(0 \leq i \leq l(w))$, $p(t) = \sum_{i=0}^{l(w)} w_i t^i$, $j = 0$.

(2) *Set* $\hat{p}(t) = p(t) = p(t)\lambda g(t)$ *with* $\lambda = \max\left\{\left[-\min_{0\leq i\leq n-1} w_i/g_i\right], 0\right\}$.
If $\lambda \neq 0$ *then set* $l(w) = l(w) + 1$. *Set* $\lambda = 0$.

(3) *Repeat until* $j > l(w)$

if $w_j \notin \mathcal{N}$ *then set* $\hat{p}(t) = p(t) + \lambda g(t)t^j$, *and determine*

$y_j \in \mathcal{N}$ *such that* $y_j \equiv w_j \pmod{g_0}$ *and set* $\lambda = (w_j - y_j)/g_0$,

$p(t) = \hat{p}(t) - \lambda g(t)t^j$, $l(w) = n + j$.

Set $j = j + 1$.

(4) *While* $l(w) > 0$ *and* $w_{l(w)} = 0$ *do* $l(w) = l(w) - 1$.

Remark. We used in Algorithm 7.1 the polynomial $\hat{p}(t)$ only for the convenience of the proof of the next theorem. One can easily modify Step 3 without using $\hat{p}(t)$.

Theorem 7.1. *Algorithm 7.1 terminates successfully i.e* $\{g,\mathcal{N}\}$ *is a NS. If* $g(t)$ *is squarefree then it takes* $O(\mu(w))$ *operations on numbers if size* $O(\mu(w))$ *where*

the O-constant depends only on the choice of the polynomial $g(t)$. More precisely, we have on every stage of the algorithm

$$\sum_{j=0}^{l(w)} |\hat{w}_j| < (n+1)(1+g_0)m(w), \tag{7.1}$$

where $\hat{p}(t) = \sum_{j=0}^{l(w)} \hat{w}_j t^j$.

Proof. After stage 2. all the coefficients of $p(t)$ and $\hat{p}(t)$ are non–negative and an easy computation shows that for the coefficients of $\hat{p}(t)$ (7.1) holds. We shall show that in the repeat loop the coefficients of $\hat{p}(t)$ remain always non–negative and their sum can not increase. In the sequel we set $g_n = 1$ and $g_{n+1} = g_{n+2} = \ldots = 0$.

Of course if $w_j \in \mathcal{N}$ then nothing else as an increment of j is done. Assume that $w_j \notin \mathcal{N}$. At the first occurrence of this event $\lambda = 0$, hence $\hat{p}(t)$ remains unchanged and the assertion is true. Otherwise, assume that $j_1 < j$ was the last index when $w_{j_1} \notin \mathcal{N}$ occurred. At this stage the algorithm sets $\lambda \neq 0$ and

$$p(t) = \hat{p}(t) - \lambda g(t)t^{j_1} = \sum_{i=0}^{j_1-1} \hat{w}_i t^i + \sum_{i=0}^{n} \left(\hat{w}_{j_1+i} - \lambda g_i\right) t^{j_1+i}. \tag{7.2}$$

Between the occurrence $w_{j_1} \notin \mathcal{N}$ and $w_j \notin \mathcal{N}$ the data $p(t)$ and λ do not change, which means that $w_i = \hat{w}_i$, $i = 0, \ldots, j_1$ and $w_{j_1+1} = \hat{w}_{j_1+1} - \lambda g_1, \ldots, w_{j-1} = \hat{w}_{j-1} - \lambda g_{j-1-j_1}$ are all non–negative. When $w_j \notin \mathcal{N}$ occurs then the algorithm sets

$$\hat{p}(t) = p(t) + \lambda g(t)t^j = \sum_{i=0}^{j} w_i t^i + \sum_{i=0}^{n} \left(w_{j+i} + \lambda g_i\right) t^{j+i}.$$

Using (7.2) we get

$$w_{j+s} + \lambda g_s = \hat{w}_{j+s} - \lambda g_{j-j_1+s} + \lambda g_s \geq \hat{w}_{j+s} \geq 0$$

for $s = 0, \ldots, n$ by the monoton decreasing property of the coefficients of $g(t)$.

The sum of the coefficients of the modified $\hat{p}(t)$ is

$$\sum_{i=0}^{j+n} w_i + \lambda \sum_{i=0}^{n} g_i = \sum_{i=0}^{j+n} \hat{w}_i - \lambda \sum_{i=0}^{n} g_i + \lambda \sum_{i=0}^{n} g_i = \sum_{i=0}^{j+n} \hat{w}_i.$$

So we proved (7.1).

Take $\hat{p}(t) = \hat{w}_0 + \ldots + \hat{w}_{l(w)} t^{l(w)}$ and $S_j(\hat{p}(t)) = \hat{w}_j + \ldots + \hat{w}_{l(w)} t^{l(w)}$. Then

$$\hat{p}(t) = \hat{w}_0 + \ldots + \hat{w}_{j-1} t^{j-1} + t^j S_j(\hat{p}(t))$$

holds. The coefficients of $S_j(\hat{p}(t))$ are non–negative and their sum is bounded, there exists so only finitely many possibilities for $S_j(\hat{p}(t))$. If Algorithm 7.1 would be non–finite, then there would exist integers $0 \leq j_1 \leq j_2$ such that $S_{j_1}(\hat{p}(t)) = S_{j_2}(\hat{p}(t))$. This means that there exists a polynomial $q(t) \in \mathcal{N}[t]$ with

$$S_{j_1}(\hat{p}(t)) = q(t) + t^{j_2-j_1} S_{j_2}(\hat{p}(t)). \tag{7.3}$$

We have $q(t) = 0$ because it has non–negative coefficients and the sum of the coefficients on both sides of (7.3) is equal. Hence

$$S_{j_1}(\hat{p}(t))\ (t^{j_2-j_1} - 1) = 0$$

holds in $\mathbb{Z}[t]/g(t)\mathbb{Z}[t]$. The assumption that none of the roots of $g(t)$ are roots of unity and the last equality implies $g(t)|S_{j_1}(\hat{p}(t))$. Hence $p(t) = w_0 + \ldots + w_{j_1-1}t^{j_1-1} + 0t^{j_1} + \ldots + 0t^{l(w)}$ holds after $j = j_1$ in the repeat loop, we have therefore in the remaining steps always $w_j \in \mathcal{N}$ and the algorithm terminates.

The second statement follows immediately from Theorem 6.2 and (7.1). □

References

[1] P. Horster, *Kryptographie*, Reihe Informatik **47** (1985), Bibliographisches Institut, Mannheim, Wien, Zürich.

[2] D.E. Knuth, *The art of computer programming. Vol. 2 Seminumerical algorithms*, Second edition, Addison-Wesley Publ. Co., 1981.

[3] B. Kovács, *Canonical number system in algebraic number fields*, Acta Math. Acad. Sci. Hungar. **37** (1981), 405-407.

[4] B. Kovács and A. Pethö, *Canonical systems in the ring of integers*, Publ. Math Debrecen **30** (1983), 39-45.

[5] B. Kovács and A. Pethö, *Number systems in integral domains especially in orders of algebraic number fields*, to appear (Acta Sci. Math. Szeged).

[6] R.C. Merkle and M.E. Hellman, *Hiding information and signatures in trapdoor knapsacks*, IEEE IT **24** (1978), 525-530.

[7] A. Shamir, *A polynomial time algorithm for breaking the basic Merkle-Hellman cryptosystem*, FOCS **23** (1982), 145-152.

Number–theoretic transforms and a theorem of Sylvester - Kronecker - Zsigmondy

M. Tasche, G. Steidl and R. Creutzburg

Abstract. The main drawback of the number-theoretic transform is the rigid relationship between the obtainable transform length and the possible computer word length. In this survey paper, the problem of the parameter determination for number-theoretic transforms is solved by studying cyclotomic polynomials. The result is closely connected with a theorem of Sylvester, Kronecker and Zsigmondy concerning prime factorizations for values of cyclotomic polynomials.

1 Introduction

The number-theoretic transform (NTT) introduced as a generalization of the discrete Fourier transform over a residue class ring of rational or Gaussian integers in order to perform fast cyclic convolutions without round-off errors and to obtain high-speed algorithms, allows an implementation with better efficiency and accuracy than the fast Fourier transform. However, the main drawback of the NTT is the rigid relationship between the obtainable transform length and the possible computer word length [7], pp.211–216. In recent papers [2], [3], [4], [6], [9], [10], the authors have discussed the important problem of parameter determination for NTT's by studying cyclotomic polynomials. In this survey paper we sketch our results. We show that our procedure is closely connected with a famous result of Sylvester, Kronecker and Zsigmondy concerning prime factorizations for values of cyclotomic polynomials.

2 Number–theoretic transforms

Let R be a finite commutative ring with identity 1 and let $N > 2$ be a fixed integer. Then $e \in R$ is called a primitive N–th root of unity if $e^N = 1$ and $e^n - 1$

Computational Number Theory

$(n = 1, \ldots, N-1)$ are invertible in R. The following theorem gives a criterion for an element to be a primitive N–th root of unity. We denote the N–th cyclotomic polynomial by Φ_N.

Theorem 2.1. [2]. *Let R be a finite commutative ring with identity 1 and let $N > 2$. Then $e \in R$ is a primitive N-th root of unity if and only if $\Phi_N(e) = 0$ and $N = N \cdot 1 \in R$ is invertible in R.*

Assume that there exists a primitive N–th root of unity $e \in R$. The number–theoretic transform (NTT) of length N is defined to be the following mapping between N–dimensional vectors $\underline{x} = (x(0), \ldots, x(N-1)) \in R^N$ and $\underline{\hat{x}} = (\hat{x}(0), \ldots, \hat{x}(N-1)) = F\underline{x} \in R^N$ with

$$\hat{x}(k) := \sum_{n=0}^{N-1} x(n)e^{nk} \quad (k = 0, \ldots, N-1).$$

The NTT $F : R^N \longrightarrow R^N$ is a linear bijective mapping. Its inverse $F^{-1} : R^N \longrightarrow R^N$ reads $\underline{x} = F^{-1}\underline{\hat{x}}$ with

$$x(n) = N^{-1} \sum_{k=0}^{N-1} \hat{x}(k)e^{-nk} \quad (n = 0, \ldots, N-1),$$

where $N^{-1} \in R$ denotes the inverse of N in R (see Theorem 2.1). The NTT has a structure and properties resembling those of the discrete Fourier transform [7], pp.211–216, particularly the cyclic convolution property

$$F(\underline{x} * \underline{y}) = (F\underline{x}) \circ (F\underline{y}) = \underline{\hat{x}} \circ \underline{\hat{y}} \quad (\underline{x}, \underline{y} \in R^N).$$

Here $\underline{x} * \underline{y} = (z(0), \ldots, z(N-1))$ with

$$z(n) := \sum_{k=0}^{N-1} x((n-k) \bmod N)y(k) \quad (n = 0, \ldots, N-1)$$

denotes the cyclic convolution and

$$\underline{\hat{x}} \circ \underline{\hat{y}} := (\hat{x}(0)\hat{y}(0), \ldots, \hat{x}(N-1)\hat{y}(N-1))$$

signifies the componentwise Hadamard product. From the cyclic convolution property it follows that

$$\underline{x} * \underline{y} = F^{-1}(\underline{\hat{x}} \circ \underline{\hat{y}}) \quad (\underline{x}, \underline{y} \in R^N).$$

This is the basic idea of a fast convolution algorithm. If the NTT and its inverse can be computed by a fast algorithm, then the computation of $F^{-1}(\underline{\hat{x}} \circ \underline{\hat{y}})$ requires considerably less arithmetical operations than the direct computation of $\underline{x} * \underline{y}$. First we consider the important special case $R = \mathbb{Z} / m\mathbb{Z}$ with a fixed odd integer $m > 1$. We call m the modulus. Assume that m possesses the prime factorization

$$m = p_1^{r_1} \ldots p_s^{r_s}$$

with distinct primes p_j and $r_j > 0 \quad (j = 1, \ldots, s)$. By the Chinese Remainder Theorem, $\mathbb{Z}\,/\,m\mathbb{Z}$ can be decomposed into the direct sum of local rings:

$$\mathbb{Z}\,/\,m\mathbb{Z} \cong \mathbb{Z}\,/\,p_1^{r_1}\mathbb{Z} \oplus \ldots \oplus \mathbb{Z}\,/\,p_s^{r_s}\mathbb{Z}.$$

Hence there exists a primitive N–th root of unity in $\mathbb{Z}\,/\,m\mathbb{Z}$ if and only if there exist primitive N–th roots of unity in every local ring $\mathbb{Z}/p_j^{r_j}\mathbb{Z}$ $(j = 1, \ldots, s)$. Thus

$$N \mid \gcd(p_1 - 1, \ldots, p_s - 1)$$

is necessary and sufficient for the existence of a primitive N–th root of unity in $\mathbb{Z}\,/\,m\mathbb{Z}$ [2].

The main drawback of the NTT in $(\mathbb{Z}\,/\,m\mathbb{Z})^N$ is the rigid relationship between the three parameters m, e and N. By Theorem 2.1 and from the numerical point of view, the following four essential conditions on a NTT are required:

- The modulus m must be a divisor of $\Phi_N(e)$ and $\gcd(N, m) = 1$.
- The length N has to be large enough and highly factorizable in order to implement fast algorithms.
- The primitive N–th root of unity e should have a simple binary representation ($e = 2$, for example), so that the arithmetic modulo m is easy to perform.
- The modulus m has to be large enough to avoid overflow, but on the other hand small enough, so that the machine word length is not exceeded. Furthermore, m should have a simple binary representation.

For instance the Fermat number transform with $N = 2^{d+1} \quad (d > 0)$, $e = 2$ and $m = 2^{2^d} + 1$ is a good compromise between these conditions. In the following, we sketch the determination of all possible moduli m for given $N, e \in \mathbb{Z} \quad (N > 2, \mid e \mid \geq 2)$ such that e is a primitive N–th root of unity in $\mathbb{Z}\,/\,m\mathbb{Z}$. By the first condition, we are interested in the prime factorization of $\Phi_N(e)$.

3 Prime factorization for values of cyclotomic polynomials

The following result originates with Sylvester, Kronecker and Zsigmondy and was developed in detail by Richter [8].

Theorem 3.1. **[8].** *Let $N, e \in \mathbb{Z} \quad (N > 2, |e| \geq 2)$ be given. Let p denote the greatest prime factor of N with $p^t \| N \quad (t \geq 1) \quad$ (that means $p^t | N$ and $p^{t+1} \nmid N$). The integer q is defined as follows:*

$$q := \begin{cases} p & \textit{if } ord_p(e) = N/p^t, \\ 1 & \textit{otherwise.} \end{cases}$$

Then $\Phi_N(e)$ possesses the prime factorization

$$\Phi_N(e) = q \prod z^{s(z)},$$

where $\prod$ is defined as the product of all primes $z > 2$ with $ord_z(e) = N$ and $s(z)$ denotes that positive integer with $z^{s(z)} \| e^N - 1$. If further $(N, e) \neq (3, -2)$,$(6, 2)$, then there exists at least one prime $z > 2$ with $ord_z(e) = N$.

By the help of the Theorems 2.1 and 3.1, the following construction of suitable moduli is obtained.

Theorem 3.2. *Let $N, e \in \mathbb{Z}$ $(N > 2,\ |e| \geq 2)$ be given, where $(N, e) \neq (3, -2)$, $(6, 2)$. Then e is a primitive N–th root of unity in $\mathbb{Z}\,/\,m\mathbb{Z}$ if and only if $m \in \mathbb{Z}$ $(m > 1)$ is a divisor of $M := \Phi_N(e)/q$.*

Proof. Let $m \in \mathbb{Z}$ $(m > 1)$ be a divisor of M. Then we obtain by Theorem 3.1 that $\Phi_N(e) \equiv 0 \bmod m$ and $\gcd(N, m) = 1$. Using Theorem 2.1, e is a primitive N–th root of unity in $\mathbb{Z}\,/\,m\mathbb{Z}$.

If $e \in \mathbb{Z}$ $(|e| \geq 2)$ is a primitive N–th root of unity in $\mathbb{Z}\,/\,m\mathbb{Z}$, then $m|\Phi_N(e)$ and $\gcd(N, m) = 1$ by Theorem 2.1. Then Theorem 3.2 implies $m|M$. □

Example 1. If $e = 2$ and $N = 120$, then m is equal to the prime $M = \Phi_{120}(2) =$ 4 562 284 561 [1]. In this case there exists exactly one modulus m such that $e = 2$ is a primitive 120–th root of unity in $\mathbb{Z}\,/\,m\mathbb{Z}$.

Example 2. If $e = 2$ and $N = 210$, then $m > 1$ is a divisor of $M = \Phi_{210}(2)$, where the prime factorization of $\Phi_{210}(2)$ [1] reads as follows

$$\Phi_{210}(2) = 211 * 664\ 441 * 1\ 564\ 921.$$

Here there exist exactly 7 moduli $m > 1$ with $m|M$ such that $e = 2$ is a primitive 210–th root of unity in $\mathbb{Z}\,/\,m\mathbb{Z}$.

Interesting results can be obtained as simple corollaries of Theorem 3.2. All possible moduli for the Mersenne number transform read as follows:

Corollary 3.1. *Let $N = p > 2$ be a prime. Then $e = 2$ is a primitive p–th root of unity in $\mathbb{Z}\,/\,m\mathbb{Z}$ if and only if $m \in \mathbb{Z}$ $(m > 1)$ is a divisor of the Mersenne number $M = \Phi_p(2) = 2^p - 1$.*

Proof. The assertion follows from Theorem 3.2 and

$$\Phi_p(x) = x^{p-1} + \ldots + x + 1 = \frac{x^p - 1}{x - 1} \quad (x \neq 1).$$

□

All possible moduli for the Fermat number transform read as follows:

Corollary 3.2. *Let* $N = 2^{d+1}$ $(d > 0)$. *Then* $e = 2$ *is a primitive* N*-th root of unity in* $\mathbb{Z}\,/\,m\mathbb{Z}$ *if and only if* $m \in \mathbb{Z}$ $(m > 1)$ *is a divisor of the Fermat number* $M = \Phi_N(2) = 2^{2^d} + 1$.

Proof. The assertion follows directly from Theorem 3.2 and $\Phi_N(x) = x^{2^d} + 1$. □

Further results can be found in [3]. Applying number-theoretic transforms, fast algorithms for cyclic convolutions and deconvolutions can be described [5]. These algorithms work without round-off errors and yield the exact results under natural conditions.

4 Complex number–theoretic transforms

One drawback of the NTT in a residue class ring of rational integers results from the rigid relationship between obtainable transform length and the modulus. One aim of the complex NTT in a residue class ring of Gaussian integers is to overcome this restriction [7],pp.236-239. We consider $R = \mathbb{Z}[i]\,/\,m\mathbb{Z}[i]$, where $m \in \mathbb{Z}[i]$ is a fixed odd Gaussian integer with norm $N(m) \geq 5$. Assume that m possesses the complex prime factorization

$$m = i^h p_1^{r_1} \ldots p_s^{r_s} c_1^{l_1} \ldots c_t^{l_t},$$

where $h \in \{0,1,2,3\}$, $r_j \geq 1$, $l_k \geq 1$, $p_j \in \mathbb{Z}$ are distinct rational primes with $p_j \equiv 3 \bmod 4$ and $c_k \in \mathbb{Z}[i]$ are distinct complex primes with a rational prime $N(c_k)$ and $N(c_k) \equiv 1 \bmod 4$ $(j = 1,\ldots,s;\ k = 1,\ldots,t)$. Splitting $\mathbb{Z}[i]\,/\,m\mathbb{Z}[i]$ into the direct sum of local rings, one can see that the condition

$$N \mid \gcd(p_j^2 - 1,\ N(c_k) - 1)\ j = 1,\ldots,s;\ k = 1,\ldots,t$$

is necessary and sufficient for the existence of a primitive N–th root of unity in $\mathbb{Z}[i]\,/\,m\mathbb{Z}[i]$ [2]. The following theorem on the complex prime factorization of $\Phi_N(e)$ for $e \in \mathbb{Z}[i]$ is a generalization of the known result of Sylvester, Kronecker and Zsigmondy (see Theorem 3.1) and is proved in [6]. We see that the prime divisor of $\Phi_N(e)$ can be characterized by certain congruence relations.

Theorem 4.1. [6]. *Let* $N \in \mathbb{Z}$ $(N > 4)$ *and* $e \in \mathbb{Z}[i]$ $(N(e) \geq 2)$ *be given. Let* p *denote the greatest rational prime factor of* N *with* $p^t \| N$ $(t \geq 1)$. *The Gaussian integer* q *is defined as follows:*

$$q := \begin{cases} 2 & \text{if } N = 2^t\ (t > 2) \text{ and } e \text{ is odd}, \\ p & \text{if } p \equiv 3 \bmod 4 \text{ and } ord_p(e) = \frac{N}{p^t}, \\ c & \text{if } p \equiv 1 \bmod 4,\ N(c) = p \text{ and } ord_c(e) = \frac{N}{p^t} \neq ord_{\bar{c}}(e), \\ p & \text{if } p \equiv 1 \bmod 4,\ N(c) = p, \text{ and } ord_c(e) = \frac{N}{p^t} = ord_{\bar{c}}(e), \\ 1 & \text{otherwise.} \end{cases}$$

Then $\Phi_N(e)$ possesses the complex prime factorization

$$\Phi_N(e) = i^h q \prod z^{s(z)},$$

where $h \in \{0,1,2,3\}$ and $\prod$ is defined as the product of all that primes $z \in \mathbb{Z}[i]$ $(N(z) > 2)$ with $ord_z(e) = N$. Further $s(z)$ denotes that positive integer with $z^{s(z)} \| e^N - 1$. Except for $(N,e) = (5,-1 \pm i)$, $(6,1 \pm i)$, $(6,2)$, $(10,1 \pm i)$, there exists at least one prime $z \in \mathbb{Z}[i]$ $(N(z) > 2)$ with $ord_z(e) = N$.

From the above result and Theorem 2.1, the following construction of suitable moduli is obtained.

Theorem 4.2. *Let $N \in \mathbb{Z}$ $(N > 4)$ and $e \in \mathbb{Z}[i]$ $(N(e) \geq 2)$ be given, where $(N,e) \neq (5,-1 \pm i)$, $(6,1 \pm i)$, $(6,2)$, $(10,1 \pm i)$. Then e is a primitive N-th root of unity in $\mathbb{Z}[i] \,/\, m\mathbb{Z}[i]$ if and only if $m \in \mathbb{Z}[i]$ $(N(m) > 1)$ is a divisor of $M := \frac{\Phi_N(e)}{q}$.*

The following example shows the reason for this generalization. We can obtain new and smaller moduli for the same transform length N.

Example 3. For $e = 2$ and $N = 360$, the rational prime factorization of $\Phi_{360}(2)$ reads as follows [1]

$$\Phi_{360}(2) = 168\,692\,292\,721 \,*\, 469\,775\,495\,062\,434\,961$$

such that $e = 2$ is a primitive 360–th root of unity in $\mathbb{Z} \,/\, m\mathbb{Z}$ for every rational divisor $m > 1$ of $\Phi_{360}(2)$. Otherwise, for $e = 1 + i$ and $N = 360$, the value $\Phi_{360}(1+i)$ is a rational integer and its rational prime factorization reads [1]

$$\Phi_{360}(1+i) = \Phi_{45}(4) = 631 * 23\,311 * 18\,837\,001$$

such that $e = 1 + i$ is a primitive 360–th root of unity in $\mathbb{Z}[i] \,/\, m\mathbb{Z}[i]$ for every divisor $m \in \mathbb{Z}[i]$ $(N(m) > 1)$ of $\Phi_{360}(1+i)$.

Many interesting results can be obtained as simple corollaries of Theorem 4.2. All possible moduli for the complex pseudo Mersenne number transform read as follows (compare with Corollary 3.1):

Corollary 4.1. *Let $p > 3$ be a rational prime. Then $e = 2i$ is a primitive $4p$–th root of unity in $\mathbb{Z}[i] \,/\, m\mathbb{Z}[i]$ if and only if $m \in \mathbb{Z}[i]$ $(N(m) > 1)$ is a divisor of $M := \frac{(2^p-1)(2^p+1)}{3}$. Further, $e = 1 + i$ is a primitive $8p$–th root of unity in $\mathbb{Z}[i] \,/\, m\mathbb{Z}[i]$ if and only if $m \in \mathbb{Z}[i]$ $(N(m) > 1)$ is a divisor of M.*

Proof. The assertion follows from Theorem 4.2 and

$$\Phi_{8p}(1+i) = \Phi_{4p}(2i) = \Phi_{2p}(-4) = \Phi_p(4) = \Phi_p(2)\Phi_{2p}(2) = \frac{(2^p-1)(2^p+1)}{3}.$$

□

All possible moduli for the complex pseudo Fermat number transform read as follows (compare with Corollary 3.2):

Corollary 4.2. *Let* $N = 2^{d+1}$ $(d > 1)$. *Then* $e = 1 + i$ *is a primitive* $2N$*-th root of unity in* $\mathbb{Z}[i] \,/\, m\mathbb{Z}[i]$ *if and only if* $m \in \mathbb{Z}[i]$ $(N(m) > 1)$ *is a divisor of* $M := 2^{2^d} + 1$.

Proof. Theorem 4.2 and

$$\Phi_{2N}(1+i) = \Phi_N(2i) = \Phi_N(2) = 2^{2^d} + 1$$

yield the assertion. □

More results on complex NTT's can be found in [4]. Finally note that we have extended the above results to domains of algebraic integers [9], [10].

References

[1] J. Brillhart, D.H. Lehmer, J.L. Selfridge, B. Tuckerman and S.S. Wagstaff, Jr., *Factorizations of* $b^n \pm 1, b = 2, 3, 5, 6, 7, 10, 11, 12$ *up to high Powers*, Contemp. Math. **22** (1983), Amer. Math. Soc., Providence, R.I..

[2] R. Creutzburg and M. Tasche, *F-Transformation und Faltung in kommutativen Ringen*, Elektron. Informationsverarb. Kybernet. **21** (1985), 129-149.

[3] R. Creutzburg and M. Tasche, *Number-theoretic transforms of prescribed length*, Math. Comp. **47** (1986), 693-701.

[4] R. Creutzburg and M. Tasche, *Parameter determination for complex number-theoretic transforms using cyclotomic polynomials*, Math. Comp. **52** (1989), 189-200.

[5] G. Drauschke and M. Tasche, *Exact deconvolution using number-theoretic transforms*, Comput. Math. Appl. **15** (1988), 757-768.

[6] G. Drauschke and M. Tasche, *Prime factorizations for values of cyclotomic polynomials in* $\mathbb{Z}[i]$, Arch. Math. **49** (1987), 292-300.

[7] H.J. Nussbaumer, *Fast Fourier Transform and Convolution Algorithms*, Springer-Verlag Berlin 1981.

[8] B. Richter, *Die Primfaktorzerlegung der Werte der Kreisteilungspolynome*, J. Reine Angew. Math. **254** (1972), 123-132.

[9] G. Steidl, M. Hänler and M. Tasche, *On a number-theoretic result of Zsigmondy in domains of quadratic integers*, Arch. Math. **53** (1989), 30-39.

[10] G. Steidl and M. Tasche, *On a number-theoretic result of Sylvester-Kronecker-Zsigmondy*, Math. Nachr. **140** (1989), 233-247.

A probabilistic class group and regulator algorithm and its implementation

J. Buchmann and S. Düllmann

Abstract. A new probabilistic algorithm for the determination of class groups and regulators of an algebraic number field $\mathcal{F}$ is presented. Heuristic evidence is given which shows that the expected running time of the algorithm is $\exp(\sqrt{\log D \log\log D})^{c+o(1)}$ where D is the absolute discriminant of $\mathcal{F}$, where $c \in \mathbb{R}_{>0}$ is an absolute constant, and where the $o(1)$-function depends on the degree of $\mathcal{F}$.

Part 1: The algorithm

1 Introduction

Computing the class group and the regulator of an algebraic number field $\mathcal{F}$ are two major tasks of algorithmic algebraic number theory. In the last decade, several regulator and class group algorithms have been suggested (e.g. [20],[21],[22],[3]). In [2] the problem of the computational complexity of those algorithms was adressed for the first time. This question was then studied in [2] in great detail. The theoretical results and the computational experience show that computing class groups and regulators is a very difficult problem. More precisely, it turns out that even under the assumption of the generalized Riemann hypothesis (GRH) the fastest known deterministic algorithm is of complexity $D^{1/5+o(1)}$ where D is the absolute discriminant of $\mathcal{F}$. This running time is exponential in the length of the input data since this length is polynomially bounded in $\log D$.

Computationally hard problems in number theory are a very good source for secure cryptosystems as is well known for the factoring problem. Therefore, in [8], [5], [9] and [10] the use of quadratic fields in cryptology was suggested and discussed. Stimulated by this new application, Hafner and McCurley [14] invented a new probabilistic class group algorithm for imaginary quadratic fields which was generalized to arbitrary fields in [4]. For quadratic fields, it can be shown under the assumption of the GRH that the expected running time of this new method is $L(D)^{\sqrt{2}+o(1)}$ where $L(D) = \exp\sqrt{\log D \log\log D}$. We use an additional assumption about the number of "smooth reduced ideals", then we obtain the

Computational Number Theory

computational complexity $L(D)^{1.7+o(1)}$ where the $o(1)$-function depends on the degree of the field $\mathcal{F}$.

In this paper we will describe the theory for the new algorithm and we will discuss its implementation for imaginary quadratic orders.

2 The idea

The algorithm of Hafner and McCurley makes use of the fact that under the GRH the class group Cl of an imaginary quadratic field $\mathcal{F}$ is generated by the prime ideals of norm below $6(\log D)^2$. By means of a probabilistic method the algorithm finds a basis for the lattice $L \subseteq \mathbb{Z}^k$ of relations for a generating system for Cl containing k elements. Since $\mathbb{Z}^k/L \cong Cl$, the knowledge of this basis implies the knowledge of the structure and the basis of Cl. The fact that a basis for the full lattice L has been found can be verified by means of the analytic class number formula by which an approximation h^* to the class number h of $\mathcal{F}$ with

$$h < h^* < 2h$$

can be determined in time polynomial in $\log D$ ([7]): A linearly independent system of k lattice vectors is a basis of L if its determinant is less than h^*.

For arbitrary number fields $\mathcal{F}$ the prime ideals of $\mathcal{F}$ whose norms do not exceed $12(\log D)^2$ will also generate Cl (see [7]) (assuming again the GRH). As we will show below, it is still possible to generate random relations among a small generating system for Cl. It is, however, not known how to verify in polynomial time that the proper class number has been found because in the general case the analytic class number formula only yields a polynomial time method for calculating a number h^* with

$$hR < h^* < 2hR$$

where h is the class number and where R is the regulator of $\mathcal{F}$.

Moreover, it seems to be more appropriate to calculate the regulator and the class number of $\mathcal{F}$ simultaneously because those two invariants are closely related.

We will, therefore, introduce a lattice L whose determinant $d(L)$ is hR. We will then explain how to find random vectors in L, how to verify that a basis for L has been found, and how to determine Cl and R if such a basis is known.

For $c \in \mathbb{R}_{>0}$ we define

$$F_c = \{\wp | \wp \text{ first degree prime ideal of } \mathcal{F}, N\wp \leq c\}$$

where for any fractional ideal A of $\mathcal{F}$ we denote by NA the absolute norm of A. F_c is a finite set. We write

$$F_c = \{\wp_1, \wp_2, \ldots, \wp_{k(c)}\}.$$

Furthermore, we denote by A_c the set of all the non zero numbers in $\mathcal{F}$ which, considered as principal ideals, can be represented as a power product of the ideals in the *factor base* F_c, i.e.

$$A_c = \left\{\alpha \in \mathcal{F}^\times \mid \alpha\mathcal{O}_F = \prod_{i=1}^{k(c)} \wp_i^{e_i}\right\}.$$

Every number in A_c represents a relation among the classes of the ideals in F_c. We are interested in the lattice of those relations and we, therefore, introduce two maps.

For $\alpha \in A_c$ with

$$\alpha\mathcal{O}_F = \prod_{i=1}^{k(c)} \wp_i^{e_i}$$

we define

$$\varphi_c'(\alpha) = (e_1, \ldots, e_{k(c)}).$$

Moreover, using the logarithm map

$$\begin{array}{rcccl} \mathrm{Log} & : & \mathcal{F}^\times & \longrightarrow & \mathbb{R}^r \\ & & \alpha & \longmapsto & (\log|\alpha|_1, \ldots, \log|\alpha|_r) \end{array}$$

where $r = s + t - 1$, s the number of real places of $\mathcal{F}$, t the number of complex places of $\mathcal{F}$ and where $||_i$ $(1 \leq i \leq m = s + t)$ are the archimedian valuations on $\mathcal{F}$, we define

$$\varphi_c(\alpha) = (e_1, \ldots, e_{k(c)}, \mathrm{Log}\,\alpha).$$

Hence, we have two maps

$$\varphi_c\colon A_c \to \mathbb{Z}^{k(c)} \times \mathbb{R}^r$$

and

$$\varphi_c'\colon A_c \to \mathbb{Z}^{k(c)}.$$

Both maps are actually homomorphisms of the multiplicative group A_c into the additive groups $\mathbb{Z}^{k(c)} \times \mathbb{R}^r$ or $\mathbb{Z}^{k(c)}$, respectively.

Note that for imaginary quadratic fields we have $r = 0$ and therefore, both maps coincide.

We can now define the lattice L_c.

Theorem 2.1. *Let $c \in \mathbb{R}_{>0}, c > 12(\log D)^2$ and assume the GRH. The set $L_c = \varphi_c(A_c)$ is a $(k(c)+r)$-dimensional lattice of determinant hR. The set $L_c' = \varphi_c'(A_c)$ is a $k(c)$-dimensional lattice of determinant h and we have $\mathbb{Z}^{k(c)}/L_c' \cong Cl$.*

Proof. See [4]. □

3 Determination of random relations in L_c

We assume that $c \in \mathbb{R}_{>0}$ is chosen such that $c > 12(\log D)^2$. This implies that the classes of the ideals in the factor base generate the class group of $\mathcal{F}$.

3.1 Numbers of small norm

The most obvious way of producing vectors in L_c is to find numbers $\alpha \in \mathcal{F}$ which completely factor over the factor base F_c.

This was first proposed by Pohst and Zassenhaus [21] but their algorithm only yields the class group of $\mathcal{F}$ and not the regulator. Also they use principal ideal testing in order to confirm the result of their computation and principal ideal testing is a very difficult problem. Their method of generating numbers of small norm in $\mathcal{F}$ is to enumerate certain parallelepipeds or ellipsoids in the conjugate lattice of the maximal order of $\mathcal{F}$. The algorithm of Pohst and Zassenhaus has proven to work well as long as the discriminant is small. For large discriminants the method seems to be too time consuming.

Actually, all methods that do not assume the GRH seem to be impractical for large discriminants since their running time is exponential in the discriminant.

If we assume the GRH then we can use ideas from the factoring algorithm "number field sieve" (NFS) (see [15]) to generate lattice vectors very quickly. Here is a short description of the method. Assume that the number field $\mathcal{F}$ is given as

$$\mathcal{F} = \mathbb{Q}[x]/f(x)\mathbb{Q}[x]$$

where $f \in \mathbb{Z}[x]$ is monic and irreducible. Let α be a root of f. It is actually sufficient to use only those prime ideals of degree one whose norm does not divide the index of the equation order $\mathbb{Z}[\alpha]$ in the maximal order of $\mathcal{F}$. For all such prime numbers p we calculate the roots $c_{p,1}, \ldots, c_{p,s_p}$ of $f \mod p$. Then $p\mathcal{O}_F + (\alpha - c_{p,l_i})\mathcal{O}_F$ $(1 \leq i \leq s_p)$ are all the degree one prime ideals dividing p. This means that the generation of the factor base is quite easy in that case. Next we choose $A_i, B_i \in \mathbb{Z}$, $i = 1, 2$, $A_1 < A_2$, $B_1 < B_2$ and we test all the numbers $a + b\alpha$ with $A_1 \leq a \leq A_2$, $B_1 \leq b \leq B_2$ for smoothness. As in the NFS this can again be done by a fast sieving procedure since for each prime p the number $a + b\alpha$ can be divisible by at most one degree one prime ideal of norm p and the ideal corresponding to the pair p, c_p divides $a + b\alpha$ if and only if $a + bc_p \equiv 0 \mod p$. The efficiency of this method depends on the existence of numbers of small norm in $\mathcal{O}_F$. The heuristic results of Cohen and Lenstra [11] and Cohen and Martinet [12] indicate that except for imaginary quadratic fields the probability for the class number of $\mathcal{F}$ to be small is very high. But if the class number of $\mathcal{F}$ is small then there are many numbers of small norm. First experiments indicate that the procedure is very easy to program and also very efficient. But as in the NFS, it is not clear how to prove anything about the method even under the assumption of the GRH.

3.2 The idea of Hafner & McCurley

In order to obtain an algorithm which under the assumption of GRH can be rigorously analysed Hafner and McCurley [14] modify an algorithm of Seyen [23] in which vectors in L_c are obtained by using reduction theory.

Let $\mathcal{F}$ be a complex quadratic field, i.e. if $r = 0, L_c = L'_c$. The method of Hafner and McCurley [14] to produce random relations in L_c works as follows:

They suggest the following : fix a parameter $B \in \mathbb{Z}_{>0}$. Put

$$\mathbb{Z}_B = \{0, 1, \ldots, B-1\}.$$

After those preparations generate vectors in L_c as follows: Pick at random a vector $\underline{v} = (v_1, \ldots, v_{k(c)}) \in \mathbb{Z}_B^{k(c)}$. Determine the reduced ideal A' in the class of

$$A = \prod_{i=1}^{k(c)} \wp_i^{v_i}. \tag{3.1}$$

Try to factor A' over the factor base F_c. Suppose this factorization can be completed successfully, i.e.

$$A' = \prod_{i=1}^{k(c)} \wp_i^{v'_i} \tag{3.2}$$

where the exponent vector $\underline{v}' = (v'_1, \ldots, v'_{k(c)})$ has rational integer entries. Then, since A and A' belong to the same ideal class, the ideal

$$A(A')^{-1} = \prod_{i=1}^{k(c)} \wp_i^{v_i - v'_i} \tag{3.3}$$

is principal which means that the vector $\underline{e} = \underline{v} - \underline{v}'$ belongs to the lattice L_c.

For fields of arbitrary degree we can, in principle, adopt the method of Hafner and McCurley. We must, however, realize that reduction of an ideal is no longer unique. An ideal class can contain many reduced ideals. Nevertheless, any reduced ideal A' in the class of A can be written as

$$A' = \frac{1}{\alpha} A \tag{3.4}$$

with $\alpha \in A - \{0\}$. If A' can be written as in (3.2), then we have found

$$(v_1 - v'_1, \ldots, v_{k(c)} - v'_{k(c)}, \operatorname{Log} \alpha) \in L_c.$$

3.3 Reduction

What does *reduction* mean? A fractional ideal A' of $\mathcal{F}$ is called *reduced* if it is primitive and the least positive integer in A is a *minimum* in A'. A number α in a fractional ideal A of $\mathcal{F}$ is called a *minimum*, if the only number $\beta \in A$ satisfying $|\beta|_i < |\alpha|_i$ for $1 \le i \le m$ is $\beta = 0$. A reduced ideal in the class of A can be determined as follows. Consider the Minkowski map

$$\begin{array}{lcl} \mathcal{F} & \longrightarrow & \mathbb{R}^n \\ \alpha & \longmapsto & \underline{\alpha} = (\alpha^{(1)}, \ldots, \alpha^{(s)}, Re\ \alpha^{(s+1)}, Im\ \alpha^{(s+1)}, \ldots, Re\ \alpha^{(m)}, Im\ \alpha^{(m)}) \end{array}$$

where $\alpha^{(1)}, \ldots, \alpha^{(n)}$ are the algebraic conjugates of $\alpha \in \mathcal{F}$, numbered such that $\alpha^{(i)} \in \mathbb{R}$ for $1 \le i \le s$ and $\alpha^{(t+i)} = \overline{\alpha^{(i)}}$ for $s+1 \le i \le s+t = m$.

The image $\underline{A}$ under this map is an n-dimensional lattice in $\mathbb{R}^n$ of determinant $\sqrt{D}\ NA$. Compute a number $\alpha \in A$ such that $\underline{\alpha}$ is a shortest non zero vector in $\underline{A}$. Then clearly, $\frac{a}{\alpha}A$ is reduced where a is the least positive integer such that $\frac{a}{\alpha}A$ is integral (see [17]).

Since computing shortest vectors in lattices is polynomial time for fixed dimension, we see that the above algorithm takes polynomial time for fixed degree n. In particular, if A is the product of two reduced ideals, then the running time of the algorithm will be polynomial in $\log D$.

3.4 Finding relations using general reduction theory

In the general case we must find vectors

$$\underline{e} = (e_1, \ldots, e_{k(c)}, \operatorname{Log}\alpha) \in L_c$$

where $\operatorname{Log}\alpha$ is not uniquely determined by $e_1, \ldots, e_{k(c)}$. Hence, it is not sufficient to randomize $\underline{e}$ by choosing an exponent vector $\underline{v}' = (v_1, \ldots, v_{k(c)})$ and to reduce the corresponding c-smooth ideal but we choose a random vector

$$\underline{v} = (v_1, \ldots, v_{k(c)+r}) \in \mathbb{Z}_D^{k(c)+r}.$$

The last r coordinates of $\underline{v}$ will determine the last r coordinates of $\underline{e}$. We compute a reduced ideal $A' = \frac{1}{\alpha}A$ in the class of $A = \prod_{i=1}^{k(c)} \wp_i^{v_i}$. The choice of the particular reduced ideal depends on the parameters $v_{k(c)+1}, \ldots, v_{k(c)+r}$ in the following way: since A' is reduced, it follows that α is a minimum of A. The image of the set M_A under the logarithm map is discrete. More precisely, it has been shown in [2] that there are constants c_1, c_2 (all the constants c_i used in the sequel depend only on the degree of $\mathcal{F}$) such that every cube in $\mathbb{R}^r$ of side length $(\log D)^{-c_1}$ contains at least one logarithm vector of a minimum in A and, correspondingly, every cube in $\mathbb{R}^r$ of side length $(\log D)^{c_2}$ contains at most one logarithm vector of a minimum of A. Cover all $\mathbb{R}^r$ with cubes of side length $(\log D)^{-c_1}$. The position of each cube is uniquely determined by a r-dimensional integer vector. Our procedure finds the reduced ideal $A' = \frac{1}{\alpha}A$ such that $\operatorname{Log}\alpha$ is closest to the cube determined by $(v_{k(c)+1}, \ldots, v_{k(c)+r})$. The vector $(v_{k(c)+1}, \ldots, v_{k(c)+r})$ is called a *coordinate vector* of $\operatorname{Log}\alpha$. By means of the results in [2] this can be done in time polynomial in $\log D$.

3.5 Running time

For $\beta > 0$ define

$$L[\beta] = \exp\left(\sqrt{\log D \log\log D}\right)^{\beta+o(1)}$$

where for fixed n the notation $o(1)$ represents a function of D which tends to 0 as D tends to infinity.

In order to be able to determine the expected running time of the algorithm, we assume that the number $N_{L[\beta]}$ of $L[\beta]$-smooth reduced ideals is

$$N_{L[\beta]} = D^{\frac{1}{2}} L[-1/(4\beta)].$$

For imaginary quadratic fields this equation is a theorem due to Seysen [23] and it can be proved for real quadratic fields by the same method.

For higher degree fields the equation is still a conjecture. We will refer to it as *"smoothness assumption"*

Choose $\beta > 0$ such that $c = L[\beta]$. The following results have been proved in [4].

Theorem 3.1. *Under the GRH and the smoothness assumption a generating system for L_c consisting of $L[\beta]$ elements can be computed in expected running time $L[2\beta + \frac{1}{4\beta}]$.*

Theorem 3.2. *Under the assumption of the GRH and the smoothness condition the class number, the structure of the class group and the regulator of an algebraic number field of degree n and absolute discriminant D can be computed in expected running time $L[1.7008...]$.*

Part 2: The implementation

4 The algorithm

In the second part of this paper we describe the implementation of the algorithm of Hafner and McCurley for imaginary quadratic fields. The general idea described in section 2 leads to the following algorithm.

Algorithm 4.1.

Input: *Absolute discriminant D.*
Output: *Class number h and structure and minimal generating set of the class group Cl.*

(1) *Compute the approximation h^* with $h < h^* < 2h$.*

(2) *Compute the factor base $F_c = \{\wp_1, \ldots, \wp_{k(c)}\}$ for $c > 6(\log D)^2$.*

(3) *Find $k(c)$ linearly independent vectors generating a $k(c)$-dimensional sublattice Λ of L_c.*

(4) *Compute the determinant Δ of Λ.*

(5) *If $\Delta < h^*$ go to step (8).*

(6) *Find an additional vector of L_c to enlarge the sublattice Λ.*

(7) *Compute the determinant Δ of the enlarged lattice Λ and go to step (5).*

(8) *Compute the structure and a minimal generating set of Cl.*

The most important and the most time consuming parts of the algorithm are steps (3) and (4). For the implementation of these steps we have developed a new strategy. These steps are described in detail in sections 6 and 7.

In sections 5, 8 and 9 we explain the implementation of the remaining steps.

In section 10 we present numerical examples and timings.

5 Computation of the approximation h^* and the factor base

In **step (1)** we compute the approximation h^* with

$$h < h^* < 2h. \tag{5.1}$$

By the analytic class number formula we know that

$$h = \frac{\sqrt{D}}{\pi} \prod_{p \in \mathbb{P}} \left(1 - \left(\frac{D}{p}\right) \cdot \frac{1}{p}\right)^{-1}.$$

So we calculate h^* by evaluating the partial product

$$h^* = \frac{3}{2}\frac{\sqrt{D}}{\pi}\prod_{p\le c'}\left(1-\left(\frac{D}{p}\right)\cdot\frac{1}{p}\right)^{-1}.$$

An approximate value for c' which guarantees that (5.1) is satisfied can be obtained from [7]. In our cases we computed

$$c' = 49(\log D)^2.$$

In **step (2)** we compute the factor base

$$F_c = \{\wp|\wp \text{ prime ideal of } \mathcal{F},\ N\wp \le c\} = \{\wp_1, \dots, \wp_{k(c)}\}$$

for $c = 8(\log D)^2$.

In order to find those prime ideals we have to determine all prime numbers $p \le c$ for which $\left(\frac{D}{p}\right) = 1$, but this has already been done in step (1). For each of those primes we have to find $b \in \mathbb{Z}$ such that

$$b^2 \equiv -D \mod (4p).$$

This is done by Shanks algorithm RESSOL (see [25]).

6 Generating the relations

In this section we describe how to compute the relations among the elements of the factor base (step (3) of Algorithm 4.1). We have to find $k = k(c)$ linearly independent relation vectors $\underline{e}_1, \dots, \underline{e}_k \in L_c$.

The method suggested by Hafner and McCurley [14] is described in section 3.1. A practical algorithm for the computation of power products of ideals has been presented in [5], see also [24]. A method for factoring ideals into prime ideals can be found in [18].

Hafner and McCurley propose to take $B = D$. Moreover, while generating the i-th vector they add $2kD$ to the i-th component of the random exponent vector in order to obtain a diagonally dominant matrix which is guaranteed to be non singular.

There are two problems arising in practice:

- In order to generate a relation one has to compute k powers of ideals each with very large exponent of average size $D/2$. Moreover, one has to compute the product of those k ideal powers. Therefore, the computation of the power product is extremely time consuming.
- The matrix generated by this process is not sparse which makes the computation of the determinant in the next step very difficult. If the matrix were sparse, one could use special methods for the computation of the determinant.

For practical purposes we have developed a much more efficient strategy. We do not know, however, how to prove any complexity results for this method.

We renumber the elements of the factor base F_c such that

$$N\wp_i \leq \log D \text{ and } \wp_i \text{ not ambiguous for } 1 \leq i \leq k_0,$$
$$N\wp_i > \log D \text{ or } \wp_i \text{ ambiguous for } k_0 < i \leq k.$$

Note that k_0 is small compared to k. In Table 6.1 we show the values of k and k_0 for some examples:

Table 6.1

D	k	k_0	D	k	k_0	D	k	k_0
$4 \cdot 10^8 + 4$	223	6	$4 \cdot 10^{18} + 4$	843	7	$4 \cdot 10^{27} + 4$	1703	8
$4 \cdot 10^9 + 4$	261	5	$4 \cdot 10^{19} + 4$	928	6	$4 \cdot 10^{28} + 4$	1872	10
$4 \cdot 10^{10} + 4$	316	7	$4 \cdot 10^{20} + 4$	1042	9	$4 \cdot 10^{29} + 4$	1962	11
$4 \cdot 10^{11} + 4$	393	5	$4 \cdot 10^{21} + 4$	1162	7	$4 \cdot 10^{30} + 4$	2078	10
$4 \cdot 10^{12} + 4$	436	5	$4 \cdot 10^{22} + 4$	1228	9	$4 \cdot 10^{31} + 4$	2272	11
$4 \cdot 10^{13} + 4$	500	5	$4 \cdot 10^{23} + 4$	1305	7	$4 \cdot 10^{32} + 4$	2395	13
$4 \cdot 10^{14} + 4$	597	7	$4 \cdot 10^{24} + 4$	1464	8	$4 \cdot 10^{33} + 4$	2490	6
$4 \cdot 10^{15} + 4$	621	5	$4 \cdot 10^{25} + 4$	1535	8	$4 \cdot 10^{34} + 4$	2609	17
$4 \cdot 10^{16} + 4$	709	9	$4 \cdot 10^{26} + 4$	1702	11	$4 \cdot F_7$	3257	12
$4 \cdot 10^{17} + 4$	795	6						

$$(4 \cdot F_7 = 4 \cdot (2^{2^7} + 1) = 1\,361\,129\,467\,683\,753\,853\,853\,498\,429\,727\,072\,845\,828)$$

The random vectors $\underline{v}_i = (v_{i1}, \ldots v_{ik})$ are chosen as follows:
We pick at random

$$v_{ij} \in \{0, \ldots, 15\} \text{ for } 1 \leq j \leq k_0.$$

So only the first entries are random numbers and they are taken from a very small interval. There is no theoretical reason why to take the bound 15, it is just our practical experience that this bound works.

Moreover we set

$$v_{ij} = 0 \qquad \text{for } k_0 < j \leq k,\ j \neq i,$$
$$v_{ii} = 1 \qquad \text{if } i > k_0.$$

So the remaining entries of the vector are (with at most one exception) set to 0. For $i > k_0$ we set $v_{ii} = 1$ to make sure that each element of the factor base is represented in the relation matrix.

In this way we obtain vectors with at most $k_0 + 1$ non zero entries. The computation of the power product now becomes very easy: The powers of the ideals

$\wp_j$ $(1 \leq j \leq k_0)$ with exponents up to 15 are precomputed in the beginning and then stored. While computing the individual power products no further exponentiation of ideals and at most k_0 multiplications are necessary. So we can very rapidly compute the power product.

As a result of the factorization process, we obtain vectors $\underline{e_i}$ $(1 \leq i \leq k)$ which are very sparse. Setting $v_{ij} = e_{ij}$ $(1 \leq i, j \leq k)$ we obtain the *relation matrix* $V = (v_{ij})_{1 \leq i,j \leq k}$ where each column represents one of the relation vectors. Note that in our matrix notation the first index denotes the number of the column, the second index denotes the number of the row.

The structure of the matrix V is the following:

- The small part of the matrix containing the rows numbered $1 \ldots k_0$ is not sparse.
- The larger part of the matrix containing the rows numbered $(k_0 + 1) \ldots k$ is sparse. There are only the diagonal elements and some entries outside the diagonal which are not zero. The non zero entries outside the diagonal come from the factoring process and are very small.

If the matrix generated so far is singular, which is detected during the further computation we generate additional vectors from L_c. In practice this does not happen frequently.

In Table 6.2 we give the number k of vectors which need to be generated, the number t of trials which were necessary to find those k vectors and the average number $\bar{t}$ of trials necessary to find one relation.

Table 6.2

D	k	t	$\bar{t}$
$4 \cdot 10^{8} + 4$	223	240	1.1
$4 \cdot 10^{9} + 4$	261	314	1.2
$4 \cdot 10^{10} + 4$	316	394	1.2
$4 \cdot 10^{11} + 4$	393	676	1.7
$4 \cdot 10^{12} + 4$	436	847	1.9
$4 \cdot 10^{13} + 4$	500	1094	2.2
$4 \cdot 10^{14} + 4$	597	1139	1.9
$4 \cdot 10^{15} + 4$	621	1733	2.8
$4 \cdot 10^{16} + 4$	709	1730	2.4
$4 \cdot 10^{17} + 4$	795	2856	3.6
$4 \cdot 10^{18} + 4$	843	3736	4.4
$4 \cdot 10^{19} + 4$	928	4578	4.9
$4 \cdot 10^{20} + 4$	1042	5439	5.2
$4 \cdot 10^{21} + 4$	1162	7987	6.8

D	k	t	$\bar{t}$
$4 \cdot 10^{22} + 4$	1228	8218	6.7
$4 \cdot 10^{23} + 4$	1305	12540	9.6
$4 \cdot 10^{24} + 4$	1464	16409	11.2
$4 \cdot 10^{25} + 4$	1535	24669	16.0
$4 \cdot 10^{26} + 4$	1702	21454	12.6
$4 \cdot 10^{27} + 4$	1703	35521	20.8
$4 \cdot 10^{28} + 4$	1872	38850	20.7
$4 \cdot 10^{29} + 4$	1962	53385	27.1
$4 \cdot 10^{30} + 4$	2078	68339	32.8
$4 \cdot 10^{31} + 4$	2272	72783	32.0
$4 \cdot 10^{32} + 4$	2395	94972	39.6
$4 \cdot 10^{33} + 4$	2490	141461	56.8
$4 \cdot 10^{34} + 4$	2609	147860	56.6
$4 \cdot F_7$	3257	684002	209.1

7 Computing the determinant

In this section we describe how to compute the determinant of the lattice generated by the columns of the relation matrix (step (4) of Algorithm 4.1).

The general approach is to compute the Hermite normal form (HNF) of that matrix. Then the determinant is the absolute value of the product of the diagonal entries. Moreover, if the relation matrix is in HNF it is very easy to handle the additional relation vectors in step (7) of the algorithm.

As we have seen the relation matrix $V = (v_{ij})_{1 \le i,j \le k}$ produced by the method of the previous section is very sparse. We have developed a special procedure for the determination of the HNF of that matrix which exploits the sparsity. This procedure consists of four stages which we describe below.

In the first stage we only deal with rows in which all relevant entries are zero except for one which is 1 or -1:

Algorithm 7.1.

(1) *Initialize* $l \leftarrow 1$ *(* l *denotes the number of the row currently dealt with).*

(2) *Find a row index* i $(l \le i \le k)$ *and a column index* j $(l \le j \le k)$ *such that*

$$\begin{aligned} v_{ij} &= \pm 1, \\ v_{i\nu} &= 0 \ \textit{for}\ l \le \nu \le k,\ \nu \ne j. \end{aligned}$$

If no such indices can be found, go to step (5).

(3) *Interchange row* l *and row* i *of* V*, interchange column* l *and column* j *of* V.

(4) *Let* $l \leftarrow l + 1$ *and go back to step (2).*

(5) *Put* $l_1 \leftarrow l$ *and stop. (*l_1 *denotes the first row to be handled in the next stage.)*

Note that in Algorithm 7.1 we perform row operations although, in general, during Hermite reduction only column operations are allowed. So we use an additional vector in which we keep track of the permutation of the rows. Using this vector we are later able to find the correct element of the factor base corresponding to a certain row of the matrix.

Note further that during this algorithm there is no addition of columns. So the size of the entries remains the same.

In the next stage we handle those rows that contain only 1,-1 or 0 entries. Now we allow more than one non zero entry. For each iteration we look for the row containing the minimal number of non zero entries in the relevant positions.

Algorithm 7.2.

(1) *Initialize* $l \leftarrow l_1$.

(2) *Find row index* i $(l \leq i \leq k)$ *such that*

$$v_{i\nu} \in \{0, 1, -1\} \text{ for } l \leq \nu \leq k,$$
$$L_i := \#\{v_{i\nu} \neq 0 \ (l \leq \nu \leq k)\} \text{ is minimal.}$$

If no such index can be found, go to step (7).

(3) *Let* $j_1, \dots j_{L_i}$ *be the positions of the non zero entries in row* i.
For $\nu = 2$ *to* L_i *add or subtract column* j_1 *to column* j_ν *to get* $v_{i,j_\nu} = 0$.

(4) *Interchange row* l *and row* i *of* V, *interchange column* l *and column* j_1 *of* V.

(5) *Let* $l \leftarrow l + 1$.

(6) *If during the computations of step (3) one or more of the matrix entries exceeded half the maximum number representable by single precision integer arithmetic, go to step (7), else go back to step (2).*

(7) *Put* $l_2 \leftarrow l$ *and stop. (*l_2 *denotes the first row to be handled in the next stage.)*

Again we have to keep track of the permutation of rows. Note that during Algorithm 7.2 there are no multiplications. Therefore, the entries of the matrix do not grow as fast as in the standard HNF algorithm. As a result, this computation can be effected using single precision integer arithmetic for a long time. Nevertheless, the size of the entries has to be controlled during the computation (step (6)). Our practical experience is that, if at all, this stopping criterion will be satisfied only after many iterations.

After these two precomputation steps the number $k - l_2$ of rows, which remain to be considered is, according to our practical experience, much less than half of the original matrix size k. Some statistics are given in a table at the end of this section.

In the next stage we use standard Hermite reduction with some slight modifications which we do not describe here. We again use single precision integers to store the matrix entries, but we use double precision floating point arithmetic to be able to detect integer overflow. If integer overflow occurs, the HNF algorithm stops. We then use the LLL-algorithm (see [16], [19]) to shorten the entries of the remaining part of the matrix and we again try HNF-computation. This iteration of HNF- and LLL-computation is stopped if the size of the entries of the output of the LLL-computation does not allow any further computation in the HNF-algorithm using single precision integer storage of the matrix entries. An outline of this stage is the following:

Algorithm 7.3.

(1) *Initialize* $l \leftarrow l_2$.

(2) *Perform Hermite reduction of the matrix* $(v_{ij})_{l \le i,j \le k}$ *until integer overflow occurs. Now there is* l' $(l \le l' \le k)$ *such that the rows numbered* $l, l+1, \ldots, l'-1$ *are in HNF.*

(3) *If* $[l' = k+1]$ *or* $[l' = l$ *and step (4) has been performed at least once* $]$ *go to step (6).*

(4) *Perform LLL-reduction of the matrix* $(v_{ij})_{l' \le i,j \le k}$.

(5) *Put* $l \leftarrow l'$ *and go back to step (2).*

(6) *Put* $l_3 \leftarrow l'$ *and stop. (*l_3 *denotes the first row to be handled in the next stage.)*

Now the size of the remaining matrix $(k - l_3)$ is, according to our practical experience, less than 20% of the original matrix size (see Table 7.1.).

So we find the HNF of the remaining matrix $(v_{ij})_{l_3 \le i,j \le k}$ by means of the modular HNF algorithm (see [13]). As the module we pick the determinant of the matrix computed by means of modular Gaussian elimination.

This stage is speeded up by another slight modification to the general procedure described in section 4: In step (3) of Algorithm 4.1 we generate not only k but $k' = k+5$ linearly independent vectors of the lattice L_c and in step (4) we compute the determinant Δ of the sublattice Λ of L_c generated by those vectors.

So we have to handle an rectangular matrix, which does not effect the ideas described so far.

We now perform modular Gaussian elimination twice operating on two different square matrices built from the columns numbered $l_3 \ldots k$ and $(l_3 + 5) \ldots (k+5)$ of the rectangular matrix. So we obtain two determinants Δ_1 and Δ_2 satisfying $\Delta | \Delta_1$ and $\Delta | \Delta_2$. For the modular Hermite reduction we chose as module

$$\Delta' = \gcd(\Delta_1, \Delta_2)$$

which is much less then Δ_1 or Δ_2.

Moreover, it turns out that in most examples we have $\Lambda = L_c$ if we generate $k+5$ vectors. In this case steps (6) and (7) of Algorithm 4.1 do not have to be executed.

In Table 7.1 we show the effect of our reduction strategy for some examples.

Table 7.1

D	k	$k - l_1$	$k - l_2$	$k - l_2$	$k - l_3$
$4 \cdot 10^8 + 4$	223	29	13	8	2
$4 \cdot 10^9 + 4$	261	46	18	8	5
$4 \cdot 10^{10} + 4$	316	42	22	12	5
$4 \cdot 10^{11} + 4$	393	71	25	11	7
$4 \cdot 10^{12} + 4$	436	98	29	13	6
$4 \cdot 10^{13} + 4$	500	147	59	23	4
$4 \cdot 10^{14} + 4$	597	141	59	26	10
$4 \cdot 10^{15} + 4$	621	218	98	36	16
$4 \cdot 10^{16} + 4$	709	214	72	30	7
$4 \cdot 10^{17} + 4$	795	321	136	44	5
$4 \cdot 10^{18} + 4$	843	382	162	49	8
$4 \cdot 10^{19} + 4$	928	398	171	63	30
$4 \cdot 10^{20} + 4$	1042	438	185	65	27
$4 \cdot 10^{21} + 4$	1162	494	225	75	32
$4 \cdot 10^{22} + 4$	1228	579	255	93	34
$4 \cdot 10^{23} + 4$	1305	698	326	106	57
$4 \cdot 10^{24} + 4$	1464	838	395	132	61
$4 \cdot 10^{25} + 4$	1535	941	502	170	80
$4 \cdot 10^{26} + 4$	1702	961	478	157	75
$4 \cdot 10^{27} + 4$	1703	1036	580	180	83
$4 \cdot 10^{28} + 4$	1872	1117	621	198	122
$4 \cdot 10^{29} + 4$	1962	1277	707	225	126
$4 \cdot 10^{30} + 4$	2078	1311	725	237	122
$4 \cdot 10^{31} + 4$	2272	1464	801	263	156
$4 \cdot 10^{32} + 4$	2395	1567	919	291	170
$4 \cdot 10^{33} + 4$	2490	1889	1170	379	224
$4 \cdot 10^{34} + 4$	2609	1770	1030	347	203
$4 \cdot F_7$	3257	2552	1681	560	323

Table 7.1 contains the following data:

- D (absolute discriminant),
- k (total size of the matrix),
- $k - l_1$ (size of the remaining matrix after applying Algorithm 7.1),
- $k - l'_2$ (size of the matrix after the rows containing exactly two entries are eliminated, this is an intermediate result in Algorithm 7.2),
- $k - l_2$ (size of the remaining matrix after applying Algorithm 7.2),
- $k - l_3$ (size of the remaining matrix after applying Algorithm 7.3).

8 Finding additional relation vectors

The output of step (4) is the value of Δ and a square matrix

$$V = (v_{ij})_{1 \le i,j \le k(c)}$$

in Hermite normal form (HNF). The columns of that matrix form a basis of Λ.

If $\Delta > h^*$ we have to perform steps (6) and (7).

In **step (6)** we have to find an additional vector $\underline{e} \in L_c$. This is done by the same method as in step (3). In **step (7)** we compute the determinant Δ' of the sublattice Λ' generated by Λ and the additional vector $\underline{e}$. For this purpose we apply Hermite reduction modulo Δ (see [13]) to the matrix $(V, \underline{e})$ and we replace V by the first $k(c)$ columns of the resulting HNF. Then

$$\Delta' = \prod_{i=1}^{k(c)} v_{ii}$$

is the determinant of the enlarged lattice Λ'.

After setting $\Delta = \Delta', \Lambda = \Lambda'$ we go back to step (5).

9 Structure of the class group

If we finally achieve $\Delta < h^*$ we continue at **step (8)**. Here the Smith normal form (SNF) of the matrix V has to be computed. For details see [14].

Let m_i $(1 \le i \le k)$ be the diagonal entries of the SNF of the matrix V. Then we know that

$$\begin{aligned} m_i \quad &| \quad m_{i+1} \quad (1 \le i \le k-1), \\ h \quad &= \quad \prod_{i=1}^{k} m_i, \\ Cl \quad &\cong \quad \bigoplus_{i=1}^{k} \mathbb{Z}/m_i\mathbb{Z}. \end{aligned}$$

So we obtain the structure of the class group from the SNF of V.

Moreover, if we keep track of the unimodular transformation matrices occuring in the computation of the SNF, we obtain a minimal generating set for the class group in terms of the elements of the factor base.

10 Computational results

In Table 10.1 we give the class numbers and the class group structures computed by Algorithm 4.1 for some examples.

Table 10.1

D	h	class group structure
$4 \cdot 10^{8} + 4$	16 416	4, 4 104
$4 \cdot 10^{9} + 4$	40 960	2, 2, 2, 2, 2 560
$4 \cdot 10^{10} + 4$	193 584	2, 2, 48 396
$4 \cdot 10^{11} + 4$	454 176	2, 2, 2, 56 772
$4 \cdot 10^{12} + 4$	938 880	2, 4, 117 360
$4 \cdot 10^{13} + 4$	2 968 912	2, 2, 742 228
$4 \cdot 10^{14} + 4$	18 544 768	2, 2, 4, 1 159 048
$4 \cdot 10^{15} + 4$	32 953 344	2, 2, 2, 2, 2, 4, 257 448
$4 \cdot 10^{16} + 4$	188 953 856	2, 2, 2, 2, 11 809 616
$4 \cdot 10^{17} + 4$	374 837 568	2, 2, 2, 46 854 696
$4 \cdot 10^{18} + 4$	1 056 540 304	2, 2, 264 135 076
$4 \cdot 10^{19} + 4$	3 298 883 812	2, 1 649 441 906
$4 \cdot 10^{20} + 4$	14 849 576 832	2, 2, 2, 1 856 197 104
$4 \cdot 10^{21} + 4$	43 410 764 928	2, 2, 2, 2, 2, 2, 678 293 202
$4 \cdot 10^{22} + 4$	158 960 976 800	2, 2, 2, 19 870 122 100
$4 \cdot 10^{23} + 4$	376 171 851 136	2, 2, 2, 2, 23 510 740 696
$4 \cdot 10^{24} + 4$	1 154 987 161 920	2, 4, 144 373 395 240
$4 \cdot 10^{25} + 4$	2 990 443 065 024	2, 2, 2, 2, 186 902 691 564
$4 \cdot 10^{26} + 4$	16 503 514 325 952	2, 4, 2 062 939 290 744
$4 \cdot 10^{27} + 4$	38 172 032 669 184	2, 2, 2, 2, 2, 2, 596 438 010 456
$4 \cdot 10^{28} + 4$	159 585 440 418 304	2, 4, 4, 4 987 045 013 072
$4 \cdot 10^{29} + 4$	436 605 442 139 680	2, 2, 109 151 360 534 920
$4 \cdot 10^{30} + 4$	1 175 363 328 387 072	2, 2, 2, 2, 2, 8, 4 591 263 001 512
$4 \cdot 10^{31} + 4$	4 112 798 696 458 052	2, 2 056 399 348 229 026
$4 \cdot 10^{32} + 4$	17 455 078 066 482 816	2, 2, 4, 1 090 942 379 155 176
$4 \cdot 10^{33} + 4$	29 696 585 481 692 160	2, 2, 2, 2, 2, 2, 4, 116 002 287 037 860
$4 \cdot 10^{34} + 4$	189 652 590 177 168 096	2, 2, 2, 23 706 573 772 146 012
$4 \cdot F_7$	17787 144 930 223 461 408	2, 8 893 572 465 111 730 704

The class group structure is given in terms of the elementary divisors m_i $(1 \le i \le k)$, $m_i \neq 1$.

Now we give the running times of Algorithm 4.1 to get the results of Table 10.1.

Table 10.2

D	total	step1/2	step3	step4 stage1/2	step4 stage3	step4 stage4	step8
$4 \cdot 10^8 + 4$	8.5	0.3	7.8	0.1	0	0.1	0.2
$4 \cdot 10^9 + 4$	10.9	0.3	10.2	0.1	0	0.1	0.1
$4 \cdot 10^{10} + 4$	17.4	0.4	16.6	0.1	0	0.1	0.2
$4 \cdot 10^{11} + 4$	29.7	0.5	28.2	0.2	0	0.2	0.6
$4 \cdot 10^{12} + 4$	32.7	0.6	30.6	0.2	0.1	0.2	0.4
$4 \cdot 10^{13} + 4$	60.4	0.8	59.0	0.3	0.1	0.1	0.1
$4 \cdot 10^{14} + 4$	83.6	0.9	84.1	0.4	0.1	0.8	0.5
$4 \cdot 10^{15} + 4$	117.5	0.6	112.4	0.7	0.3	2.5	1.0
$4 \cdot 10^{16} + 4$	190.4	0.7	188.3	0.6	0.2	0.3	0.3
$4 \cdot 10^{17} + 4$	257.5	0.9	251.4	1.5	3.4	0.2	0.1
$4 \cdot 10^{18} + 4$	415.1	1.0	408.9	2.1	1.8	0.4	0.9
$4 \cdot 10^{19} + 4$	509.6	1.1	477.0	2.5	1.8	27.1	0.1
$4 \cdot 10^{20} + 4$	882.8	1.2	855.1	2.2	2.3	17.6	0.4
$4 \cdot 10^{21} + 4$	1215	1.3	1160	5.6	4.6	43.0	0.7
$4 \cdot 10^{22} + 4$	1466	1.4	1414	7.6	20.5	22.7	0.2
$4 \cdot 10^{23} + 4$	2378	1.6	2093	14.2	16.3	252.7	0.4
$4 \cdot 10^{24} + 4$	3418	1.8	3191	25.4	42.7	156.9	0.2
$4 \cdot 10^{25} + 4$	5250	1.9	4687	45.9	88.6	426.7	0.2
$4 \cdot 10^{26} + 4$	7093	2.1	6321	42.2	104.5	622.9	0.2
$4 \cdot 10^{27} + 4$	9806	2.1	8757	66.6	287.9	690.4	1.7
$4 \cdot 10^{28} + 4$	13396	2.4	11231	86.6	260.0	1816	0.3
$4 \cdot 10^{29} + 4$	18587	2.6	15369	128.7	343.4	2743	0.1
$4 \cdot 10^{30} + 4$	22332	2.7	19561	135.4	471.0	2161	0.8
$4 \cdot 10^{31} + 4$	30086	2.9	23831	183.6	1764	4304	0.9
$4 \cdot 10^{32} + 4$	54138	3.2	40901	274.7	1365	11594	0.3
$4 \cdot 10^{33} + 4$	99223	5.5	69357	548.8	2556	26754	1.7
$4 \cdot 10^{34} + 4$	123404	3.6	100362	374.2	1202	21461	0.2
$4 \cdot F_7$	511978	4.7	415128	1742	14494	80609	0.2

All times are CPU seconds on a Sun-4/60 SPARC-station 1. We have listed
- absolute discriminant D,
- total running time of Algorithm 4.1,
- running time for steps (1) and (2),
- running time for step (3),
- running time for step (4), stages 1 and 2 (see section 7),
- running time for step (4), stage 3 (see section 7),
- running time for step (4), stage 4 (see section 7),
- running time for step (8).

We do not list the running time for the steps (6) and (7) because - as we mentioned in section 7 - for most examples these steps do not have to be performed.

References

[1] E. Bach, *What to do until the witness comes*, preprint 1988.

[2] J. Buchmann, *Zur Komplexität der Berechnung von Einheiten und Klassenzahlen algebraischer Zahlkörper*, Habilitationsschrift, Düsseldorf, 1987.

[3] J. Buchmann, *On the computation of units and class numbers by a generalization of Lagranges' algorithm*, J. Number Theory **26** (1987), 8-30.

[4] J. Buchmann, *A subexponential algorithm for the determination of class groups and regulators of algebraic number fields*, to appear in: Proceedings of Seminaire de Theorie des Nombres, Paris.

[5] J. Buchmann, S. Düllmann and H.C. Williams, *On the complexity and efficiency of a new key exchange system*, Proceedings EUROCRYPT'89, to appear.

[6] J. Buchmann and V. Kessler, *Computing a reduced lattice basis from a generating system with applications*, Preprint 1990.

[7] J. Buchmann and H.C. Williams, *On the computation of the class number of an algebraic number field*, Math. comp. **53** (1989), 679-688.

[8] J. Buchmann and H.C. Williams, *A key exchange system based on imaginary quadratic fields*, J. Cryptology **1** (1988), 107-118.

[9] J. Buchmann and H.C. Williams, *A key exchange system based on real quadratic fields*, Proceedings CRYPTO 89.

[10] J. Buchmann and H.C. Williams, *Quadratic fields and cryptology*, Number Theory and Cryptography, London Math. Society Lecture Notes Series **154** (1990), 9-26.

[11] H. Cohen and H.W. Lenstra Jr., Heuristics on class groups of number fields, Lect. Notes in Math. **1068** (1984), 33-62.

[12] H. Cohen and J. Martinet, Etude heuristique des groupes de classes des corps de nombres, J. reine angew. Math. **404** (1990), 39-76.

[13] P.D. Domich, R. Kannan and L.E. Trotter, *Hermite normal form computation using modulo determinant arithmetic*, Math. Op. Res. **12** (1987), 50-59.

[14] J.L. Hafner and K.S. McCurley, *A rigorous subexponential algorithm for computation of class groups*, Journal AMS, to appear.

[15] A.K. Lenstra, H.W. Lenstra Jr., M.S. Manasse and J.M. Pollard, *The number field sieve*, Proceedings FOCS 90, to appear.

[16] A.K. Lenstra, H.W. Lenstra Jr. and L. Lowasz, *Factoring polynomials with rational coefficients*, Math. Ann. **261** (1982), 513-534.

[17] H.W. Lenstra Jr.. *Integer programming with a fixed number of variables*, Math. Op. Res. **8** (1983), 538-548.

[18] K.S. McCurley, *Cryptographic key distribution and computation in class groups*, Proceedings of NATO ASI Number Theory and Applications, Kluwer Academic Publishers 1989, 459-479.

[19] M. Pohst, *A modification of the LLL reduction algorithm*, J. Symbolic Computation **4** (1987), 123-127.

[20] M. Pohst, P. Weiler and H. Zassenhaus. *On effective computation of fundamental units I,II*, Math. Comp. **38** (1982), 275-329.

[21] M. Pohst and H. Zassenhaus, *Über die Berechnung von Klassenzahlen und Klassengruppen algebraischer Zahlkörper*, J. Reine angew. Math. **361** (1985), 50-72.

[22] M. Pohst and H. Zassenhaus. *Algorithmic algebraic number theory*, Cambridge University Press. 1989.

[23] M. Seysen, *A Probabilistic Factorization algorithm with quadratic Forms of Negative Discriminant*. Math. Comp **48** (1987), 737-780.

[24] D. Shanks. *Class number, a theory of factorization and genera*, Proc. Symp. Pure Math. **20** (1969 Institute on number theory), Amer. Math. Soc., Providence, 1971, 415-440.

[25] D. Shanks. *Five number theoretic algorithms*, Proc. Second Manitoba Conference on Numerical Math., 1972, 51-70.

Prime–producing quadratic polynomials and class numbers of quadratic orders

F. Halter-Koch

Abstract. It is the aim of this paper to present a unified theory of prime producing quadratic polynomials in connection with quadratic orders of class number one or two valid both in the real and in the imaginary case. Thereby it is possible to generalize, strengthen and complete most of the results occuring in the literature, esp. for positive discriminants.

1 Introduction

It is well known that Euler's polynomial $x^2 + x + 41$ represents primes for $0 \leq x \leq 39$. Frobenius [3] observed that the source of this phenomenon lies in the fact, that all positive definite binary quadratic forms of discriminant $D = (-163) = 1 - 4 \cdot 41$ are equivalent, i. e. $h_{-163} = 1$; he considered also the other discriminants $D < 0$ with $h_D = 1$, some discriminants $D < 0$ with $h_D = 2$ and some discriminants $D > 0$ with $h_D = 1$ and showed that they also deliver quadratic polynomials representing many primes. Rabinowitsch [15] proved, that conversely, for $D < 0$ there is a criterion for $h_D = 1$ in terms of prime-producing quadratic polynomials. Hendy [7] extended the results of Frobenius and Rabinowitsch to all discriminants $D < 0$ satisfying $h_D \leq 2$. For a more complete history of the problem and the connection with the class-number problem confer [16] ch.3.II.

Kutsuna [10] was the first who showed that also for certain positive discriminants there is a criterion for class number one in terms of prime–producing quadratic polynomials. Recently, Mollin and Williams showed in a series of papers how to extend and complete these results for positive discriminants, especially of Richaud-Degert-Type, cf. [12], [14] and the literature cited there.

2 Preliminaries on quadratic orders

In this section I collect some facts on quadratic orders and present them in a form which is suitable for my purpose; for proofs I refer to [1], [8] or [9].

Computational Number Theory

An integer $D \in \mathbb{Z}$ is called a *discriminant*, if D is not a square and either $D \equiv 0 \bmod 4$ or $D \equiv 1 \bmod 4$; let in the sequel D always be a discriminant. Put

$$\omega_D = \begin{cases} \frac{1}{2}\sqrt{D}\,, & \text{if } D \equiv 0 \bmod 4, \\ \frac{1}{2}(1+\sqrt{D})\,, & \text{if } D \equiv 1 \bmod 4, \end{cases}$$

and

$$\mathcal{R}_D = \mathbb{Z} \oplus \mathbb{Z}\omega_D\,.$$

$\mathcal{R}_D$ is an order in the quadratic number field $\mathbb{Q}(\sqrt{D})$; if D_0 is the discriminant of $\mathbb{Q}(\sqrt{D})$ and f is the conductor of $\mathcal{R}_D$, then $D = D_0 f^2$. I call D_0 the *fundamental discriminant* associated with D.

For $\alpha \in \mathbb{Q}(\sqrt{D})$ let $\bar{\alpha}$ be the conjugate of α and $N(\alpha) = \alpha\bar{\alpha}$. For an ideal $(0) \neq \mathcal{I} \lhd \mathcal{R}_D$ set $N(\mathcal{I}) = (\mathcal{R}_D : \mathcal{I})$; then $|N(\alpha)| = N(\alpha\mathcal{R}_D)$ for $0 \neq \alpha \in \mathcal{R}_D$.

An ideal $(0) \neq \mathcal{I} \lhd \mathcal{R}_D$ is called *primitive*, if $m^{-1}\mathcal{I} \not\subset \mathcal{R}_D$ for all $m \in \mathbb{N}$, $m \geq 2$. An element $\alpha \in \mathcal{R}_D$ is called *primitive*, if $\alpha \notin \mathcal{R}_D^\times \cup \{0\}$ and the principal ideal $\alpha\mathcal{R}_D$ is primitive. Every primitive ideal $\mathcal{I} \lhd \mathcal{R}_D$ is of the form

$$\mathcal{I} = a\mathbb{Z} \oplus (b + \omega_D)\mathbb{Z}$$

with $a = N(\mathcal{I}) \in \mathbb{N}_+$, $b \in \mathbb{Z}$, $-\frac{a}{2} < b \leq \frac{a}{2}$, and $a \mid N(b + \omega_D)$.

An ideal $\mathcal{I} \lhd \mathcal{R}_D$ is called *regular*, if $\mathcal{R}_D = \{\alpha \in \mathbb{Q}(\sqrt{D}) \mid \alpha\mathcal{I} \subset \mathcal{I}\}$. Every regular ideal is invertible. Every non-zero principal ideal of $\mathcal{R}_D$ is regular, and every ideal $(0) \neq \mathcal{I} \lhd \mathcal{R}_D$ with $(N(\mathcal{I}), f) = 1$ is regular.

Two ideals $\mathcal{I}_1$, $\mathcal{I}_2$ of $\mathcal{R}_D$ are called *equivalent*, if $\beta_1\mathcal{I}_1 = \beta_2\mathcal{I}_2$ for some $\beta_1, \beta_2 \in \mathcal{R}_D \backslash \{0\}$. Let $\mathcal{C}_D$ be the set of equivalence classes of invertible ideals of $\mathcal{R}_D$. Ideal multiplication induces a group structure on $\mathcal{C}_D$ and makes $\mathcal{C}_D$ canonically isomorphic to $\mathrm{Pic}\mathcal{R}_D$. Set $h_D = \#\mathcal{C}_D$ and

$$M_D = \begin{cases} \sqrt{\frac{-D}{3}}\,, & \text{if } D < 0, \\ \sqrt{\frac{D}{5}}\,, & \text{if } D > 0; \end{cases}$$

then every class in $\mathcal{C}_D$ contains a regular primitive ideal $\mathcal{I}$ with $N(\mathcal{I}) < M_D$ (for $D > 0$ this follows from the theory of minima of indefinite forms, cf. [2] ch. XI). For an invertible ideal $\mathcal{I} \lhd \mathcal{R}_D$ let $[\mathcal{I}] \in \mathcal{C}_D$ be the class containing $\mathcal{I}$.

3 Prime–producing quadratic polynomials: necessary conditions

Definition. *Let D be a discriminant, $D = D_0 f^2$ with associated fundamental discriminant D_0 and conductor f. Let $q \in \mathbb{N}_+$ be squarefree with $q \mid D_0$, $q \nmid f$.*

a) *Define the quadratic polynomial* $F_{D,q} \in \mathbb{Z}[X]$ *by*

$$F_{D,q}(X) = \begin{cases} qX^2 - \frac{D}{4q}\,, & \text{if } 4q \mid D, \\ qX^2 + qX + \frac{q^2-D}{4q}\,, & \text{if } 4q \nmid D. \end{cases}$$

b) *Let* $\mathcal{J}_q \lhd \mathcal{R}_D$ *be the unique ideal with* $\mathcal{J}_q^2 = q\mathcal{R}_D$, *explicitly:*

$$\mathcal{J}_q = \begin{cases} q\mathbb{Z} \oplus \frac{\sqrt{D}}{2}\mathbb{Z}\,, & \text{if } 4q \mid D, \\ q\mathbb{Z} \oplus \frac{q+\sqrt{D}}{2}\mathbb{Z}\,, & \text{if } 4q \nmid D. \end{cases}$$

Theorem 3.1. *Let D be a discriminant, $D = D_0 f^2$ with associated fundamental discriminant D_0 and conductor f. Let $q \in \mathbb{N}_+$ be squarefree with $q \mid D_0$, $q \nmid f$, and suppose that $|F_{D,q}(x)|$ is either 1 or a prime for all integers x satisfying $0 \le x < \frac{1}{2}(M_D - 1)$. Then*

$$\mathcal{C}_D \;=\; \{[\mathcal{R}_D], [\mathcal{J}_q]\}\,.$$

Proof. Assume $\mathcal{C}_D \neq \{[\mathcal{R}_D], [\mathcal{J}_q]\}$. Then there is a primitive regular ideal $\mathcal{I} \lhd \mathcal{R}_D$ which is neither equivalent to $\mathcal{R}_D$ nor $\mathcal{J}_q$ and satisfies $N(\mathcal{I}) = r < M_D$. If $\mathcal{I} = \mathcal{P}_1 \cdot \ldots \cdot \mathcal{P}_n$ with prime ideals $\mathcal{P}_i$ of $\mathcal{R}_D$, then all $\mathcal{P}_i$'s are primitive and regular and at least one $\mathcal{P}_i$ is neither equivalent to $\mathcal{R}_D$ nor $\mathcal{J}_q$. Therefore I may suppose that $\mathcal{I}$ is a prime ideal, and thus r is a prime. The ideal $\mathcal{I}\mathcal{J}_q$ is primitive and $N(\mathcal{I}\mathcal{J}_q) = rq$; therefore

$$\mathcal{I}\mathcal{J}_q \;=\; rq\mathbb{Z} \;\oplus\; (t + \omega_D)\mathbb{Z}$$

with $t \in \mathbb{Z}, -\frac{rq}{2} < t \le \frac{rq}{2}$ and

$$rq \mid N(t + \omega_D) \;=\; \begin{cases} t^2 - \frac{D}{4}\,, & \text{if } D \equiv 0 \bmod 4, \\ t^2 + t + \frac{1-D}{4}\,, & \text{if } D \equiv 1 \bmod 4. \end{cases}$$

From $t + \omega_D \in \mathcal{I}\mathcal{J}_q$ I infer $(t + \omega_D)\mathcal{R}_D = \mathcal{I}\mathcal{J}_q\mathcal{Q}$ with an invertible ideal $\mathcal{Q} \lhd \mathcal{R}_D$; as $\mathcal{Q}$ is not a principal ideal, I obtain

$$1 \;<\; u \;=\; N(\mathcal{Q}) \;=\; \frac{|N(t + \omega_D)|}{rq}.$$

Now I have to distinguish 3 cases.

CASE 1: $4q \mid D$. From

$$|N(t + \omega_D)| \;=\; \left|t^2 - \frac{D}{4}\right| \;=\; urq$$

I infer $q \mid t$; putting $x = \frac{t}{q}$, it follows that

$$|F_{D,q}(x)| \;=\; ur$$

is neither 1 nor a prime, and $|x| = \frac{|t|}{q} \le \frac{r}{2}$. If $r = 2$, then $|x| \le 1$ and thus $\frac{1}{2}(M_D - 1) < 1$, $M_D < 3$; but there are only finitely many such D and for these the theorem is true. Therefore I may suppose that r is odd, and then

$$|x| \;\le\; \frac{r-1}{2} \;<\; \frac{M_D - 1}{2},$$

as asserted.

CASE 2: $D \equiv 4 \bmod 8,\ 2 \mid q$. From

$$|N(t+\omega_D)| = \left|t^2 - \frac{D}{4}\right| = urq$$

and $\frac{q}{2} \mid \frac{D}{4}$ I infer $\frac{q}{2} \mid t$, and as $\frac{D}{4}$ is odd, t is odd, too. Putting $t = (2x+1)\frac{q}{2}$ with $x \in \mathbb{Z}$, I obtain

$$|F_{D,q}(x)| = ur,$$

which is neither 1 nor a prime. x satisfies the inequalities

$$-\frac{r}{2} - \frac{1}{2} < x = \frac{t}{q} - \frac{1}{2} \le \frac{r}{2} - \frac{1}{2} < \frac{M_D - 1}{2};$$

if $x \ge 0$, I am done; if $x < 0$, observe that $F_{D,q}(x) = F_{D,q}(-x-1)$ and $0 \le -x-1 < \frac{p-1}{2} < \frac{M_D-1}{2}$.

CASE 3: $D \equiv 1 \bmod 4$. From

$$|N(t+\omega_D)| = \left|t^2 + t - \frac{D-1}{4}\right| = urq$$

I infer $|(2t+1)^2 - D| = 4urq$ and thus $q \mid (2t+1)$; putting $2t+1 = q(2x+1)$ with $x \in \mathbb{Z}$, I obtain

$$|F_{D,q}(x)| = ur,$$

which is neither 1 nor a prime. x satisfies the inequalities

$$\frac{-r-1}{2} + \frac{1}{2q} < x = \frac{2t+1-q}{2q} \le \frac{p-1}{2} + \frac{1}{2q};$$

if $r = 2$ and $q = 1$, then $|x| \le 1$ and thus $\frac{1}{2}(M_D - 1) < 1$, $M_D < 3$; but there are only finitely many such D, and for these the theorem is true. If $(p,q) \ne (2,1)$, then

$$x \le \frac{p-1}{2} < \frac{M_D - 1}{2},$$

and, if $x \ge 0$, the assertion follows. If $x < 0$, observe that $F_{D,q}(-x-1) = F_{D,q}(x)$ and $0 \le -x-1 < \frac{p-1}{2} < \frac{M_D-1}{2}$. □

Remark. It is not known whether there exist positive discriminants satisfying the hypothesis of Theorem 3.1 and not being of Richaud-Degert-Type.

Corollary 3.1. *Let $D \equiv 1 \bmod 4$ be a discriminant.*

a) *If $|x^2 + x + \frac{1-D}{4}|$ is either 1 or a prime for all integers x satisfying $0 \le x < \frac{1}{2}(M_D - 1)$, then $h_D = 1$.*

b) *If $|Dx^2 + Dx + \frac{D-1}{4}|$ is either 1 or a prime for all integers x satisfying $0 \le x < \frac{1}{2}(M_D - 1)$, then $h_D = 1$.*

Proof. Obvious from Theorem 3.1 with $q = 1$ resp. $q = D$. □

Remark. For $D < 0$ a) strengthens the classical result of Rabinowitsch [15], while b) seems to be new. For $D > 0$ a) strengthens Kutsuna's Theorem [10] (see also [14] Remark 3.2).

Corollary 3.2. *Let D be a discriminant and suppose that either*

a) *$D \equiv 12 \bmod 16$, and $|2x^2 + 2x + \frac{4-D}{8}|$ is 1 or a prime for all integers x satisfying $0 \leq x < \frac{1}{2}(M_D - 1)$*

or

b) *$D \equiv 0 \bmod 8$, and $|2x^2 - \frac{D}{8}|$ is 1 or a prime for all integers x satisfying $0 \leq x < \frac{1}{2}(M_D - 1)$.*

Then

$$h_D = \begin{cases} 1, & \text{if } |u^2 - \frac{D}{4}v^2| = 2 \text{ has a solution } (u,v) \in \mathbb{Z}^2 , \\ 2, & \text{otherwise} . \end{cases}$$

Proof. The equation $|u^2 - \frac{D}{4}v^2| = 2$ has a solution $(u,v) \in \mathbb{Z}^2$ if and only if $\mathcal{J}_2$ is a principal ideal of $\mathcal{R}_D$. Thus the assertion follows from Theorem 3.1 with $q = 2$. □

Remark. For $D < 0$ Corollary 3.2 strengthens a result of Hendy [7]. For $D > 0$ it is known that h_D is odd if and only if $D = 4p$ with a prime $p \equiv 3 \bmod 4$ [5] ch. 29; thus Corollary 3.2 strengthens [14] Theorems 3.2 and 3.3.

Corollary 3.3. *Let $D \equiv 1 \bmod 4$ be a discriminant and q a prime with $q \mid D$, $q^2 \nmid D$. Suppose that $|qx^2 + qx + \frac{q^2-D}{4q}|$ is either 1 or a prime for all integers x satisfying $0 \leq x < \frac{1}{2}(M_D - 1)$. Then*

$$h_D = \begin{cases} 1, & \text{if } |u^2 - Dv^2| = 4q \text{ has a solution } (u,v) \in \mathbb{Z}^2 , \\ 2, & \text{otherwise.} \end{cases}$$

Proof. The equation $|u^2 - Dv^2| = 4q$ has a solution $(u,v) \in \mathbb{Z}^2$ if and only if $\mathcal{J}_q$ is a principal ideal of $\mathcal{R}_D$. Thus the assertion follows from Theorem 3.1. □

Remark. For $D < 0$ Corollary 3.3 strengthens a result of Hendy [7]. For $D > 0$ it is known that h_D is odd if and only if $D = pq$ with primes $p \equiv q \equiv 3 \bmod 4$ [5] ch. 29; thus Corollary 3.2 strengthens [14] Theorem 3.4.

4 Prime–producing quadratic polynomials: sufficient conditions

Theorem 4.1. *Let D be a discriminant, $D = D_0 f^2$ with associated fundamental discriminant D_0 and conductor f. Let $q \in \mathbb{N}_+$ be squarefree with $q \mid D_0$, $q \nmid f$ and $A > 0$ such that $|N(\alpha)| \geq A$ for all primitive $\alpha \in \mathcal{R}_D$. If $\mathcal{C}_D = \{[\mathcal{R}_D], [\mathcal{J}_q]\}$, then $|F_{D,q}(x)|$ is either 1 or a prime for all integers x satisfying*

$$\begin{array}{ccccl} \dfrac{D-4A^2}{4q^2} & < & x^2 & < & \dfrac{D+4A^2}{4q^2}\,, \quad \text{if } 4q \mid D, \\ \dfrac{D-4A^2}{4q^2} & < & (x+\frac{1}{2})^2 & < & \dfrac{D+4A^2}{4q^2}\,, \quad \text{if } 4q \nmid D. \end{array}$$

Proof. Suppose that, for some $x \in \mathbb{Z}$, $|F_{D,q}(x)|$ is composite, $F_{D,q}(x) = ru$ with $r, u \in \mathbb{Z}$, $|r| \geq 2$, $|u| \geq 2$. Then $qru = N(\alpha)$ with

$$\alpha = \begin{cases} qx + \frac{\sqrt{D}}{2}\,, & \text{if } 4q \mid D, \\ q(x+\frac{1}{2}) + \frac{\sqrt{D}}{2}\,, & \text{if } 4q \nmid D. \end{cases}$$

Obviously, $\alpha \in \mathcal{R}_D$ is primitive, and the principal ideal $\alpha\mathcal{R}_D$ has a prime decomposition

$$\alpha\mathcal{R}_D = \mathcal{J}_q \mathcal{P}_1 \cdot \ldots \cdot \mathcal{P}_n$$

with prime ideals $\mathcal{P}_\nu$ and $n \geq 2$. As $h_D \leq 2, \alpha$ is reducible in $\mathcal{R}_D$, i. e., $\alpha = \alpha_1\alpha_2$ with $\alpha_1, \alpha_2 \in \mathcal{R}_D \backslash (\mathcal{R}_D^\times \cup \{0\})$. Since α_1, α_2 are also primitive, this implies

$$A^2 \leq |N(\alpha)| = q \cdot |F_{D,q}(x)|.$$

If $4q \mid D$, then $|F_{D,q}(x)| = |qx^2 - \frac{D}{4q}| \geq \frac{A^2}{q}$, i. e., either $x^2 \geq \frac{D+4A^2}{4q^2}$ or $x^2 \leq \frac{D-4A^2}{4q^2}$;

if $4q \nmid D$, then $|F_{D,q}(x)| = |q(x+\frac{1}{2})^2 - \frac{D}{4q}| \geq \frac{A^2}{q}$, i. e., either $(x+\frac{1}{2})^2 \geq \frac{D+4A^2}{4q^2}$ or $(x+\frac{1}{2})^2 \leq \frac{D-4A^2}{4q^2}$. □

In order to apply Theorem 4.1 in concrete situations it is necessary to know a reasonable value for A. The following Lemma describes A in terms of diophantine equations:

Lemma 4.1. *Let D be a discriminant. Then, for $A > 0$, the following assertions are equivalent:*

a) *$|N(\alpha)| \geq A$ for all primitive $\alpha \in \mathcal{R}_D$.*

b) *If $m \in \mathbb{N}$ satisfies $1 < m < A$ and $(u, v) \in \mathbb{Z}^2$ is a solution of the diophantine equation*

$$\begin{array}{rcll} \left| u^2 - \dfrac{D}{4}v^2 \right| & = & m\,, & \text{if } D \equiv 0 \bmod 4, \\ \left| u^2 - Dv^2 \right| & = & 4m\,, & \text{if } D \equiv 1 \bmod 4, \end{array}$$

then $m = t^2$ with $t \in \mathbb{N}_+$ and $(u, v) = t$.

Proof. Suppose first, that $|N(\alpha)| \geq A$ for all primitive $\alpha \in \mathcal{R}_D$, and let $(u,v) \in \mathbb{Z}^2$ be a solution of the diophantine equation in b) for some $1 < m < A$; set $t = (u,v) \in \mathbb{N}_+$. If $D \equiv 0 \bmod 4$, I obtain

$$\left| \left(\frac{u}{t}\right)^2 - \frac{D}{4}\left(\frac{v}{t}\right)^2 \right| = |\, N(\alpha)\, | = \frac{m}{t^2}$$

with $\alpha = \frac{u}{t} + \frac{v}{t} \cdot \frac{\sqrt{D}}{2} \in \mathcal{R}_D$. As $(\frac{u}{t}, \frac{v}{t}) = 1$, α is either primitive or $\alpha \in \mathcal{R}_D^\times$. But $\frac{m}{t^2} \leq m < A$, and thus $\alpha \in \mathcal{R}_D^\times$, $|N(\alpha)| = 1$, $m = t^2$. If $D \equiv 1 \bmod 4$, I proceed as above, this time with $\alpha = \frac{u+v\sqrt{D}}{2t}$.

If b) is satisfied and $\alpha \in \mathcal{R}_D$ is primitive, I set

$$\begin{aligned} \alpha &= u + v\tfrac{\sqrt{D}}{2}\,, && \text{if } D \equiv 0 \bmod 4, \\ \alpha &= \tfrac{1}{2}(u + v\sqrt{D})\,, && \text{if } D \equiv 1 \bmod 4, \end{aligned}$$

and obtain

$$\begin{aligned} \left| u^2 - \frac{D}{4}v^2 \right| &= |N(\alpha)|\,, && \text{if } D \equiv 0 \bmod 4, \\ |\, u^2 - Dv^2 \,| &= 4|N(\alpha)|\,, && \text{if } D \equiv 1 \bmod 4, \end{aligned}$$

and thus $|N(\alpha)| \geq A$. □

For $D > 0$ the diophantine equation of Lemma 4.1 has been studied in [13] and [4].

For $D < 0$ I may set

$$A = \begin{cases} \frac{-D}{4}\,, & \text{if } D \equiv 0 \bmod 4, \\ \frac{1-D}{4}\,, & \text{if } D \equiv 1 \bmod 4. \end{cases}$$

There are only finitely many discriminants $D < 0$ satisfying $h_D \leq 2$, and these are well known [16] ch. 3.II. The results obtained from Theorem 4.1 for these discriminants using the above A are not the best possible ones.

For $D > 0$ I obtain bounds for A from [4] Lemma 1 or from the continued fraction expansion of ω_D, cf. [4]. The results obtained in this manner for the discriminants of Richaud-Degert-Type

$$D = 4a^2 + 1, \quad D = a^2 + 4, \quad D = a^2 + 4a$$

with $h_D = 1$ using the constants

$$A = a, \quad A = \left[\frac{\sqrt{D}}{2}\right] + 1, \quad A = a$$

are stronger than those of Mollin [12] Cor. 1 and 2 but far from beeing best possible. For $D = a^2 + 4$ and $D = a^2 + 4a$ stronger results have been proved by Frobenius [3] using reduction of binary forms. I shall return to these discriminants in section 5. Here I consider a Richaud-Degert-Type discriminant with class number two:

Corollary 4.1. *Let a be an odd positive integer and $D = 4(a^2 + 1)$ a discriminant with $h_D = 2$. Then $|2x^2 - \frac{a^2+1}{2}|$ is a prime for all integers x satisfying $|x| < \sqrt{a^2+1}$.*

Proof. By [4] there are no reduced integers $\alpha \in \mathcal{R}_D$ with $|N(\alpha)| < \frac{\sqrt{D}}{2} = \sqrt{a^2+1}$. Taking $A = \sqrt{a^2+1}$ in Theorem 4.1 and observing that $|2x^2 - \frac{a^2+1}{2}|$ cannot be 1 gives the assertion. □

5 Discriminants of Richaud-Degert-Type with class number one

Theorem 5.1. *Let D be a discriminant with $h_D = 1$.*

i) *If $D = 4p^2+1$ with an integer $p \geq 2$, then p is a prime, and $|x^2+x-p^2|$ is a prime for all integers x satisfying*

$$\left|x + \frac{1}{2}\right| < \sqrt{8p^2 - 12p + \frac{15}{2}}$$

except $x = -2p-1, -2p, -1, 0, 2p-1$ and $2p$.

ii) *If $D = p^2+4$ with an odd integer $p \geq 3$, then p is a prime, and $|x^2+x-\frac{p^2+3}{4}|$ is a prime for all integers x satisfying*

$$\left|x + \frac{1}{2}\right| < \sqrt{\frac{17}{4}p^2 - 12p + 10}$$

except $x = \frac{-3p-3}{2}, \frac{-3p+1}{2}, \frac{-p-1}{2}, \frac{p-1}{2}, \frac{3p-3}{2}$, and $\frac{3p+1}{2}$.

iii) *If $D = p^2+4p$ with an odd integer $p \geq 3$, then p is a prime, and $|x^2+x+\frac{1-4p-p^2}{4}|$ is a prime for all integers x satisfying*

$$\left|x + \frac{1}{2}\right| < \sqrt{\frac{17}{4}p^2 - 3p + 1}$$

except $x = \frac{-3p-13}{2}, \frac{-3p-1}{2}, \frac{-p-3}{2}, \frac{p+1}{2}, \frac{3p-1}{2}$, and $\frac{3p+11}{2}$.

Proof. i) It follows from [6], that p is a prime. If $\alpha \in \mathcal{R}_D$ is primitive, $\alpha = x + y\omega_D$ with $x, y \in \mathbb{Z}$, $(x,y) = 1$, then, putting

$$x = -(p+1)u + pv, \qquad y = u - v,$$

I obtain

$$N(\alpha) = x^2 + xy - p^2y^2 = p(u^2 - v^2) + uv.$$

The form

$$\psi(u,v) = |p(u^2 - v^2) + uv|$$

has the minimal values $\psi(1,1) = 1, \psi(1,0) = p$ and $\psi(2,-1) = 3p-2$. Thus, if $\alpha \in \mathcal{R}_D$ is primitive with $|N(\alpha)| < (3p-2)^2$ then either $|N(\alpha)| = 1$ or $|N(\alpha)|$ is a prime or $p | N(\alpha)$.

Now suppose that $\alpha \in \mathcal{R}_D$ is of the form $\alpha = x + \omega_D$. Then $N(\alpha) = x^2 + x - p^2$, and $|N(\alpha)| < (3p-2)^2$ if and only if $|x + \frac{1}{2}| < \sqrt{8p^2 - 12p + \frac{15}{4}}$. As $1 + 2p + 2\omega$ is the fundamental unit of $\mathcal{R}_D$, $|x^2 + x - p^2| \neq 1$ for all x, and $x^2 + x - p^2 \equiv 0 \bmod p$ if and only if $x \equiv 0$ or $-1 \bmod p$; from these observations the assertion follows.

ii) and iii) are proved similarly; for details see Frobenius [3] who proved an a little bit weaker statement. Observe however, that Frobenius overlooked some exceptional values of x. □

Remark. Weaker forms of assertions i) and ii) of Theorem 3.1 have been proved by Mollin [12] Cor. 1 and 2 and Louboutin [11] p. 171.

Theorem 5.1 depends on the explicit knowledge of the first few minima of the quadratic form belonging to the principal class of discriminant D. There are several further discriminants for which these first few minima are known and for which results as in Theorem 5.1 can be obtained, e. g. $D = 4(a^2 \pm 2)$ (cf. [11]).

Note. (added in proof, 2-10-1990). In a recent paper, S. Louboutin [Prime Producing Quadratic Polynomials and Class-numbers of Real Quadratic Fields, Canadian J. Math. **42** (1990), 315 – 341] proved that any positive fundamental discriminant satisfying the assumption of Theorem 3.1 with $q = 1$ or $q = 2$ must be of Richaud-Degert-Type; he also obtained some partial results for general q.

References

[1] P.G.L. Dirichlet, *Vorlesungen über Zahlentheorie*, Braunschweig 1983, Chelsea Reprint 1968.

[2] L.E. Dickson, *Introduction to the theory of Numbers*, Dover Publ. 1957.

[3] F.G. Frobenius, *Über quadratische Formen, die viele Primzahlen darstellen*, Sitzungsber. d. Kgl. Preuß. Akad. Wiss. Berlin (1912), 966-980.

[4] F. Halter-Koch, *Quadratische Ordnungen mit großer Klassenzahl*, J. Number Theory **14** (1990), 82-94.

[5] H. Hasse, *Number Theory*, Springer 1980.

[6] H. Hasse, *Über mehrklassige, aber eingeschlechtige reell-quadratische Zahlkörper*, Elem. Math. **20** (1965), 49-59.

[7] M.D. Hendy, *Prime Quadratics Associated with Complex Quadratic Fields of Class Number Two*, Proc. AMS **43** (1974), 253-260.

[8] E.L. Ince, *Cycles of reduced ideals in quadratic fields*, Math. Tables vol. IV, Cambridge 1968.

[9] B.W. Jones, *The Arithmetic Theory of Quadratic Forms*, Carus Math. Monographs 1967.

[10] M. Kutsuna, *On a Criterion for the Class Number of a Quadratic Number Field to be One*, Nagoya Math. J. **79** (1980), 123-129.

[11] S. Louboutin, *Continued Fractions and Real Quadratic Fields*, J. Number Theory **30** (1988), 167-176.

[12] R.A. Mollin, *Class Number One Criteria for Real Quadratic Fields I*, Proc. Japan Acad. **63** Ser. A (1987), 121-125.

[13] R.A. Mollin, *On the Insolubility of a Class of Diophantine Equations and the Non-triviality of the Class Numbers of Related Real Quadratic Fields of Richaud-Degert Type*, Nagoya Math. J. **105** (1987), 39-47.

[14] R.A. Mollin and H.C. Williams, *Class number one for real quadratic fields, continued fractions and reduced ideals*, In R. A. Mollin (Ed.): Number Theory and Applications, Kluwer Acad. Publ. 1989.

[15] G. Rabinowitsch, *Eindeutigkeit der Zerlegung in Primfaktoren in quadratischen Zahlkörpern*, J. reine u. angew. Math. **142** (1913), 153-164.

[16] P. Ribenboim, *The Book of Prime Number Records*, Springer 1988.

[17] R. Sasaki, *Generalized Ono invariants and Rabinowitch's Theorem for real quadratic fields*, Nagoya Math. J. **109** (1988), 117-124.

[18] H. Yokoi, *Class-number one problem for certain kind of real quadratic fields*, Proc. Int. Conf. on Class Numbers and Fundamental Units, Katata 1986, 125-137.

Applications of a new class number two criterion for real quadratic fields

R.A. Mollin *

Abstract. The primary purpose of this paper is to provide a real quadratic field analogue of the well-known Hendy criterion for class number 2 of complex quadratic fields in terms of prime-producing quadratic polynomials. We do this for real quadratic fields of Extended-Richaud-Degert (ERD)-type. Examples are provided as well as an explicit determination of all ERD-types $d \equiv 1 \bmod 8$ with class number $h(d) = 2$. We also provide. for what are called narrow ERD-types, a general class number criterion which allows us to determine an algorithm for listing all $d = 4m^2 + 1$ of a given class number. (for even m). Finally, we provide a general condition on arbitrary square-free d for $h(d)$ to be less than or equal to 2 in terms of prime producing quadratic polynomials. This continues work in [3] - [13].

1 Notation and preliminaries

Throughout d will be a positive square-free integer. Let $[\alpha, \beta]$ be the module $\{\alpha x + \beta y : x, y \in \mathbb{Z}\}$. Note that the ring of integers $\mathcal{O}_K$ of a quadratic field $K = \mathbb{Q}(\sqrt{d})$ is $[1, w]$ where

$$w = (\sigma - 1 + \sqrt{d})/\sigma \text{ with } \sigma = \begin{cases} 2, & \text{if } d \equiv 1 \pmod 4), \\ 1, & \text{if } d \not\equiv 1 \pmod 4). \end{cases}$$

The discriminant Δ of K is $(w - \overline{w})^2 = 4d/\sigma^2$, and the absolute norm of α is $N(\alpha) = \alpha\overline{\alpha}$ where $\overline{x}$ is the algebraic conjugate of x. Details and proofs of the following remarks can be found in [14], (also see the elucidation in [13]).

Remark 1.1. If I is an ideal of $\mathcal{O}_K$ (not contained in $\mathbb{Z}$), then $I = [a, b+cw]$ where $a, b, c \in \mathbb{Z}, a > 0, c > 0, c \mid b$, $c \mid a$ and $ac \mid N(b+cw)$. For a given I in $\mathcal{O}_K$ a and c are unique and a is the least positive rational integer in I, denoted $L(I) = a$. Also let $ca = N(I) =$ norm of I. If $I = (\alpha)$, principal, then $N(I) = |N(\alpha)|$. If $I \sim J$ then there is a $\gamma \in I$ such that $(\gamma)J = (L(J))I$. Here $\sim$ denotes equivalence of ideals in the class group.

*The author gratefully acknowledges the financial support of N.S.E.R.C. Canada grant number A8484.

Computational Number Theory

An ideal I is said to be *primitive* if $L(I) = N(I)$; i.e., $c = 1$. I is called a *reduced ideal* in $\mathcal{O}_K$ if I is primitive and there does *not* exist a non-zero $\alpha \in I$ such that *both* $|\alpha| < L(I)$ and $|\overline{\alpha}| < L(I)$ hold.

Remark 1.2. If I is an ideal of $\mathcal{O}_K$ then there is at least one *reduced* ideal J with $I \sim J$. If I is an ideal of $\mathcal{O}_K$ then there is at least one *primitive* ideal J with $I \sim J$.

If I is a reduced ideal in $\mathcal{O}_K$ then $L(I) < \sqrt{\Delta}$. If I is a primitive ideal in $\mathcal{O}_K$ and $L(I) < \sqrt{\Delta}/2$ then I is reduced in $\mathcal{O}_K$.

Remark 1.3. Let $I = [N(I), b+w]$ be primitive then the expansion of $(b+w)/N(I)$ as a continued fraction proceeds as follows

$$(P_0, Q_0) = \begin{cases} (b, N(I)), & \text{if } d \not\equiv 1 \pmod 4, \\ (2b+1, 2N(I)), & \text{if } d \equiv 1 \pmod 4. \end{cases}$$

and, recursively for $i = 0, 1, \ldots$ $d = P_{i+1}^2 + Q_i Q_{i+1}$, $P_{i+1} = a_i Q_i - P_i$ where $a_i = \lfloor (P_i + \sqrt{d})/Q_i \rfloor$ with $\lfloor\ \rfloor$ being the greatest integer function.

Thus $1 \leq Q_i < 2\sqrt{d}$ and $1 \leq P_i < \sqrt{d}$ for $1 \leq i \leq k$. Let $a = a_0$ then $(b+w)/N(I) = < a, \overline{a_1, a_2, \ldots, a_k} >$ as a continued fraction of *period* length k.

Remark 1.4. Let $I = [N(I), b+w_0]$ be reduced then the expansion of $(b+w)/N(I)$ into a continued fraction yields *all* of the reduced ideals in $\mathcal{O}_K$ equivalent to I; i.e.,

$$I_1 = [Q_0/\sigma, (P_0 + \sqrt{d})/\sigma] = I \sim I_2 = [Q_1/\sigma, (P_1 + \sqrt{d})/\sigma] \sim \ldots \sim$$
$$I_k = [Q_{k-1}/\sigma, (P_{k-1} + \sqrt{d})/\sigma].$$

Thus the $(P_i + \sqrt{d})/Q_i$ are complete quotients in the continued fraction expansion of $(b+w)/N(I)$ and the Q_i/σ's represent *norms* of all reduced ideals equivalent to I.

2 Reduced ideals, continued fractions and $h(d) = 2$

The development in Section 1 suggests the following generalization of [2] Proposition 2, p.169.

Proposition 2.1. *Let $I = [N(I), b+w]$ be a reduced ideal in $\mathcal{O}_K$.*

(a) *If J is also reduced and $I \sim J$ then $N(J) = Q_i/\sigma$ for some i with $1 \leq i \leq k$, where the Q_i's appear in the continued fraction expansion of $(b+w)/N(I)$.*

(b) *If J and $\overline{J}$ are the only ideals of Norm $N(J)$, where J is reduced, and $N(J) = Q_i/\sigma$ for some i with $1 \leq i \leq k$ in the continued fraction expansion of $(b+w)/N(I)$ then either $J \sim I$ or $\overline{J} \sim I$.*

Remark 2.1. To get Louboutin's result [ibid] we merely take I to be in the principal class in which case $N(I) = 1$ and $b = 0$; whence, if J and $\overline{J}$ are the only ideals of norms $N(J)$ then $J \sim 1$ if and only if $N(J) = Q_i/\sigma$ for some i with $1 \leq i \leq k$ in the continued fraction expansion of w.

Now we would like to find criteria for $h(d) = 2$ in terms of the factorization of certain quadratic polynomials much in the same way as Hendy [1] accomplished the task for complex quadratic fields of class number 2. It turns out, that we can use Proposition 2.1 to explicitely do this for ERD types.

3 Quadratic polynomials, class number 2 and ERD types

In [1] Hendy provided necessary and sufficient conditions for $h(-d) = 2$ in terms of prime-producing quadratic polynomials. It is the goal of this section to provide such criteria for $h(d) = 2$ when d is of ERD-type; i.e., $d = \ell^2 + r$ where $4\ell \equiv 0 \pmod{r}$. Let

$$f_d(x) = \begin{cases} -x^2 + x + \dfrac{d-1}{4}, & \text{if } d \equiv 1 \pmod 4, \\ -x^2 + d, & \text{if } d \equiv 2, 3 \pmod 4. \end{cases}$$

Consider a reduced ideal $I = [N(I), -b + w]$ then we have $f_d(b) = -N(w - b)$. Thus the $f_d(x)$ is our canonical choice of a *norm-induced* polynomial which we will use to give an analogue of the aforementioned result of Hendy [1]. First we deal with the case where $d \equiv 2, 3 \pmod 4$.

Theorem 3.1. *Let $d \not\equiv 1 \pmod 4$, $d > 3$ be of ERD type $d = \ell^2 + r$ and assume that d cannot be written in the form $d = m^2 \pm 2$. Then $h(d) = 2$ if and only if (for $1 \leq x \leq \ell$) $d - x^2$ is*

(i) *(for $|r|$ odd) prime, twice a prime, $|r|$ times a prime, $2|r|$ times a prime, or the product of two primes $(\ell + (r-1)/2)(\ell - (r-1)/2)$ which occurs at $|r+1|/2$, and $|r|$ is prime or 1.*

(ii) *(for $|r|$ even) prime, twice a prime, $|r|/2$ times a prime or $|r|$ times a prime and $|r|/2$ is prime. Moreover, if $r < 0$ then additionaly $(r + 4\ell - 4)/2$ times a prime where $(r + 4\ell - 4)$ is prime.*

Proof. First assume that $h(d) = 2$. Let

$$\alpha = \begin{cases} 1, & \text{if } d \equiv 3 \pmod 4, \\ 0, & \text{if } d \equiv 2 \pmod 4. \end{cases}$$

We will first calculate, for each case, the continued fraction expansion of both $\sqrt{d}$ and $(\sqrt{d} + \alpha)/2$, and then analyze the factorization of $d - x^2$ for each case. Note that the ideal $J = [2, \alpha + \sqrt{d}]$ above 2 is not principal since $|r| \neq 2$ (where we invoke Proposition 2.1 and an examination of the continued fraction expansion of $\sqrt{d}$ given below). Observe that $\lfloor \sqrt{d} \rfloor = \ell$ or $\ell - 1$. Thus we have

CASE I. $|r|$ odd.

CASE I(A). $\lfloor\sqrt{d}\rfloor = \ell$, (whence $r > 0$). The continued fraction expansion of $\sqrt{d}$ is then represented by

i	0	1	2
P_i	0	ℓ	ℓ
Q_i	1	r	1
a_i	ℓ	$2\ell/r$	2ℓ

That of $(\sqrt{d}+\alpha)/2$ is

i	0	1	2
P_i	α	$\ell-1$	$(r+1)/2$
Q_i	2	$(2\ell+r-1)/2$	$(2\ell-r+1)/2$
a_i	$(\ell+\alpha-1)/2$	1	$\begin{cases} 1 & \text{if } r<\ell \\ 2 & \text{if } r=\ell \end{cases}$

i	3	4
P_i	$\begin{cases} \ell-r & \text{if } r<\ell \\ (r+1)/2 & \text{if } r=\ell \end{cases}$	$\begin{cases} \ell-r & \text{if } r<\ell \\ \ell-1 & \text{if } r=\ell \end{cases}$
Q_i	$\begin{cases} 2r & \text{if } r<\ell \\ (2\ell+r-1)/2 & \text{if } r=\ell \end{cases}$	$\vdots$
a_i	$\begin{cases} (\ell-r)/r & \text{if } r<\ell \\ 1 & \text{if } r=\ell \end{cases}$	$\vdots$

CASE I(B). $\lfloor\sqrt{d}\rfloor = \ell-1$, (whence $r<0$), then for $\sqrt{d}$

i	0	1	2	3
P_i	0	$\ell-1$	$r+\ell$	$r+\ell$
Q_i	1	$r+2\ell-1$	$-r$	$\vdots$
a_i	$\ell-1$	1	$-2(\ell+r)/r$	$\vdots$

Observe that if $r = -1$ then the period length k is 2. The expansion for $(\sqrt{d}+\alpha)/2$ is

i	0	1	2	3
P_i	α	$\ell-1$	$\ell+2$	$\ell+r$
Q_i	2	$(2\ell+r-1)/2$	$-2r$	$\vdots$
a_i	$(\ell+\alpha-1)/2$	2	$-(\ell+r)/r$	$\vdots$

CASE II. $|r|$ even.

CASE II(A). $\lfloor\sqrt{d}\rfloor = \ell$ (whence $r > 0$). The expansion for $\sqrt{d}$ in this case is the same as in I(a) whereas the expansion of $(\sqrt{d}+\alpha)/2$ is

i	0	1	2
P_i	α	ℓ	ℓ
Q_i	2	$r/2$	$\vdots$
a_i	$(\ell+\alpha)/2$	$4\ell/r$	$\vdots$

CASE II(B). $\lfloor\sqrt{d}\rfloor = \ell-1$ (whence $r < 0$). The expansion of $\sqrt{d}$ is the same as in I(b), while the expansion for $(\sqrt{d}+\alpha)/2$ is

i	0	1	2	3
P_i	α	$\ell-2$	$(r+2\ell)/2$	$(r+2\ell)/2$
Q_i	2	$(r+4\ell-4)/2$	$-r/2$	$\vdots$
a_i	$(\ell+\alpha-2)/2$	1	$-2(2\ell+r)/r$	$\vdots$

Now we examine $d-x^2$ for each case. Let p and q be odd primes dividing d such that $p < q$.

We *first assume* that $\wp \sim 1$ where $\wp$ lies over p, then in Case I(a) $p = r$. If $q > \sqrt{d}$ then no other odd prime $s \neq p, q$ can divide $d - x^2$. The reason is that $s < \sqrt{d}$ in that case and so by Proposition 2.1, $s = (2\ell+r-1)/2$ or $s = (2\ell-r+1)/2$. In either instance we get $pqs > d-1$. However, $d-1 \geq d-x^2 > pqs$, a contradiction. We have shown that if $q > \sqrt{d}$ then $d-x^2$ is r times a prime or $2r$ times a prime, (since $d-x^2 \not\equiv 0 \pmod 4$).

Now suppose that $q < \sqrt{d}$. Therefore by Proposition 2.1 we must have $q = (2\ell+r-1)/2$ or $q = (2\ell-r+1)/2$. In this case, if there is a third odd prime dividing d and $s > \sqrt{d}$ then we get a contradiction as above. If $s < \sqrt{d}$ then $p = r$, $q = (2\ell+r-1)/2$ and $s = (2\ell-r+1)/2$ (without loss of generality) by Proposition 2.1. However, we then get $pqs \geq d-1$, a contradiction as above. We have shown, that if $q < \sqrt{d}$ then $d-x^2$ is r times a prime or $2r$ times a prime.

If I(b) holds then $p = -r$, (observing that $r + 2\ell - 1$ is even). As in I(a) above, if $q > \sqrt{d}$, then no other odd prime divides $d - x^2$, and the result follows. If $q < \sqrt{d}$ then $q = (2\ell + r - 1)/2$, whence no third odd prime divides $d - x^2$ and the result follows.

In the case where II(a) holds Proposition 2.1 does not allow $\wp \sim 1$ since $|r|$ is even. If II(b) holds then $p = r + 2\ell - 1$. Thus $q < \sqrt{d}$ is forced; whence, $q = -r/2$ or $q = (r + 4\ell - 4)/2$. As above there cannot be a third odd prime dividing $d - x^2$. Hence, we have shown in this case, that $d - x^2$ is $-r/2$ times a prime or $-r$ times a prime. This completes the case where $\wp \sim 1$ and two odd primes divide $d - x^2$.

Now assume $\wp \not\sim 1$ and two odd primes divide $d - x^2$. Therefore, by Proposition 2.1, if Case I(a) holds then $p = (2\ell + r - 1)/2$ or $p = (2\ell - r + 1)/2$. If $q > \sqrt{d}$ then $pq \geq d - 1$, a contradiction, unless $r = 1$ and $\ell = p$ in which case we are done. Therefore $q < \sqrt{d}$ so either $q = r$ or $q = (2\ell + r - 1)/2$ (when $p = (2\ell - r + 1)/2$), or $q = (2\ell - r + 1)/2$ (when $p = (2\ell + r - 1)/2$). As above no third odd prime can divide $d - x^2$. We have shown that $d - x^2$ is r times a prime, $2r$ times a prime or $(\ell + (r-1)/2)(\ell - (r-1)/2)$, a product of two primes at $x = (r+1)/2$.

In Case I(b), $p = (2\ell + r - 1)/2$. Again $q > \sqrt{d}$ leads to a contradiction so $q < \sqrt{d}$; whence $q = -r$. For similar reasons to the previous cases no third odd prime can divide $d - x^2$. Thus, in this case $d - x^2$ is $-r$ times a prime or $-2r$ times a prime.

If Case II(a) holds then $p = r/2$. If $q < \sqrt{d}$ then Proposition 2.1 leads to $q = r/2$, a contradiction. Thus, $q > \sqrt{d}$ and no third odd prime can divide $d - x^2$. Therefore $d - x^2$ is $r/2$ times a prime or r times a prime. In Case II(b) we get $p = -r/2$ or $p = (r + 4\ell - 4)/2$. Again $q < \sqrt{d}$ is forced and no third odd prime can divide $d - x^2$. Thus $d - x^2$ is $-r/2$ times a prime or $-r$ times a prime.

This completes the case where two odd primes divide $d - x^2$. If only one odd prime divides $d - x^2$ then it is prime or twice a prime. Thus we have a complete analysis of the factorization of $d - x^2$.

It remains to show that $|r|/2$ and $(r + 4\ell - 4)/2$ are primes when $|r|$ is even, and $\ell + (r-1)/2, \ell - (r-1)/2$ and $|r|$ are primes (or $|r| = 1$) when $|r|$ is odd. The arguments for each of these is essentially the same so we only present the proof for $r > 0$ odd. If an odd prime p divides r then by Proposition 2.1, $p = r$, $p = (2\ell + r - 1)/2$ or $p = (2\ell - r + 1)/2$. However, in the latter two cases we get a contradiction.

Conversely, assume that (i) and (ii) hold. It suffices to show that a given reduced ideal $I = [N(I), b + w]$, either $I \sim 1$ or $I \sim J$.

CASE 1. $|r|$ is odd.

If $r > 0$ then from the above calculations for $\sqrt{d}$ and $(\sqrt{d} + \alpha)/2$ we can get that $J \sim L$ or L' where L is a prime above $(2\ell + r - 1)/2$ and L' is a prime above $(2\ell - r + 1)/2$. If $r < 0$ then $J \sim L$. In either instance $R \sim 1$ where R is the prime over $|r|$.

We have that $d - P_{i+1}^2 = Q_iQ_{i+1}$ in the continued fraction expansion of $(b + w)/N(I)$. Since $1 \le P_{i+1} < \sqrt{d}$ then $1 \le P_{i+1} \le \ell$. Then the hypothesis holds so that one of the following occurs

Q_iQ_{i+1} is prime, forcing $I \sim 1$.
Q_iQ_{i+1} is twice a prime, forcing either $I \sim 1$ or $I \sim J$.
Q_iQ_{i+1} is $|r|$ times a prime, forcing $I \sim R \sim 1$.
Q_iQ_{i+1} is $2|r|$ times a prime, forcing $I \sim 1$, $I \sim J$, or $I \sim JR \sim J$.
Q_iQ_{i+1} is $((2\ell + r - 1)/2)((2\ell - r + 1)/2)$, forcing $I \sim LL' \sim J^2 \sim 1$.

CASE 2. $|r|$ is even has essentially the same argument. □

Example 3.1. Let $d = 122 = 2 \cdot 61 = 11^2 + 1$ with $h(122) = 2$.

x	122	$-$	x^2
1	121	$=$	11^2
2	118	$=$	$2 \cdot 59$
3	113		
4	106	$=$	$2 \cdot 53$
5	97		
6	86	$=$	$2 \cdot 43$
7	73		
8	58	$=$	$2 \cdot 29$
9	41		
10	22	$=$	$2 \cdot 11$
11	1		

Remark 3.1. Example 3.1 shows that when $d = \ell^2 + 1 \equiv 2 (\text{mod } 4)$, $h(d) = 2$ if and only if $d - x^2$ is a prime or twice a prime for all x with $1 \le x \le \ell$. This phenomenon has a generalization.

Proposition 3.1. *Let p a fixed prime dividing d. If $f_d(x)$ is prime or p times a prime for all x with $1 \le x \le \omega$ then $h(d) \le 2$.*

Proof. It suffices to show that for a given reduced ideal $I = [N(I), b + w] \sim 1$ or $I \sim P$ where P is the ideal above p. Expand $(b + w)/N(I)$ and consider $d - P_{i+1}^2 = Q_iQ_{i+1}$ for $i \ge 1$.

CASE 1. $d \equiv 2, 3 (\text{mod } 4)$. Since $P_{i+1} < \sqrt{d}$ then by hypothesis Q_iQ_{i+1} is prime or p times a prime. Therefore $I \sim P$ or $I \sim 1$.

CASE 2. $d \equiv 1 (\text{mod } 4)$. Since Q_i is even then we may set $P_{i+1} = 2x - 1$ to get $f_d(x) = -x^2 + x + (d-1)/4 = Q_iQ_{i+1}/4$ where $x = (P_{i+1}+1)/2 < (\sqrt{d}+1)/2 = \omega$. Thus by hypothesis $Q_iQ_{i+1}/4$ is prime or p times a prime; i.e., $I \sim 1$ or $I \sim P$. □

The following examples illustrate remaining cases of Theorem 3.1.

Example 3.2. ($r > 0$ even). Let $d = 447 = 21^2 + 6$ with $h(d) = 2$. Thus for $\sqrt{d}$

i	0	1	2
P_i	0	21	21
Q_i	1	6	1
a_i	21	7	42

and for $(1 + \sqrt{d})/2$

i	0	1	2
P_i	1	21	21
Q_i	2	3	2
a_i	11	14	22

and,

x	d	$-$	x^2
1	446	=	$2 \cdot 223$
2	443		
3	438	=	$2 \cdot 3 \cdot 73$
4	431		
5	422	=	$2 \cdot 211$
6	411	=	$3 \cdot 137$
7	398	=	$2 \cdot 199$
8	383		
9	366	=	$2 \cdot 3 \cdot 61$
10	347		
11	326	=	$2 \cdot 163$
12	303	=	$3 \cdot 101$
13	278	=	$2 \cdot 139$
14	251		
15	222	=	$2 \cdot 111$
16	191		
17	158	=	$2 \cdot 79$
18	128	=	$3 \cdot 41$
19	86	=	$2 \cdot 43$
20	47		
21	$3 \cdot 7$		

Example 3.3. ($r < 0$ even). Let $d = 215 = 15^2 - 10$ with $h(d) = 2$.

x	$215 - x^2$	
1	214	$= 2 \cdot 107$
2	211	
3	206	$= 2 \cdot 103$
4	199	
5	190	$= 2 \cdot \cdot 5 \cdot 19$
6	179	
7	166	$= 2 \cdot 83$
8	151	
9	134	$= 2 \cdot 67$
10	115	$= 5 \cdot 23$
11	94	$= 2 \cdot 47$
12	71	
13	46	$= 2 \cdot 23$
14	19	

Note that $(r + 4\ell - 4)/2 = 23$.

Example 3.4. ($r < 0$ odd). Let $d = 143 = 12^2 - 1$ with $h(d) = 2$.

x	$143 - x^2$	
1	142	$= 2 \cdot 71$
2	139	
3	134	$= 2 \cdot 67$
4	127	
5	118	$= 2 \cdot 59$
6	107	
7	94	$= 2 \cdot 47$
8	79	
9	62	$= 2 \cdot 31$
10	43	
11	22	$= 2 \cdot 11$

Remark 3.2. Theorem 3.1 characterizes $h(d) = 2$ for $d \equiv 2, 3 \pmod 4$ of ERD types (except for the troublesome forms $d = \ell^2 \pm 2$) in terms of the factorization of $d - x^2$. To solve the problem for $d \equiv 1 \pmod 4$ in terms of $-x^2 + x + (d-1)/4$ remains open although Proposition 3.1 provides some evidence.

We now explicitly determine all ERD types $d \equiv 1 \pmod 8$ with $h(d) = 2$.

Theorem 3.2. *If $d \equiv 1 \pmod 8$ and d is of ERD type then $h(d) = 2$ if and only if $d = 65$ or 105.*

Proof. Let $h(d) = 2$ and $\wp = [2, w]$. It can be shown in a fashion similar to that used in the proof of Theorem 3.1 that $\wp$ is not principal. However $\wp^2 \sim 1$ and, by Proposition 3.1 this forces $Q_i = 8$ for some i with $1 \le i \le k$, whenever $4 < \sqrt{d}/2$. Since $d \equiv 1 \pmod 8$ then clearly $d = \ell^2 + r$ with ℓ even. Thus, by an analysis similar to that given in the proof of Theorem 3.1 we reduce to only two cases for the continued fraction expansion of w.

Case 1. $\lfloor\sqrt{d}\rfloor = \ell = 2a$ with $\ell/2 \ge r \ge 1$. Hence $Q_1 = (2\ell + r - 1)/2$, $Q_2 = (2\ell - r + 1)/2$ and $Q_3 = 2r$ with $k = 6$ unless $r = 1$ in which case $k = 3$. If $Q_1/2 = 4$ then $2\ell + r = 17$ forcing $r = 1$ and $\ell = 8$, i.e., $d = 65$. If $Q_2/2 = 4$ then $2\ell - r = 15$ forcing $r = 5$ so $d = 105$ or $r = 1$ where $d = 65$. Clearly $Q_3/2 \ne 4$, so that completes this case.

Case 2. $\lfloor\sqrt{d}\rfloor = \ell - 1 = 2a - 1$ with $-\ell/2 \le r < -1$. Here $Q_1 = (2\ell + r - 1)/2$ and $Q_2 = -2r$ with $k = 4$. If $Q_1/2 = 4$ then $r = -1$ and $d = 80$, a contradiction. If $Q_2/2 = 4$ then r is even, a contradiction.

Hence if $d > 64$ then $h(d) = 2$ if and only if $d = 65$ or 105. A quick check of $d < 64$ shows no d with d of ERD-type and $h(d) = 2$. □

Remark 3.3. We may not employ the same technique as in Theorem 3.2 to investigate the $d \not\equiv 1 \pmod 4$ of ERD type. The reason is that although $4 < \sqrt{d}$ may hold we do *not* have $\wp^2$ primitive for $\wp$ above 2 so we may not employ Proposition 2.1.

Remark 3.4. The $d \equiv 5 \pmod 8$ case has not been handled in terms of prime producing quadratic polynomial criteria for $h(d) = 2$ when d is of ERD type because there is no convenient prime in general to investigate, as there is with $p = 2$ in the $d \not\equiv 5 \pmod 8$ case. Nevertheless, even in the latter case Theorem 3.1 shows that d's of the form $\ell^2 \pm 2$ are problematic because in that case the ideals above 2 are principal, so we again do not have a convenient prime to investigate. Nevertheless restricting to $d = \ell^2 + r$ with $|r| \in \{1, 4\}$ does allow more to be said. Let $h = h(d)$ in what follows.

Theorem 3.3. *Let $d = \ell^2 + r$ with $|r| \in \{1, 4\}$. Then p is inert for all primes p with $p^h < \omega$ unless $h \equiv 0 \pmod 2$ in which case p may be ramified.*

Proof. Let $p^h < \omega$ and let $\wp$ be an $\mathcal{O}_K$-prime above p. If $\wp$ is not inert then $N(\wp^h) = \pm p^h = (x - dy^2)/\sigma^2$.

The following shows that we may assume $y \ne 0$.

CLAIM 1. If $y = 0$ then p is ramified and h is even.

We have $\wp^h = \left(\frac{x+y\sqrt{d}}{\sigma}\right)$. If $y = 0$ then $\frac{x+y\sqrt{d}}{\sigma} = \lambda$, a rational integer, and $\wp^h = \overline{\wp}^h$; i.e., $\wp^h$ is ambiguous. However the only non-trivial (primitive) ambiguous ideals are those whose ideal prime factors are ramified. (To see this set $\mathcal{A} = [N(\mathcal{A}), b+w] = \overline{\mathcal{A}} = [N(\mathcal{A}), b+\overline{w}]$. Then $w - \overline{w} \in \mathcal{A}$ forcing $N(d) \mid N(w-\overline{w})$; i.e., primes dividing $\mathcal{A}$ must ramify). Also $|N(\wp^h)| = p^h = \lambda^2$, so h is even.

CLAIM 2. If p splits in K then $p^h \geq \omega$.

By [3] Lemma 1.1, p.40 since $\pm p^h\sigma^2 = x^2 - dy^2$ has a nontrivial solution (x, y) then $p^h \geq ((2t_d)/\sigma - N(\varepsilon_d) - 1)/u_d^2 = B$ where the fundamental unit ε_d of K is given by $\varepsilon_d = (t_d + u_d\sqrt{d})/\sigma$.

It suffices now to check that this bound satisfies $B \geq \lfloor\omega\rfloor$.

If $r = 1$ and $\sigma = 2$ then $N(\varepsilon_d) = -1$ so $B = t_d/u_d^2 = \ell/2 = \lfloor\omega\rfloor$. If $r = 1$ and $\sigma = 1$ then $B = 2t_d/u_d^2 = 2\ell = 2\lfloor\omega\rfloor$. If $r = -1$ then $\sigma = 1$ and $N(\varepsilon_d) = 1$ so $B = (2t_d - 2)/u_d^2 = 2(\ell - 1) = 2\lfloor\omega\rfloor$. If $r = 4$ then $\sigma = 2$ and $N(\varepsilon_d) = -1$ so $B = t_d/u_d^2 = \ell/2 = \lfloor\sqrt{d}\rfloor/2$. If $r = -4$ then $\sigma = 2$ and $N(\varepsilon_d) = 1$ so $B = \ell - 2 \geq \lfloor\omega\rfloor$. □

Corollary 3.1. *If p splits then $h(d) \geq \log\omega/\log p$.*

Applications.

(1) Let $d = 4m^2 + 1$. By Theorem 3.3 $d \equiv 1 \pmod 8$ forces $m \leq 2^{h(d)}$ whence

$$\begin{aligned} h(d) = 1 &\quad \text{if and only if} \quad d = 17\ (m = 2),\\ h(d) = 2 &\quad \text{if and only if} \quad d = 65\ (m = 4),\\ h(d) = 3 &\quad \text{if and only if} \quad d = 257\ (m = 8),\\ h(d) = 4 &\quad \text{if and only if} \quad d = 145\ (m = 6). \end{aligned}$$

We may continue the use of Theorem 3.3 as an algorithm for finding all such $h(d)$.

(2) $d \equiv 5 \pmod 8), d = 4m^2 + 1$. Let p be the least prime quadratic residue modulo d then $d \leq 4p^{2h(d)} + 1$ by Theorem 3.3.

(3) $d \equiv 2 \pmod 4$ and $d = (2m + 1)^2 + 1$. Here we could take the least odd prime quadratic residue and get a bound as in (2). However, we may invoke Theorem 3.3, and get that $2m + 1$ is prime if $h(d) = 2$. Moreover, if there is a prime $p \mid m$ with $(2/p) = 1$ then $m = p$ if $h(d) = 1$.

(4) Finally note that if $d \not\equiv 5 \pmod 8$ $d = \ell^2 + r$ with $|r| \in \{1, 4\}$ and $h(d)$ is odd then $4^{h(d)} \geq d$.

References

[1] M.D. Hendy, *Prime Quadratics Associated With Complex Quadratic Fields of Class number Two*, Proc. Amer. Math. Soc. **43** (1974), 253–260.

[2] S. Louboutin, *Continued Fraction and Real Quadratic Fields*, J. Number Theory **30** (1988), 167–176.

[3] R.A. Mollin, *On the Insolubility of a Class of Diophantine Equations and the Non-triviality of the Class Numbers of Related Real Quadratic Fields of Richaud-Degert Type*, Nagoya Math. J. **105** (1987), 39–47.

[4] R.A. Mollin, *Diophantine Equations and Class Numbers*, J. Number Theory **24** (1986), 7–19.

[5] R.A. Mollin, *Necessary and Sufficient Conditions for the Class Number of a Real Quadratic Field to be One and a Conjecture of S.Chowla*, Proc. Amer. Math. Soc. **102** (1988), 17–21.

[6] R.A. Mollin, *Class Number One Criteria for Real Quadratic Fields I*, Proc. Japan. Acad **63** Ser. A. (1987), 121–125.

[7] R.A. Mollin, *Class Number One Criteria for Real Quadratic Fields II*, Proc. Japan. Acad **63** Ser. A. (1987), 162–164.

[8] R.A. Mollin, *Class numbers and the Divisor Function* (to appear).

[9] R.A. Mollin and H.C. Williams, *Continued Fractions of Period Five and Real Quadratic Fields of Class Number One*, Acta Arith. **LVI** (1990), 55–63.

[10] R.A. Mollin and H.C. Williams, *Quadratic Non-Residues and Prime-Producing Polynomials*, Canad. Math. Bull. **32** (1989), 474–478.

[11] R.A. Mollin and H.C. Williams, *Class Number One for Real Quadratic Fields, Continued Fractions and Reduced Ideals*, in "Number Theory and Applications", Kluwer Academic Publishers (R.A. Mollin (ed.)) (NATO ASI), vol.C 265, 1989, 481–496.

[12] R.A. Mollin and H.C. Williams, *Solution of the Class Number One Problem for Real Quadratic Fields of Extended Richaud-Degert type (with one possible exception)*, in "Number Theory" (R.A. Mollin (ed.)) Walter de Gruyter, Berlin, 1990, 417–425.

[13] R.A. Mollin and H.C. Williams, *Computation of the Class Number of a Real Quadratic Field* (to appear: Advances in the Theory of Computing and Comp. Math.).

[14] H.C. Williams and M.C. Wunderlich, *On the Parallel Generation of the Residues for the Continued Fraction Factoring Algorithm*, Math. Comp. **177** (1987), 405–423.

On a solution of a class number two problem for a family of real quadratic fields *

R.A. Mollin and H.C. Williams

Abstract. The main result of this paper is a solution of the class number 2 problem for real quadratic fields of Extended Richaud-Degert Type, (with one possible exception remaining). We also elucidate a consequence of this result as the refinement of a class number 2 criterion found in earlier work. We also describe the solution of the class number 1 problem for ERD-types (and its consequences) which provide a basis for the work herein.

1 Notation and preliminaries

A real quadratic field $\mathbb{Q}(\sqrt{d})$, (d positive square-free) is said to be of *Extended Richaud-Degert* (ERD)-type if $d = \ell^2 + r$ where $4\ell \equiv 0 \pmod{r}$. Let

$$\omega = (\sigma - 1 + \sqrt{d})/\sigma \ , \quad \text{where } \sigma = \begin{cases} 1 \ , & \text{if } d \equiv 2,3 \pmod 4, \\ 2 \ , & \text{if } d \equiv 1 \pmod 4, \end{cases}$$

and let $\mathcal{O}_K$ be the maximal order of $\mathbb{Q}(\sqrt{d})$. The discriminant Δ of K is given by $\Delta = (2/\sigma)^2 d$.

Also if by $[\alpha, \beta]$ we denote the module $\alpha\mathbb{Z} + \beta\mathbb{Z} = \{x\alpha + y\beta : x, y \in \mathbb{Z}\}$ then $\mathcal{O}_K = [1, \omega]$.

An ideal I of $\mathcal{O}_K$ can be written as $I = [a, b + c\omega]$; with $a, b, c \in \mathbb{Z}$, $a > 0$, $c \mid b$, $c \mid a$ and $ac \mid N(b + c\omega)$ where N denotes the norm from K to $\mathbb{Q}$. The ideal *conjugate* to I, denoted by $\overline{I}$, is given by $\overline{I} = [a, b + c\overline{\omega}]$ where $\overline{\omega}$ is the algebraic conjugate of ω. Furthermore if $a, b, c, \in \mathbb{Z}$, $c \mid b$, $c \mid a$, $ac \mid N(b + c\omega)$ then $[a, b + c\omega]$ is an ideal of $\mathcal{O}_K$. If $|c| = 1$; i.e., $I = [a, b + \omega]$ then I is a *primitive ideal.* Henceforth all ideals are considered to be primitive. For such $I, N(I) = a > 0$.

---

*Mathematics Subject Classification (1985 revision) 12A50 12A25
Key Words and Phrases, Real Quadratic Field, Class Number, Continued Fractions.
The first author's research is supported by NSERC Canada grant # A8484.
The second author's research is supported by NSERC Canada grant # A7649.

Computational Number Theory

A primitive ideal is said to be *reduced* if it does *not* contain any non-zero element α satisfying

$$|\alpha| < N(I), \quad |\overline{\alpha}| < N(I).$$

Proofs of the following fundamental results on reduced ideals can be found in [10].

Theorem 1.1. *I is a reduced ideal of $\mathcal{O}_K$ if and only if there exists some $\beta \in I$ such that $I = [N(I), \beta]$ with $\beta > N(I)$ and $-N(I) < \overline{\beta} < 0$.*

From Theorem 1.1 we get:

Corollary 1.1. *If I is a reduced ideal of $\mathcal{O}_K$ then $N(I) < \sqrt{\Delta}$.*

Corollary 1.1 is a necessary but not sufficient condition. For sufficiency,

Corollary 1.2. *If I is a primitive ideal of $\mathcal{O}_K$ and $N(I) < \sqrt{\Delta}/2$ then I is a reduced ideal of $\mathcal{O}_K$.*

For convenience sake we gather together some basic facts concerning continued fractions which we will be using throughout the remainder of the paper. Let ω be as above and let $\omega = < a, \overline{a_1, \ldots, a_k} >$ denote the continued fraction expansion of ω. Then $a_0 = a = \lfloor \omega \rfloor$, where $\lfloor \ \rfloor$ denotes the greatest integer function, and $a_i = \lfloor (P_i + \sqrt{d})/Q_i \rfloor$ for $i \geq 1$ where

$$P_0 = \begin{cases} 1\,, \\ 0\,, \end{cases} \quad Q_0 = \begin{cases} 2\,, & \text{if } d \equiv 1 \pmod 4, \\ 1\,, & \text{if } d \equiv 2,3 \pmod 4, \end{cases}$$

and

$$P_{i+1} = a_i Q_i - P_i\,, \quad Q_{i+1} Q_i = d - P_{i+1}^2 \quad \text{for} \quad i \geq 0\,.$$

Finally we will need some data concerning some basic analytic results concerning the class number $h(d)$. If ϵ (> 1) is the fundamental unit of $K = \mathbb{Q}(\sqrt{d})$ then $R = \log \epsilon$ is the *regulator* of K and the analytic class number formula asserts that $2h(d)R = \sqrt{\Delta} L(1, \chi_\Delta)$, where $L(1, \chi_\Delta) = \sum_{n=1}^{\infty} \left(\frac{\Delta}{n}\right) \frac{1}{n}$ and $\left(\frac{\cdot}{n}\right) = \chi_\Delta(n)$ is the Kronecker symbol.

2 Introduction

We begin with an overview of the class number one problem for ERD types which provides the foundation. In [8] we established the following result.

Theorem 2.1. *If d is of ERD-type then $h(d) = 1$ if and only if $d \in \{$ 2, 3, 5, 6, 7, 11, 13, 14, 17, 21, 23, 29, 33, 37, 38, 47, 53, 62, 69, 77, 83, 93, 101, 141, 167, 173, 197, 213, 227, 237, 293, 398, 413, 437, 453, 573, 677, 717, 1077, 1133, 1253, 1293, 1757 $\}$ with possibly only one more value remaining.*

Since the authors established in [6] - [7], under the assumption of the GRH (generalized Riemann hypothesis), that these are *exactly* the values of d for which $h(d) = 1$ then the existence of the exceptional value would be a counterexample to the Riemann hypothesis. Moreover, recent evidence communicated to the authors shows that if the exceptional value exists then it must exceed 10^{500}.

It is worth mentioning that Theorem 2.1 is highly relevant to several conjectures in the literature. They are

Conjecture 2.1. (S. Chowla [1]) *If* $p = m^2 + 1$ *is prime and* $m > 26$ *then* $h(p) > 1$.

Conjecture 2.2. (H. Yokoi [11]) *If* $d = m^2 + 4$ *is square-free and* $m > 17$ *then* $h(d) > 1$.

Conjecture 2.3. (R.A. Mollin [3]) *If* $d = m^2 - 4$ *is square-free and* $m > 21$ *then* $h(d) > 1$.

Conjecture 2.4. (R.A. Mollin and H.C. Williams [7]) *If* $d = m^2 \pm 2$ *is square-free and* $m > 20$ *then* $h(d) > 1$.

Conjecture 2.5. (R.A. Mollin and H.C. Williams [7]) *If* $d = m^2 - p$ *is square-free where* p *is an odd prime dividing* m *with* $m > 42$ *then* $h(d) > 1$.

Conjecture 2.6. (R.A. Mollin and H.C. Williams [7]) *If* $d = m^2 \pm 4\ell$ *is square-free where* $\ell > 1$ *is an odd integer dividing* m *with* $m > 39$ *then* $h(d) > 1$.

Remark 2.1. From results of the first author [3] - [4] it follows that if $h(d) = 1$ and d is of ERD type then d must be one of the forms in conjectures 2.1-2.6.

Remark 2.2. It follows from Theorem 2.1 and Remark 2.1 that five of the six conjectures 2.1-2.6 are true (but we don't know which ones) and the remaining one fails for at most one value, whose existence would provide a counterexample to the GRH. There is no mystery to this beyond the exeptional value which Tatuzawa [9] could not eliminate.

Remark 2.3. As a result of Remark 2.2 the situation for real quadratic fields of ERD type having class number one is at exactly the same position as was the general class number one problem for complex quadratic fields before the Baker-Stark-Heegner solution. However, to eliminate the possibility of one more value in Theorem 2.1 appears to be extremely difficult at this juncture. The results in the following section (for $h(d) = 2$), improve upon the techniques used to prove Theorem 2.1.

3 ERD-types of class number 2

Since we are concerned with determining all ERD-types of class number 2 then we need to know the maximum period length k of ω, which they can take on. The following table gives us the classification.

Table 3.1. Periods of all ERD-types

k	$\lfloor\sqrt{d}\rfloor$	d	
1	2	5	
1	$2a-1$	$(2a-1)^2+4$;	$a>1$
1	a	a^2+1;	a odd ; $a \geq 1$
2	$2a-1$	$(2a-1)^2+r \equiv 1 (\bmod 4)$;	$r>0$; $r \neq 4$; $r \mid (8a-4)$
2	$2a$	$(2a+1)^2-4$;	$a>1$
2	a	$a^2+r \not\equiv 1 (\bmod 4)$;	$r \mid 2a$; $r>0$
2	a	$(a+1)^2-1$;	a odd
3	$2a$	$(2a)^2+1$;	$a \geq 2$
4	$2a-1$	$(2a)^2+r \equiv 1 (\bmod 4)$;	$r<-1$; $-r \mid a$
4	$2a$	$(2a+1)^2+r \equiv 1 (\bmod 4)$;	$r<0$; $r \neq -4$; $-r \mid 4a+2$
4	a	$(a+1)^2+r \not\equiv 1 (\bmod 4)$;	$r<0$; $r \neq -1$; $-r \mid a+1$
6	$2a$	$(2a)^2+r \equiv 1 (\bmod 4)$;	$r>1$; $r \mid a$.

Now we are in a position to begin our classification.

Theorem 3.1. *If d is of ERD-type and $\Delta > 2 \times 10^9$ then, with at most one possible exception, $h(d) > 2$.*

Proof. By Tatuzawa [9] we have that (with at most one possible exception) $L(1,\chi_\Delta) > .655\eta\Delta^{-\eta}$ for $0<\eta<1/2$ and $\Delta \geq \max(e^{1/\eta}, e^{11\cdot 2})$. From well-known analytic class number formulae we have (see section 1):

$$2h(d)R = \sqrt{\Delta}L(1,\chi_\Delta).$$

Moreover by a long and tedious calculation one can verify that for each of the d's in Table 3.1 we have that

$$R < 5\log\sqrt{\Delta}.$$

Hence if we choose $\eta = 0.048$ then for $d > 2 \times 10^9$ we have with one possible exception that:

$$2h(d) > \frac{(.655)(.048)\Delta^{0.452}}{5\log\Delta} > 4.$$

□

Now we describe the technique of dealing with the remaining $\Delta < 2 \times 10^9$. Let S be the set of all rational primes $p < \sqrt{\Delta}/2$ such that $(\Delta/p) \neq -1$ and let

$$L_1 = \{Q_i/Q_0 : \ i = 1,2,\ldots,k\} \cap \mathbb{P}$$

where the Q_i and k are defined in section 1, and $\mathbb{P}$ is the set of primes.

Proposition 3.1. *If $h(d) = 1$ then $L_1 \supseteq S$.*

Proof. If $(\Delta/p) \neq -1$ then there must exist a prime ideal $\wp$ lying over p with $N(\wp) = p < \sqrt{\Delta}/2$. Hence $\wp$ is reduced. If $h(d) = 1$ then by [10], $N(\wp) \in L_1$, hence $L_1 \supseteq S$. □

Suppose that there exists some $p \in S$ such that $p \notin L_1$. Let x be a positive solution of the congruence

$$x^2 \equiv d (\text{mod } p)$$

(such a solution must exist). Put

$$P_0' = \begin{cases} p - x \, , & \text{when } x \text{ is even and } \sigma = 2 \, , \\ x \, , & \text{otherwise} \, , \end{cases}$$

and $Q_0' = \sigma p$. The ideal $\wp = (Q_0/\sigma)\mathbb{Z} + ((P_0 + \sqrt{d})/\sigma)\mathbb{Z}$ lies over p in $\mathbb{Q}(\sqrt{d})$ and is not principal since $N(\wp) < \sqrt{\Delta}/2$ and $p \notin L_1$, (ibid.).

By expanding $\phi' = (P_0' + \sqrt{d})/Q_0'$ into a continued fraction we can compute the set

$$L_2 = \{Q_i'/\sigma : \ i = 0, 1, 2, \ldots, k'\} \cap \mathbb{P}$$

where k' is the period of the continued fraction expansion of ϕ'. We now have

Proposition 3.2. *If $h(d) = 2$ then $L_1 \cup L_2 \supseteq S$.*

Proof. Let q be any prime of S. If $h(d) = 2$ there can be only two classes of reduced ideals in $\mathbb{Q}(\sqrt{d})$, the principal class and the class of ideals equivalent to $\wp$ (above). Since $q < \sqrt{\Delta}/2$ we must have $q \in L_1$ or $q \in L_2$. □

Now we can describe our algorithm for eliminating most of the d's less than 2×10^9. For each d with $k \leq 6$ we determined L_1 and L_2. We then attempted to find a prime q such that $q \in S$ but $q \notin L_1 \cup L_2$. If this occured we could, in view of Proposition 3.2 eliminate d since $h(d) \geq 3$ in this case. The few remaining numbers were then treated more carefully for the value of $h(d)$. In this final step we eliminated all the values except those for which $h(d) = 2$. Thus we have shown

Theorem 3.2. *If $d > 0$ is of ERD type then $h(d) = 2$ if and only if (with possibly only one more value remaining) d is an element of the set* {10, 15, 26, 30, 34, 35, 39, 42, 51, 65, 66, 85, 87, 95, 102, 105, 110, 119, 122, 123, 138, 143, 146, 165, 182, 194, 203, 205, 215, 221, 230, 258, 285, 287, 318, 327, 357, 362, 365, 395, 402, 429, 447, 482, 485, 527, 533, 623, 629, 635, 645, 678, 741, 782, 843, 902, 957, 965, 1022, 1085, 1157, 1173, 1205, 1245, 1298, 1469, 1517, 1533, 1605, 1685, 1853, 1965, 2013, 2037, 2045, 2085, 2093, 2117, 2301, 2373, 2397, 2613, 2717, 2813, 3005, 3237, 3365, 3597, 3605, 3813, 4245, 4277, 4773, 4893, 5645, 5757, 5885, 5957, 6573, 7157, 7733, 8333, 9005, 14405}.

This completes our desired classification of ERD-types of class number 2. We conclude with a refinement of a class number 2 criterion made possible by the above.

Remark 3.1. Theorem 3.2 completes the task of $h(d) = 2$ for ERD-types begun in [2] where only the narrow ERD-types $m^2 + 1$ and $m^2 + 4$ were discussed. Moreover our techniques for the general case are far simpler than those for the aforementioned special case in [2].

In [5] established the following criterion for $h(d) = 2$. Let $f_d(x) = d - x^2$.

Theorem 3.3. *Let d be of ERD type with $d \not\equiv 1 (mod\ 4)$ and $d \neq x^2 \pm 2$ for any $x \geq 0$. Then $h(d) = 2$ if and only if whenever $q | f_d(x)$ for any prime $q < \sqrt{\Delta}/2$ and $1 \leq x \leq \sqrt{\Delta}/2$ then $q = Q_i/\sigma$ in the continued fraction expansion of either ω or $(b_2 + \omega)/2$ where $\wp = [2, b_2 + \omega]$ is an $\mathcal{O}_K$-ideal above 2.*

Remark 3.2. We observe that in view of Theorem 3.2 we may now modify Theorem 3.3.

Theorem 3.4. *If $d \not\equiv 1 (mod\ 4)$, is of ERD Type and $d > 1298$ then with the exception of one possible value of d of the form $\ell^2 \pm 2$ the following criterion holds. $h(d) = 2$ if and only if whenever $q | f_d(x)$ for any prime $q < \sqrt{\Delta}/2$ and $1 \leq x \leq \sqrt{\Delta}/2$ then $q = Q_i/\sigma$ in the continued fraction expansion of either ω or $(b_2 + \omega)/2$ where $\wp = [2, b_2 + \omega]$ is an $\mathcal{O}_K$-ideal above 2.*

Remark 3.3. Theorem 3.4 is (despite the possible existence of the exceptional value of form $\ell^2 \pm 2$) far more valuable in terms of applicability than was Theorem 3.3, because we may exclude the ERD-forms $m^2 \pm 2$. This is a much more palatable situation.

References

[1] S. Chowla and J. Friedlander, *Class Numbers and Quadratic Residues*, Glasgow Math. J. **17** (1976), 47–52.

[2] M.-G. Leu, *On a Determination of Certain Real Quadratic Fields of Class Number Two*, J. Number Theory **33** (1989), 101–106.

[3] R.A. Mollin, *Class Number One Criterion for Real Quadratic Fields I*, Proceed. Japan. Acad **63** Ser. A. (1987), 121–125.

[4] R.A. Mollin, *Class Number One Criterion for Real Quadratic Fields II*, Proceed. Japan. Acad **63** Ser. A. (1987), 162–164.

[5] R.A. Mollin and H.C. Williams, *On Real Quadratic Fields of Class Number Two* (to appear: Math. Comp.).

[6] R.A. Mollin and H.C. Williams, *On Prime-Valued Polynomials and Class Numbers of Real Quadratic Fields*, Nagoya Math. J. **112** (1988), 143–151.

[7] R.A. Mollin and H.C. Williams, *Prime-Producing Quadratic Polynomials and Real Quadratic Fields of Class Number One*, in "Number Theory" (J. M. DeKoninck and C.Levesque (ed.)) Walter de Gruyter, Berlin, 1989, 654–663.

[8] R.A. Mollin and H.C. Williams, *Solution of the Class Number One Problem for Real Quadratic Fields of Extended Richaud-Degert Type (with One Possible Exception)*, in "Number Theory" (R.A. Mollin (ed.)) Walter de Gruyter, Berlin, 1990, 417–425.

[9] T. Tatuzawa, *On a Theorem of Siegel*, Japan J. Math. **21** (1951), 163–178.

[10] H.C. Williams and M.C. Wunderlich, *On the Parallel Generation of the Residues of the Continued Fraction Factoring Algorithm*, Math. Comp. **48** (1987), 405–423.

[11] H. Yokoi, *Class Number One Problems for Certain Kinds of Real Quadratic Fields*, Proceed. Internat. Conf. on Class Numbers and Fundamental Units, Katata, Japan, 1986.

Cubic number fields with exceptional units

V. Ennola

1 Introduction

A unit η of an algebraic number field K is called *exceptional* if $\eta - 1$ is also a unit. The term has been introduced by Nagell [20] who was led to such units already in 1928 when studying representations by cubic forms. The name "exceptional" comes from "exceptional point", i.e., a point of finite order on an elliptic curve, a term brought into use by Nagell in 1935. From the existence of such a torsion point over K one can namely derive in certain cases the existence of exceptional units in a relatively straightforward manner. For all this see the short historical survey in [20] and the further literature cited there. For later developments, see, e.g., the paper of Mestre [18]. He also refers to the interesting connection between exceptional units and the existence of the Euclidean algorithm due to Lenstra. A thorough account on this is contained in [14] and [16].

Let $\mathcal{E}_K$ denote the set of exceptional units in K. It was implicitly proved by Siegel [22] that $\mathcal{E}_K$ is a finite set for any K. For the first explicit proof, see Lang [12] and Chowla [2]. If $\eta \in \mathcal{E}_K$, let

$$\mathcal{A}_\eta = \{\eta, 1-\eta, \eta^{-1}, (1-\eta)^{-1}, 1-\eta^{-1}, \eta(\eta-1)^{-1}\}.$$

Clearly, $\mathcal{A}_\eta$ is contained in $\mathcal{E}_K$ and the six units in $\mathcal{A}_\eta$ are different, unless η satisfies the equation $\eta^2 - \eta + 1 = 0$ in which case $\mathcal{A}_\eta$ contains only two elements.

Let us assume from now on that K is a cubic field of discriminant $D = D(K)$. One can immediately verify that there are exactly two families of cubic fields containing exceptional units, namely $\mathcal{F} = \{F_l\}$ and $\mathcal{L} = \{L_k\}$, where

$$\begin{aligned} F_l &= \mathbb{Q}(\epsilon), & \operatorname{Irr}(\epsilon, \mathbb{Q}) = f_l(x) &= x^3 + (l-1)x^2 - lx - 1, \\ L_k &= \mathbb{Q}(\tau), & \operatorname{Irr}(\tau, \mathbb{Q}) = g_k(x) &= x^3 - kx^2 - (k+3)x - 1, \end{aligned}$$

for some integers l, k. Note that ϵ and $-\tau$ are exceptional units but τ is not if $k \neq -1, -2$. The discriminants of the polynomials f_l and g_k are

$$\begin{aligned} D(f_l) &= l^4 + 6l^3 + 7l^2 - 6l - 31 = (l^2 + 3l - 1)^2 - 32\ , \\ D(g_k) &= (k^2 + 3k + 9)^2\ . \end{aligned}$$

Computational Number Theory

Since $D(g_k)$ is a square in $\mathbb{Z}$, the extension $L_k/\mathbb{Q}$ is cyclic. We have

$$\begin{aligned} F_{-4} &= F_{-2} = F_{-1} = F_1 = \text{ the field of discriminant } -23 \ , \\ F_{-3} &= F_0 = \text{ the field of discriminant } -31 \ . \end{aligned}$$

These two fields are the only cubic fields of negative discriminant containing exceptional units. Here equality of fields is of course meant in a "wide sense" no distinction being made between conjugate fields. We shall exclude these cases and confine ourselves to totally real fields. Further

$$f_{-l-3}(x) = -(x-1)^3 f_l(x/(x-1)), \quad g_{-k-3}(x) = -x^3 g_k(1/x),$$

whence

$$F_{-l-3} = F_l, \quad L_{-k-3} = L_k. \tag{1.1}$$

The number $D(f_l)$ is a square in $\mathbb{Z}$ only when $l = 2$ or -5, and then

$$F_2 = F_{-5} = L_{-1} = L_{-2} = \text{ the field of discriminant } 49.$$

Hence we do not lose anything by making the assumptions

$$l \geq 3, \quad k \geq -1. \tag{1.2}$$

Fields containing exceptional units sometimes emerge incidentally. One instance of this is the isolation process in [5].

A natural question to ask is whether or not the fields F_l or L_k are distinct for different values of the parameters l and k under the assumptions (1.2). As regards the family $\mathcal{L}$, the answer is negative, there are several counter-examples. We shall say more about this in Section 2. In [20] p.168 Nagell conjectured that in fact F_l and F_m are not conjugate if $3 \leq l < m$, or in other words that $\mathcal{E}_{F_l} = \mathcal{A}_\epsilon$ if $\mathrm{Irr}(\epsilon, \mathbb{Q}) = f_l(x)$ and $l \geq 3$. This plausible hypothesis has been posed as a question in [15]. In any case we know from Nagell [19] that both $\mathcal{F}$ and $\mathcal{L}$ are infinite families.

There is a copious literature on $\mathcal{L}$. Apparently, the pioneer was Harvey Cohn [3] who discovered that the class number of L_k is even for certain values of k. The basic arithmetic can be found in M.-N. Gras [7], and interesting results and conjectures about units in [8]. Emma Lehmer [13] gives a unified method which produces particular families of cyclic units of different degrees and leads to $\mathcal{L}$ in the cubic case. Further results mainly on class numbers are contained in [21], [25], [26].

Except for some numerical observations about coincidences in $\mathcal{L}$, we shall focus our attention on $\mathcal{F}$ only. First we shall prove some results analogous to those in [8] to the effect that certain units are not nontrivial powers. By a theorem of Thomas [23], $\{\epsilon, \epsilon - 1\}$ is a fundamental pair of units for the order $\mathbb{Z}[\epsilon]$. For $3 \leq l \leq 500$, we have verified that the same is true for the maximal order of the field F_l. It seems reasonable to expect this result to hold for all values of $l \geq 3$. The truth of this conjecture would imply Nagell's conjecture mentioned above.

Further, we shall give a few class fields, and, in view of a remark in [14], we show that there are values of $l \equiv 2 \bmod 4$, $l \geq 3$ such that the class group in the ordinary sense is or is not a direct factor of the narrow class group, the former alternative occurring infinitely often.

Applying the algorithm developed in [6], we have computed the class numbers and class group structures for all fields F_l with $3 \leq l \leq 100$ and in 32 additional cases with $l \geq 107$. These results should be of particular interest because there are many noncyclic class groups in spite of our observation in [6] that most of the class groups tend to be cyclic. In particular, we have found two cases where the class group requires three generators.

In conclusion, I would like to express my gratitude to Jacques Martinet for stimulating correspondence concerning cubic fields in general. I am further much obliged to the referees for pointing out several improvements.

2 The cyclic case

We have verified that there are exactly four cyclic cubic fields L such that $L = L_k$ for more than one value of k satisfying $-1 \leq k \leq 10000$. These fields are given in Table 2.1.

Table 2.1

Conductor	k	j	j_u	Reference
7	-1	1	1	
	5	7	7	[20],[8]
	12	3^3	13	[20]
	1259	61^3	97	[20]
9	0	1	1	
	3	3	3	[20],[8]
	54	7^3	19	[20]
13	1	1	1	
	66	$3^3 \cdot 13$	13	[8]
19	2	1	1	
	2389	67^3	31	

For each k, j is the index of the order $\mathbb{Z}[\tau]$ in the maximal order of L_k, and j_u is the index of the subgroup U_τ generated by -1, τ, $\tau+1$ in the unit group U of L_k. Since $\tau' = -(\tau+1)^{-1}$ is a conjugate of τ, U_τ is also the $\mathbb{Z}[\mathrm{Gal}(L_k/\mathbb{Q})]$-submodule of U generated by -1 and τ. It seems to be evident that the values of k in the table for which $j_u > 1$ are the only ones between -1 and 10000 such that $U_\tau \neq U$.

but the proof of this is not yet quite complete. This fact would imply a conjecture of M.-N. Gras ([8] Conjeture 2) for all fields L_k with $-1 \le k \le 10000$. Our first proposition summarizes some simple properties of the units τ, $\tau+1$.

Proposition 2.1. *For any $k \ge -1$, the following statements are true:*

(i) *τ and $\tau+1$ are not nontrivial powers in L_k.*

(ii) *$\tau(\tau+1) = -\tau/\tau'$ is an nth power in L_k for $n > 1$ if and only if $k = n = 3$.*

(iii) *$j_u \equiv 0$ mod 3 if and only if $k = 3$.*

(iv) *If $k \ne 3$ then every prime factor of j_u is $\equiv 1$ mod 6.*

Proof. Note that none of the units τ, $\tau+1$, $\tau(\tau+1)$ can be a square or the opposite of a square in L_k, because τ is not totally positive or totally negative.

(i) is merely a restatement of [8] Proposition 1. Cf. also the proof of Theorem 4.1 (i) below.

(ii) Suppose that $\psi^n = -\tau\tau'^{-1}$ for some $\psi \in L_k$ and $n > 1$. Then $\psi'^n = -\tau\tau'^2$ and $(\psi'/\psi)^n = \tau'^3$. By (i), we have $n = 3$ so that $\psi'/\psi = \tau'$. From the proof of [8] Proposition 2 it follows that either $-\psi$ or $-\psi^{-1}$ is a zero of the polynomial $g_0(x)$, which leads to $k = 3$.

(iii) follows immediately from (i) and (ii).

(iv) Let p be a prime dividing j_u. Then for some integers a, b such that $ab \not\equiv 0 \bmod p$, there is a $\psi \in L_k$ such that $\psi^p = \tau^a\tau'^b$. It follows that $\psi'^p = \tau^{-b}\tau'^{a-b}$ and $(\psi^{a-b}\psi'^{-b})^p = \tau^{a^2-ab+b^2}$. Hence, by (i), $a^2 - ab + b^2 \equiv 0 \bmod p$, and the assertion follows. □

It is easy to verify (cf. [17]) that L_{-1} is the only totally real cubic field containing an exceptional unit which is a square or the opposite of a square, and that the only such units are the three zeros of $f_{-5}(x)$ and their inverses.

3 Arithmetic in F_l

From now on we shall always assume that the cubic field we are concerned with is F_l for some $l \ge 3$. Let $\mathcal{O}$, U, U_ϵ, H, H^+ denote the maximal order of F_l, the unit group of F_l, the subgroup of U generated by $\{-1, \epsilon, \epsilon - 1\}$, the class group in the ordinary sense and in the narrow sense. Further, put $h = [H:1]$, $h^+ = [H^+:1]$, and

$$j = [\mathcal{O} : \mathbb{Z}[\epsilon]] = \left(D(f_l)/D(F_l)\right)^{1/2}, \quad j_u = [U : U_\epsilon].$$

The notation $p^\nu \parallel c$ indicates that $p^\nu \mid c$, $p^{\nu+1} \nmid c$ for p prime and $c \in \mathbb{Z}$. Let $C(n)$ be the cyclic group of order n. For any $\gamma \in F_l$, let γ, γ', γ'' denote the conjugates of γ, and write $S(\gamma) = \gamma + \gamma' + \gamma''$, $N(\gamma) = \gamma\gamma'\gamma''$.

In the following three propositions we have gathered together basic arithmetic properties of F_l. Many of these results are contained in [14], [15], [17], but for the convenience of the reader we sketch the proofs. Note that in [14], [15], [17] the minimal polynomial $p(x) = x^3 - tx^2 + (t+1)x - 1$ is used. However, we have

$$x^3\, p(1 - 1/x) = f_{t-3}(x),$$

so that a zero of $p(x)$ generates the field F_l for $l = t - 3$.

Proposition 3.1. *Let p be any prime divisor of the polynomial discriminant $D(f_l)$ and let $p^\lambda \parallel D(f_l)$. Then $p \equiv \pm 1$ mod 8.*

(i) *If $p \neq 7$, then p is ramified in F_l if and only if λ is odd. In this case $p^1 \parallel D(F_l)$, and $p\mathcal{O} = \mathbf{p}^2\mathbf{q}$ where $\mathbf{p}$ and $\mathbf{q}$ are different prime ideals of $\mathcal{O}$.*

(ii) *If $p = 7$, then $l \equiv 2$ mod 7.*
For $l \equiv 23$ mod 49, we have $7^3 \parallel D(f_l)$, $7^1 \parallel D(F_l)$, $7\mathcal{O} = \mathbf{p}^2\mathbf{q}$.
For $l \not\equiv 23$ mod 49, we have $7^2 \parallel D(f_l)$, $7^2 \parallel D(F_l)$, $7\mathcal{O} = \mathbf{p}^3$.

Proof. From the expression of $D(f_l)$ it follows immediately that $p \equiv \pm 1$ mod 8 because $(32/p) = 1$. Put $a = 2(l^2 + l + 1)$, $b = l^2 - l - 9$. The following identities are true:

$$9f_l(x) - (3x + l - 1)f_l'(x) = -ax + b\,, \tag{3.1}$$

$$a^2 f_l'(b/a) = -9D(f_l)\,, \tag{3.2}$$

$$a^3 - 2(a(l-1) + 3b)^2 = 54D(f_l)\,. \tag{3.3}$$

A simple computation shows that (a, b) is a divisor of 21. If $p \neq 7$ then $p \nmid a$, by (3.3). Choosing t to satisfy $at \equiv b \bmod p^\lambda$ we have $f_l(t) \equiv f_l'(t) \equiv 0 \bmod p^\lambda$, by (3.2) and (3.1). It follows from a theorem of Voronoi ([4] p.111, Theorem I) that there exists an order $\mathcal{O}'$ of the field F_l such that $[\mathcal{O}' : \mathbb{Z}[\epsilon]] = p^{[\lambda/2]}$. The assertion now follows from well known classical results (see, e.g., [10]).

Suppose that $p = 7$. It is easily seen that $l \equiv 2$ mod 7, and writing $l = 2 + 7n$ we have

$$a \equiv 14(1 + 5n),\ b \equiv 7(-1 + 3n),\ a(l-1) + 3b \equiv 7(-1 + 5n) \bmod 49.$$

Hence, from (3.3), $7^3 \parallel D(f_l)$ if $l \equiv 23$ mod 49, and $7^2 \parallel D(f_l)$ otherwise. We have

$$\mathrm{Irr}(\epsilon - 2, \mathbb{Q}) = f_l(x+2) = x^3 + (l+5)x^2 + (3l+8)x + 2l + 3 \equiv x^3 \bmod 7\,. \tag{3.4}$$

Therefore, if $l \equiv 23$ mod 49,

$$f_l(2) \equiv 0 \bmod 49, \quad f_l'(2) \equiv 0 \bmod 7, \tag{3.5}$$

and the result follows as above. For $l \not\equiv 23$ mod 49, $f_l(x+2)$ is an Eisenstein polynomial. Hence also the last assertion is true. □

If $l \not\equiv 2$ mod 7 or $l \equiv 23$ mod 49, then $\mathbb{Q}(\epsilon, \sqrt{D})/\mathbb{Q}(\sqrt{D})$ is an unramified cyclic extension so that the class number of $\mathbb{Q}(\sqrt{D})$ is divisible by 3.

Proposition 3.2. **(i)** *The ring $\mathcal{O}$ has an integral basis of the form* 1, ϵ, $(u+v\epsilon+\epsilon^2)/j$, *where u and v are determined by*

$$28u \equiv -2l^3 - 9l^2 - 11l + 11,\ 28v \equiv l^3 + l^2 + 9l - 30 \ mod\ (7,j)j. \tag{3.6}$$

(ii) *Let $l \equiv 23 \bmod 49$ and denote $\omega = (\epsilon-2)^2/7$, $\mathcal{O}_7 = \mathbb{Z}[\epsilon,\omega]$. Then $\mathcal{O}_7 \subseteq \mathcal{O}$. The largest ideal of $\mathcal{O}_7$ which is contained in $\mathbb{Z}[\epsilon]$ is*

$$\mathbf{f} = 7\mathbb{Z}[\epsilon] + (\epsilon-2)\mathbb{Z}[\epsilon].$$

The unit group of the residue class ring $\mathcal{O}_7/\mathbf{f}$ is of order 36. In particular, if $\mathcal{O}_7 = \mathcal{O}$, then $\mathbf{f}$ is the conductor of $\mathbb{Z}[\epsilon]$ in $\mathcal{O}$ and $\mathbf{f} = \mathbf{pq}$ where $\mathbf{p}$ and $\mathbf{q}$ are defined in Proposition 3.1 (ii).

Proof. We shall use the results in the proof of the preceding proposition and the theorem of Voronoi cited there. First of all we must show that the simultaneous congruences

$$f_l(x) \equiv 0 \bmod k^3,\quad f_l'(x) \equiv 0 \bmod k^2,\quad \frac{1}{2}f_l''(x) \equiv 0 \bmod k$$

do not have a common solution for any $k > 1$. Suppose the contrary. It follows from the identity

$$-12f_l'(x) + f_l''(x)^2 = 2a. \tag{3.7}$$

and (3.1) that $k^2 \mid (a,b)$, which contradicts the fact that (a,b) is a divisor of 21. Hence $\mathcal{O}$ has an integral basis of the required form where u and v have to be determined. Put $j' = j/(7,j)$ so that $7 \nmid j'$. By Voronoi's theorem, we have

$$u \equiv (b/a)^2 + (l-1)(b/a) - l,\quad v \equiv b/a + l - 1 \bmod j'.$$

Using the identity

$$a(3l^3 + 10l^2 - 22l - 41) + 392 = (6l-10)D(f_l)$$

to remove the denominators we obtain after a short computation the congruences (3.6) modulo j'.

Suppose now that $7 \mid j$, i.e., $l \equiv 23 \bmod 49$. Using (3.5) we get from Voronoi's theorem

$$u \equiv 4 + 2(l-1) - l \equiv 4,\quad v \equiv 2 + l - 1 \equiv 3 \bmod 7$$

in accordance with (3.6) modulo 49. This proves (i).

We still suppose that $l \equiv 23 \bmod 49$. Putting $x = \epsilon - 2$ in (3.4) we find that $\epsilon - 2 \equiv 0 \bmod \mathbf{pq}$, so that ω is an integer. Since $(\epsilon-2)\omega \in \mathbb{Z}[\epsilon]$ and $(\epsilon-2)\omega \equiv 3(\epsilon-1)(\epsilon-2) \bmod 7\mathbb{Z}[\epsilon]$, we have $\omega\mathbf{f} \subseteq \mathbf{f}$ and $\mathcal{O}_7\mathbf{f} = \mathbf{f}$, so that $\mathbf{f}$ is an ideal of $\mathcal{O}_7$. Plainly

$$\mathbf{f} = 7\mathbb{Z} \oplus (5+\epsilon)\mathbb{Z} \oplus (3+\epsilon^2)\mathbb{Z},$$

and the numbers $\{a + b\omega \mid a,b \in \{0,1,\ldots,6\}\}$ form a system of representatives of $\mathcal{O}_7/\mathbf{f}$. One can easily verify that $\mathcal{O}_7/\mathbf{f} \cong \mathbb{Z}/7\mathbb{Z} \oplus \mathbb{Z}/7\mathbb{Z}$, by using the fact that $\omega^2 \equiv 3\omega \bmod \mathbf{f}$. The rest of the assertion is clearly true. □

Proposition 3.3. **(i)** *We have* $[U : U^2] = 8$, *and*

$$\{(-1)^i \epsilon^j (\epsilon - 1)^k \mid i, j, k \in \{0, 1\}\} \tag{3.8}$$

is a system of representatives of U/U^2.

(ii) *The image of* U *in* $C(2) \oplus C(2) \oplus C(2)$ *under the signature mapping has 2-rank* 2.

(iii) *We have* $h^+ = 2h$.

Proof. (i) The conjugates of ϵ lie in the intervals

$$1 < \epsilon < 1 + l^{-1}, \ -l^{-1} < \epsilon' < 0, \ -l < \epsilon'' < -l + l^{-2}. \tag{3.9}$$

Hence the only totally positive units in (3.8) are 1 and $\epsilon(\epsilon - 1)$. Suppose that $\epsilon(\epsilon - 1) = \eta^2$ for some $\eta \in F_l$. We may suppose that $N(\eta) = 1$. Denote $\mathrm{Irr}(\eta, \mathbb{Q}) = x^3 - sx^2 + tx - 1$. We have

$$\mathrm{Irr}(\epsilon(\epsilon - 1), \mathbb{Q}) = x^3 - (l^2 + l)x^2 + (2l + 1)x - 1, \tag{3.10}$$

so that $s^2 - 2t = l^2 + l, t^2 - 2s = 2l + 1$. Eliminating s we obtain

$$t^4 - (4l + 2)t^2 - 8t + 1 = 0.$$

However, this equation does not have integral solutions for $l \geq 3$. This proves (i), and (ii) and (iii) are immediate consequences. □

4 The unit group

Our first theorem contains a collection of results analogous with those of Proposition 2.1 in the cyclic case.

Theorem 4.1. *For any* $l \geq 3$, *the following statements are true:*

(i) *None of the units* ϵ, $\epsilon - 1$, $\epsilon(\epsilon - 1)$, $\epsilon/(\epsilon - 1)$ *is a nontrivial power in* F_l *or the opposite of such a power.*

(ii) *We have* $(j_u, 30) = 1$.

(iii) *For* $l \equiv 23$ *mod* 49, $\{\epsilon, \epsilon - 1\}$ *is a fundamental pair of units for the order* $\mathcal{O}_7$.

(iv) *For* $3 \leq l \leq 500$, $\{\epsilon, \epsilon - 1\}$ *is a fundamental pair of units for the maximal order* $\mathcal{O}$.

Proof. It follows from Proposition 3.3 that none of the units in (i) is a square or the opposite of a square, and it also follows that j_u is odd.

Suppose that (i) has already been proved. If $p = 3$ or 5 divides j_u, there is a unit $\eta \in U \setminus \{1\}$ such that $\eta^p = \epsilon^a(\epsilon - 1)^b$, $a \in \{0, \ldots, p - 1\}, b \in \{0, 1\}$. By (i), this could only be possible for $p = 5, (a, b) = (2, 1)$ or $(3, 1)$, the latter case being equivalent to $(a,b) = (1,2)$. At the end of the proof we shall show that these two possibilities do not occur. Hence (ii) is true.

If (iii) does not hold, then there is a unit $\eta \in \mathcal{O}_7 \setminus \mathbb{Z}[\epsilon]$, because the result does hold for the order $\mathbb{Z}[\epsilon]$ by [23]. However, Proposition 3.2 (ii) implies that $\eta^{36} \in 1 + \mathbf{f} \subseteq \mathbb{Z}[\epsilon]$, which contradicts the fact that $(j_u, 6) = 1$.

To prove (iv) we have computed the prime factorization of $D(f_l)$ for $3 \leq l \leq 500$ and have verified that the only values of l such that $D(f_l)$ is divisible by the square of a prime $p \neq 7$ are the ones given in Table 4.1.

Table 4.1

l	90	123	155	163	231	280	295	412	436	452
p	23	17	31	17	23	41	23	17	23	17

In these cases we have applied our program developed in [6], and have verified that $\{\epsilon, \epsilon - 1\}$ is in fact a fundamental pair of units for the maximal order $\mathcal{O}$. For other values of l we obtain the result from [23] or from the part (iii) of our theorem.

It remains to prove (i) and to show that the two special cases mentioned above do not occur. In each case we suppose that the unit in question is of the form η^p where p is an odd prime and $\eta \in U$, and deduce a contradiction. We denote

$$\mathrm{Irr}(\eta, \mathbb{Q}) = x^3 - sx^2 + tx - 1.$$

CASE $\eta^p = \delta$, where $\delta \in \{\epsilon, \epsilon - 1, \epsilon/(\epsilon - 1)\}$. As pointed out by the referee, the assertion is an immediate consequence of the following more general result:

Lemma 4.1. *Let ξ be an exceptional unit in an arbitrary totally real number field K and let p be an odd integer > 1. Then ξ is not a pth power in K.*

Proof. Put $p = 2n + 1$. It is easy to see in many ways that

$$(x^{2n+1} - 1)/(x - 1) > |x|^n \qquad \text{for } x \in \mathbb{R},\ x \neq \pm 1.$$

E.g., one can verify that, for $e = \pm 1$,

$$(x^{2n+1} - 1)/(x - 1) - ex^n = (x^{n+1} + e)(x^n - e)/(x - 1) > 0.$$

Suppose that $\eta^{2n+1} = \xi$ and apply the inequality to the conjugates of η. Since $\xi - 1$ is a unit by assumption, the norms from K to $\mathbb{Q}$ would satisfy the inequality

$$1/|N_{K/\mathbb{Q}}(\eta - 1)| \; > \; |N_{K/\mathbb{Q}}(\eta)|^n ,$$

which is impossible. ☐

CASE $\eta^p = \epsilon(\epsilon - 1)$. From (3.10) we infer

$$1 = S(\eta^{-p})^2 - 4S(\eta^p) = \eta^{-2p} + \eta'^{-2p} + \eta''^{-2p} - 2(\eta^p + \eta'^p + \eta''^p). \tag{4.1}$$

Put $a = \eta^{-p}$, $b = \eta'^{-p}$ so that $a > 1$, $b > 1$, and consider the function

$$h(x) = a^{2x} + b^{2x} + (ab)^{-2x} - 2(a^{-x} + b^{-x} + (ab)^x)$$

in the interval $0 \le x \le 1$. We have $h(0) = -3$, and $h(1) = 1$ by (4.1). Further, for $x > 0$,

$$h'(x) = 2(a^x - b^x)(a^x \ln a - b^x \ln b) + 2a^{-x} \ln a(1 - a^{-x}b^{-2x}) + 2b^{-x} \ln b(1 - a^{-2x}b^{-x}) > 0,$$

so that $h(x)$ is monotonically increasing. Then

$$t^2 - 4s = S(\eta^{-1})^2 - 4S(\eta) = h(1/p)$$

is an integer $\equiv 0, 1 \bmod 4$ in the open interval $(-3, 1)$, whence $t^2 - 4s = 0$. But (4.1) implies that $t^2 - 4s \equiv 1 \bmod p$, which is impossible.

CASE $\eta^5 = \epsilon^2(\epsilon - 1)$. We have

$$\operatorname{Irr}(\epsilon^2(\epsilon - 1), \mathbb{Q}) = x^3 + (l^3 + l^2 - 3)x^2 + (-l^2 + 3)x - 1. \tag{4.2}$$

We contend that the following estimates are true for any $l \ge 3$:

$$\begin{array}{ccccc}
l^{-1} - 3l^{-3} & < & \epsilon^2(\epsilon - 1) & < & l^{-1}, \\
-l^{-2} & < & \epsilon'^2(\epsilon' - 1) & < & -l^{-2} + l^{-3}, \\
-l^3 - l^2 & < & \epsilon''^2(\epsilon'' - 1) & < & -l^3 - l^2 + 3, \\
l & < & \epsilon^{-2}(\epsilon - 1)^{-1} & < & l + 3l^{-1}, \\
-l^2 - l & < & \epsilon'^{-2}(\epsilon' - 1)^{-1} & < & -l^2 - l + 3, \\
-l^{-3} & < & \epsilon''^{-2}(\epsilon'' - 1)^{-1} & < & -l^{-3} + l^{-4}.
\end{array} \tag{4.3}$$

Consider, e.g., the bounds of $\epsilon^2(\epsilon-1)$. On substituting $x = l^{-1}$ and $x = l^{-1} - 3l^{-3}$ in (4.2) we find by simple computations that the values of the polynomial have opposite signs, so that there is a zero of (4.2) between these values of x. This zero is a conjugate of $\epsilon^2(\epsilon - 1)$, and (3.9) shows that it must be equal to $\epsilon^2(\epsilon - 1)$. The other cases are dealt with similarly.

By the part (iv) of our theorem, we have $l > 500$. Put $w = \sqrt[5]{l}$. Using the classical inequalities

$$1 + \frac{ru}{1+u} \le (1+u)^r \le 1 + ru \quad \text{for } u > -1,\ 0 < r < 1,$$

we easily obtain from (4.3) for

$$s = \sum_{i=0}^{2} (\epsilon^{(i)2}(\epsilon^{(i)} - 1))^{1/5}$$

and

$$t = \sum_{i=0}^{2} (\epsilon^{(i)-2}(\epsilon^{(i)} - 1)^{-1})^{1/5}$$

the following expressions:

$$s = -w^3 + w^{-1} - \left(\frac{6}{5} + \frac{\alpha}{l}\right)w^{-2} \quad \text{where } |\alpha| < 1,$$

$$t = -w^2 + w - \left(\frac{6}{5} - \frac{\beta}{l}\right)w^{-3} \quad \text{where } 0 < \beta < 2.$$

Define

$$q = t^2 + st + t - 2s - l - 3.$$

On substituting the expressions of s and t in q we obtain after a short computation the estimate $0.1 < q < 0.5$. This is a contradiction, because $q \in \mathbb{Z}$.

CASE $\eta^5 = \epsilon(\epsilon - 1)^2$. We have

$$\mathrm{Irr}(\epsilon(\epsilon-1)^2, \mathbb{Q}) = x^3 + (l^3 + 2l^2 + l - 3)x^2 + (l^2 + 2l + 4)x - 1.$$

In exactly the same way as in the preceding case we show that the following inequalities hold for any $l \geq 3$:

$$\begin{array}{rcccl} l^{-2} - 3l^{-3} & < & \epsilon(\epsilon-1)^2 & < & l^{-2}\,, \\ -l^{-1} - l^{-2} - 3l^{-4} & < & \epsilon'(\epsilon'-1)^2 & < & -l^{-1}\,, \\ -l^3 - 2l^2 - l & < & \epsilon''(\epsilon''-1)^2 & < & -l^3 - 2l^2 - l + 3 + l^{-1} + 3l^{-3}\,, \\ l^2 + 3l & < & \epsilon^{-1}(\epsilon-1)^{-2} & < & l^2 + 3l + 3 + l^{-1}\,, \\ -l + 1 - l^{-1} & < & \epsilon'^{-1}(\epsilon'-1)^{-2} & < & -l + 1\,, \\ -l^{-3} & < & \epsilon''^{-1}(\epsilon''-1)^{-2} & < & -l^{-3} + 2l^{-4}\,. \end{array}$$

This time we get

$$s = \sum_{i=0}^{2} (\epsilon^{(i)}(\epsilon^{(i)} - 1)^2)^{1/5} = -w^3 - \left(1 + \frac{\alpha}{l}\right)w^{-1} + \frac{3}{5}w^{-2}\,,$$

$$t = \sum_{i=0}^{2} (\epsilon^{(i)-1}(\epsilon^{(i)} - 1)^{-2})^{1/5} = w^2 - w - \left(\frac{2}{5} + \frac{\beta}{l}\right)w^{-3} + \frac{1}{5}w^{-4}\,,$$

where

$$-0.174 < \alpha < 0.433, \quad -1.061 < \beta < 1.858,$$

and $w = \sqrt[5]{l}$ as before. Define now

$$q = -2t^3 + 2s^2 + 5st + t^2 - 5t - l - 4.$$

Here we need more accurate estimates than in the preceding case, but after a certain amount of computation we find that $0 < w^2 q < 4$, which is impossible. □

Conceivably, the method used to deal with the last two cases in the above proof is capable of generalization. The only major problem here is a successful choice of the approximation polynomial q. E.g., we believe that it would not be too hard to show that $\epsilon^2(\epsilon - 1)$ and $\epsilon(\epsilon - 1)^2$ are not nontrivial powers in F_l. We hope to be able to return to this topic later. All these results seem to indicate that the following may be true:

Conjecture 4.1. *For any $l \geq 3$, $\{\epsilon, \epsilon - 1\}$ is a fundamental pair of units for the maximal order $\mathcal{O}$.*

In Section 2 we saw that the corresponding result is not valid for the cyclic family $\mathcal{L}$, but there one may ask whether there are infinitely many counter-examples. If the above conjecture is true, the conjecture of Nagell mentioned in the introduction follows, because a straightforward argument based on (3.9) shows that the regulator of the pair $\{\epsilon, \epsilon - 1\}$ is an increasing function of l.

5 Table of class groups

In the first four columns of Table 5.2 we give the value of l, the discriminant $D = D(F_l)$ of the field F_l, the index $j = [\mathcal{O} : \mathbb{Z}[\epsilon]]$, and the class number h. In the column labelled with H we give the prime power orders of the cyclic factors in a direct product decomposition of H if $h > 1$.

The last column with label *g.n.* contains generator norms, i.e., norms of ideals of $\mathcal{O}$ representing ideal classes which generate these cyclic subgroups of H. In most cases the ideals are chosen to be first degree prime ideals $\mathbf{p}$ such that no other ideal of $\mathcal{O}$ has the same norm p as $\mathbf{p}$. We have tried to select $\mathbf{p}$ so that p is small but the choice is not claimed to be optimal in any way. For each l the class group H is generated by prime ideals having very small norms, but it is not always possible to find a prime ideal with a reasonable-sized norm which is contained in a given class of prime power order, even more so if one requires the norm to determine the ideal uniquely. There are twelve exceptional cases where we have to choose $\mathbf{p}$ among three ideals having norm p. Each such $\mathbf{p}$ is contained in a class of order two. Let $\mathbf{p} = p\mathcal{O} + (\epsilon - a)\mathcal{O}$. In Table 5.1 we give the numbers l, p, a. These exceptional values of p are printed in italics in Table 5.2.

Table 5.1

l	28	59	62	69	77	80	81	85	93	191	205	226
p	13	11	19	89	41	13	31	127	839	181	503	2953
a	8	2	9	82	9	7	22	57	58	15	65	2571

The computation employs the adaptation of the Voronoi algorithm developed in [6]. We have examined all fields F_l such that (i) $3 \le l \le 100$; (ii) $101 \le l \le 239$ and the polynomial discriminant $D(f_l)$ is divisible by the square of a prime ; (iii) $240 \le l \le 500$ and $j > 1$. Many noncyclic class groups occur, especially for $l \equiv 2 \bmod 7$ when $7 \mid D$. In particular, for $l = 86$ and 198, H requires three generators.

Table 5.2

l	D	j	h	H	$g.n.$
3	257	1	1		
4	697	1	1		
5	1489	1	1		
6	2777	1	2	2	3
7	4729	1	1		
8	7537	1	2	2	5
9	11417	1	3	3	3
10	16609	1	2	2	7
11	23377	1	2	2	5

Table 5.2. (continued)

l	D	j	h	H	$g.n.$
12	32009	1	3	3	3
13	42817	1	3	3	5
14	56137	1	2	2	5
15	72329	1	6	2,3	19,3
16	91777	1	3	3	5
17	114889	1	3	3	7
18	142097	1	6	2,3	17,5
19	173857	1	5	5	5
20	210649	1	4	2,2	7,11
21	252977	1	9	9	3
22	301369	1	4	4	7
23	7273	7	1		
24	418577	1	10	2,5	31,5
25	488569	1	4	4	19
26	566977	1	8	8	5
27	654449	1	7	7	3
28	751657	1	6	2,3	*13*,23
29	859297	1	7	7	5
30	978089	1	12	2,2,3	13,41,7
31	1108777	1	10	2,5	19,7
32	1252129	1	5	5	11
33	1408937	1	15	3,5	11,41
34	1580017	1	12	4,3	17,29
35	1766209	1	6	2,3	31,37
36	1968377	1	16	16	7
37	2187409	1	9	3,3	7,19
38	2424217	1	10	2,5	23,5
39	2679737	1	16	16	13
40	2954929	1	9	9	7
41	3250777	1	14	2,7	31,7
42	3568289	1	12	4,3	3,17
43	3908497	1	16	16	7
44	4272457	1	18	2,9	29,5
45	4661249	1	20	4,5	13,59
46	5075977	1	12	4,3	5,109
47	5517769	1	13	13	7
48	5987777	1	28	2,2,7	17,41,29
49	6487177	1	15	3,5	31,41
50	7017169	1	12	2,2,3	17,37,19
51	7578977	1	27	3,9	7,3

Table 5.2. (continued)

l	D	j	h	H	$g.n.$
52	8173849	1	15	3,5	7,11
53	8803057	1	16	16	19
54	9467897	1	36	4,9	283,47
55	10169689	1	14	2,7	19,7
56	10909777	1	16	16	17
57	11689529	1	21	3,7	7,61
58	12510337	1	18	2,3,3	71,5,7
59	13373617	1	24	2,4,3	*11*,17,7
60	14280809	1	30	2,3,5	19,17,127
61	15233377	1	21	3,7	5,7
62	16232809	1	20	2,2,5	11,*19*,23
63	17280617	1	28	4,7	47,127
64	18378337	1	24	8,3	11,5
65	19527529	1	18	2,3,3	181,7,17
66	20729777	1	44	4,11	31,11
67	21986689	1	21	3,7	163,19
68	23299897	1	23	23	5
69	24671057	1	38	2,19	*89*,3
70	26101849	1	24	8,3	13,41
71	27593977	1	24	8,3	5,13
72	594881	7	5	5	3
73	30769177	1	35	5,7	113,23
74	32455777	1	24	8,3	71,23
75	34210769	1	42	2,3,7	263,13,257
76	36035977	1	28	2,2,7	71,127,47
77	37933249	1	18	2,9	*41*,17
78	39904457	1	64	2,32	103,3
79	41951497	1	27	3,9	7,5
80	44076289	1	28	2,2,7	*13*,83,31
81	46280777	1	62	2,31	*31*,5
82	48566929	1	28	4,7	89,13
83	50936737	1	32	2,16	13,17
84	53392217	1	50	2,25	257,3
85	55935409	1	30	2,3,5	*127*,211,11
86	58568377	1	48	2,2,4,3	43,197,71,7
87	61293209	1	45	9,5	7,181
88	64112017	1	32	32	5
89	67026937	1	46	2,23	307,5
90	132401	23	2	2	3
91	73153777	1	39	3,13	311,19

Table 5.2. (continued)

l	*D*	*j*	*h*	*H*	*g.n.*
92	76370089	1	33	3,11	29,101
93	79691297	1	90	2,9,5	*839*,3,181
94	83119657	1	36	4,9	89,19
95	86657449	1	28	4,7	37,11
96	90306977	1	58	2,29	181,5
97	94070569	1	31	31	7
98	97950577	1	36	4,9	211,29
99	101949377	1	104	2,4,13	109,5,139
100	106069369	1	45	9,5	31,43
107	138509329	1	45	3,3,5	7,11,71
114	177875537	1	108	4,3,9	1427,7,97
121	4593673	7	9	9	5
123	830993	17	5	5	3
128	281132257	1	66	2,3,11	83,7,41
135	347039609	1	78	2,3,13	29,7,13
142	423906889	1	54	2,3,9	29,43,31
149	512886577	1	81	3,27	193,5
155	624049	31	2	2	7
156	615188777	1	132	4,3,11	281,7,139
163	2533153	17	6	2,3	29,7
170	17650801	7	12	4,3	47,41
177	1014995849	1	153	3,3,17	7,23,43
184	1183841617	1	162	2,3,27	911,7,19
191	1372924777	1	162	2,3,27	*181*,7,5
198	1583801177	1	270	2,3,3,3,5	1889,7,43,61,223
205	1818084289	1	138	2,3,23	*503*,7,13
212	2077445209	1	99	3,3,11	7,179,13
219	48236993	7	39	3,13	11,5
226	2678372977	1	186	2,3,31	*2953*,7,197
231	5523113	23	13	13	3
233	3023570137	1	144	16,3,3	307,7,241
268	107646553	7	53	53	5
280	3735169	41	3	3	17
295	14608681	23	10	2,5	41,7
317	209997361	7	35	5,7	41,31
366	372230873	7	126	2,3,3,7	1483,631,787,337
412	101155081	17	25	25	7
415	614112121	7	62	2,31	409,11
436	69253513	23	18	2,9	113,5
452	146351641	17	25	25	11
464	958229713	7	88	2,4,11	911,313,19

6 Some 2-class fields

For any algebraic number field K and any prime number p, let $\mathcal{H}(K)$, $\mathcal{H}_p(K)$, $\mathcal{H}^+(K)$, $\mathcal{H}_p^+(K)$ denote the Hilbert class field of K and its p - part in the ordinary and in the narrow sense. We shall first outline a proof of a result essentially contained in [14]. It shows, in particular, that h is even for every $l \equiv 2 \bmod 4$.

Proposition 6.1. *For each $l \geq 3$ there is exactly one quadratic extension $F_l(\sqrt{\theta})/F_l$ such that*

(i) *θ is a unit of F_l;*

(ii) *$F_l(\sqrt{\theta})/F_l$ is unramified at all finite primes of F_l.*

This extension is obtained by taking

$$\theta = \begin{cases} \epsilon - 1, & \text{if } l \equiv 0, 1 \bmod 4, \\ \epsilon(\epsilon - 1), & \text{if } l \equiv 2 \bmod 4, \\ \epsilon, & \text{if } l \equiv 3 \bmod 4. \end{cases}$$

Hence $\mathcal{H}^+(F_l) = \mathcal{H}(F_l)(\sqrt{\theta})$ if $l \not\equiv 2 \bmod 4$, and $\sqrt{\theta} \in \mathcal{H}(F_l)$ if $l \equiv 2 \bmod 4$.

Proof. Let K be any algebraic number field and let α be any integer of K prime to 2. From class field theory we know ([9] Ia, Satz 9) that the extension $K(\sqrt{\alpha})/K$ is unramified at all finite primes of K if and only if the principal ideal $\alpha\mathcal{O}$ is a square in the ideal group of K and there is an integer ξ of K satisfying the congruence $\xi^2 \equiv \alpha \bmod 4$. In the present case $K = F_l$, the prime 2 is inert in $F_l/\mathbb{Q}$ and there are exactly 7 prime residue classes mod 4 containing squares, namely the ones represented by $\{\epsilon^{2n} \mid n = 0, \dots, 6\}$. It is enough to take θ to be one of the units $\neq 1$ in (3.8) and to check all possibilities. The rest of the assertion is clear because $\epsilon(\epsilon - 1)$ is totally positive, but ϵ and $\epsilon - 1$ are not. □

It is relatively easy to construct parametric families of quadratic or quartic extensions of F_l contained in $\mathcal{H}(F_l)$ or $\mathcal{H}^+(F_l)$ for certain values of l. We shall give three examples of such families.

Theorem 6.1. *Suppose that for some integer m*

$$l = 6 + 4m + 288m^2, \quad m \not\equiv 5 \bmod 7, \quad m \not\equiv 614 \bmod 743 \tag{6.1}$$

or

$$l = 10 - 68m + 288m^2, \quad m \not\equiv 6 \bmod 7, \quad m \not\equiv 707 \bmod 743. \tag{6.2}$$

Put

$$\alpha = 1 + (l + 4)\epsilon + \epsilon^2 + 4(1 + l\epsilon + \epsilon^2)\sqrt{\epsilon^2 - \epsilon}, \quad \beta = 1 + (l - 8)\epsilon + \epsilon^2 .$$

(i) *If β is a square in F_l, then $F_l(\sqrt{\alpha})$ is a cyclic quartic extension of F_l contained in $\mathcal{H}^+(F_l)$ so that H^+ contains a cyclic subgroup of order 4. This occurs (with equality) for $l = 6$ or 10.*

(ii) *If $(\epsilon^2-\epsilon)\beta$ is a square in F_l, then $F_l(\sqrt{\alpha})$ is a noncyclic quartic extension of F_l contained in $\mathcal{H}^+(F_l)$.*

(iii) *If β is not a square in $F_l(\sqrt{\epsilon^2-\epsilon})$, then we only have $F_l(\sqrt{\alpha}) \subseteq \mathcal{H}^+(F_l(\sqrt{\epsilon^2-\epsilon}))$, and $F_l(\sqrt{\epsilon^2-\epsilon}, \sqrt{\beta})$ is a noncyclic quartic extension of F_l contained in $\mathcal{H}(F_l)$ so that the 2-rank of H is ≥ 2. This occurs for $l \geq l_0$, where l_0 is an effectively computable number.*

Clearly, exactly one of the alternatives (i), (ii), (iii) is true depending on the structure of the Galois group of $\mathrm{Irr}(\sqrt{\alpha}, F_l)$. Instead of the numbers α, β one could also work with

$$(\epsilon-1)\alpha/\epsilon = -3+4\epsilon+4\sqrt{\epsilon^2-\epsilon}, \quad (\epsilon-1)\beta/\epsilon = 9-8\epsilon$$

which would perhaps render the argument slightly more transparent.

Proof. Let $N_{6/3}$ denote the norm from $F_l(\sqrt{\epsilon^2-\epsilon})$ to F_l, and N, as usual, the norm from F_l to $\mathbb{Q}$. We have

$$N_{6/3}(\alpha) = \frac{\epsilon\beta}{\epsilon-1}, \quad N(\beta) = 72l-431 = \begin{cases} (144m+1)^2 & \text{in case (6.1)}, \\ (144m-17)^2 & \text{in case (6.2)}. \end{cases}$$

Let $\mathbf{p}$ be a prime ideal of $\mathcal{O}$ dividing β. Since $\mathbf{p} \mid 72l-431$ we have $\mathbf{p} \nmid 6$. Suppose first that $\mathbf{p} \nmid D(f_l)$. Since

$$8(8\epsilon-9) = -9(\epsilon+7)\beta + (72l-431)\epsilon \equiv 0 \bmod \mathbf{p},$$

the ideal $\mathbf{p}$ is of degree one, say $N(\mathbf{p}) = p$, and $\mathbf{p}$ is the only common prime ideal divisor of p and β.

Suppose now that $\mathbf{p} \mid D(f_l)$. Computing the polynomial $D(f_l)$ for $l = 431/72$ we find that

$$\mathbf{p} \mid 7 \cdot 17 \cdot 23 \cdot 743.$$

However, the conditions imposed upon m in (6.1) and (6.2) imply that $\mathbf{p} \nmid 7 \cdot 743$.

If $\mathbf{p} \mid 17$ we have $72l-431 \equiv 0 \bmod 289$, $l \equiv 10 \bmod 289$, $D(f_l) \equiv 136 \bmod 289$, so that 17 is ramified in $F_l/\mathbb{Q}$, by Proposition 3.1. Further, $\beta \equiv (1+\epsilon)^2 \bmod 17$ and $17\mathcal{O} = \mathbf{p}\mathbf{q}^2$ where $\mathbf{p} = (17, 1+\epsilon)$, $\mathbf{q} = (17, 4+\epsilon)$, so that $\mathbf{q} \nmid \beta$.

If $\mathbf{p} \mid 23$ we have similarly $l \equiv 366 \bmod 529$, $D(f_l) \equiv 69 \bmod 529$, $\beta \equiv (17+\epsilon)(19+\epsilon) \bmod 23$, $23\mathcal{O} = \mathbf{p}\mathbf{q}^2$ where $\mathbf{p} = (23, 19+\epsilon)$, $\mathbf{q} = (23, 12+\epsilon)$, and hence $\mathbf{q} \nmid \beta$.

It follows from these considerations that $\beta\mathcal{O}$ is a square in the ideal group of F_l because $N(\beta)$ is a square in $\mathbb{Q}$. Since $\beta \equiv (1+\epsilon)^2 \bmod 4$ it follows that $F_l(\sqrt{\beta})/F_l$ is unramified because (3.9) easily implies that β is totally positive. Hence $\sqrt{\beta} \in \mathcal{H}(F_l)$.

Let $\mathbf{P}$ be any prime divisor of α in the field $F_l(\sqrt{\epsilon^2-\epsilon})$ and take $\mathbf{p} = \mathbf{P} \cap \mathcal{O}$ so that $\mathbf{p} \mid \beta$. Since $1+l\epsilon+\epsilon^2 = \beta+8\epsilon \not\equiv 0 \bmod \mathbf{p}$, it is clear that $\mathbf{p}$ splits in $F_l(\sqrt{\epsilon^2-\epsilon})/F_l$, say $\mathbf{p} = \mathbf{P}\overline{\mathbf{P}}$, and $\overline{\mathbf{P}} \nmid \alpha$. Therefore, the principal ideal generated by α is a square in the ideal group of $F_l(\sqrt{\epsilon^2-\epsilon})$. Since $\alpha \equiv (1+\epsilon)^2 \bmod 4$, the extension $F_l(\sqrt{\alpha})/F_l(\sqrt{\epsilon^2-\epsilon})$ is unramified at all finite primes. However, some infinite primes are ramified, because for example $1+(l+4)\epsilon''+\epsilon''^2<0$, by (3.9).

Obvious arguments show that if either β or $(\epsilon^2-\epsilon)\beta$ is a square in F_l, then $F_l(\sqrt{\alpha})/F_l$ is a Galois extension with Galois group $C(4)$ or $C(2)\times C(2)$ respectively. This proves (i) and (ii).

Suppose that $\beta=\gamma^2$ where $\gamma\in F_l$. Denote $\mathrm{Irr}(\gamma,\mathbb{Q})=x^3-sx^2+tx-n$. Computing the coefficients of the polynomial $\mathrm{Irr}(\beta,\mathbb{Q})$ we obtain

$$s^2-2t=9l-4,\quad t^2-2ns=8l^2-39l+11,\quad n^2=72l-431,$$

and eliminating n and s we get the Diophantine equation $P(l,t)=0$ where

$$\begin{aligned}P(l,t) &= 64l^4-624l^3-(16t^2+895)l^2\\ &\quad +(78t^2-576t+15810)l+t^4-22t^2+3448t-6775\,.\end{aligned}$$

The curve $P(l,t)=0$ is rational, in fact, it has the parametric representation

$$\begin{aligned}56l &= 393-57b(k)c(k)/a(k),\quad 7t=-8-57b(k)d(k)/a(k)\,,\\ a(k) &= (33209732k^2+279377876k+585465233)\times\\ &\quad (117434756k^2+468447788k+40891793)\,,\\ b(k) &= 12484k^2-11808k-238081\,,\\ c(k) &= 286849020484k^2+2933755633120k+7104474746815\,,\\ d(k) &= 189337679932k^2+1392782459428k+2542856893537\,.\end{aligned}$$

With a finite number of exceptions, to a rational point (l,t) on the curve there corresponds a rational value of k. Write $k=u/v$ where u and v are coprime integers. Computing resultants one can verify that if (l,t) is an integer point, then a Thue equation $v^4a(u/v)=r$ is satisfied where r divides a certain explicitly given huge integer. By Baker's result (see [1]), this is impossible for $l\geq l_0$.

Suppose finally that $\beta/(\epsilon^2-\epsilon)=\gamma^2$ where $\gamma\in F_l$. Let $\mathrm{Irr}(\gamma,\mathbb{Q})$ be as before. Comparing the expressions of the coefficients of $\mathrm{Irr}(\beta/(\epsilon^2-\epsilon),\mathbb{Q})$ we have

$$s^2-2t=l^2-4l-3,\quad t^2-2ns=9l^2-38l-86,\quad n^2=72l-431,$$

and $Q(l,t)=0$ where

$$\begin{aligned}Q(l,t) &= 81l^4-972l^3+(-18t^2+2772)l^2\\ &\quad +(76t^2-576t+504)l+t^4+172t^2+3448t+2224\,.\end{aligned}$$

The curve $Q(l,t)=0$ is of genus 1, because its only singularities are the two ordinary double points $(l,t)=((39\pm\sqrt{5})/6,-3(3\pm\sqrt{5})/2)$. By a theorem of Baker, we cannot have $l\geq l_0$. Hence (iii) follows. □

Theorem 6.2. *Suppose that $l=16m^2-2$ for some positive integer m, and take*

$$\alpha=(l+1)\epsilon+\epsilon^2+4m(l\epsilon+\epsilon^2)\sqrt{\epsilon^2-\epsilon}.$$

Then $F_l(\sqrt{\alpha})$ is a cyclic quartic extension of F_l contained in $\mathcal{H}^+(F_l)$.

Proof. Let $N_{6/3}$ be as in the proof of Theorem 6.1. We have $N_{6/3}(\alpha) = \epsilon^3/(\epsilon - 1)$ so that α is a unit. The extension $F_l(\sqrt{\alpha})/F_l(\sqrt{\epsilon^2 - \epsilon})$ is unramified at finite primes, because $\alpha \equiv \epsilon^2 - \epsilon \equiv \epsilon^6 \bmod 4$. Since α is not totally positive, the same is not true at all infinite primes. The rest follows on the same lines as before. □

Theorem 6.3. *Suppose that* $56l = m^2 + 279$ *where* $m \equiv \pm 1 \bmod 14$, $m \not\equiv 0 \bmod 23$, $m \not\equiv 0 \bmod 137$. *Let* θ *be the unit defined in Proposition 6.1.*

(i) *For each such* l *we have* $\mathcal{H}^+(F_l) = \mathcal{H}(F_l)(\sqrt{8\epsilon - 7\epsilon^2})$.

(ii) *If* $l \not\equiv 2 \bmod 4$ *and* $l \geq l_0$, *where* l_0 *is an effectively computable number, then* $F_l(\sqrt{(8\epsilon - 7\epsilon^2)\theta})$ *is a quadratic extension of* F_l *contained in* $\mathcal{H}(F_l)$ *so that* h *is even.*

Proof. Using exactly the same argument as in the proof of Theorem 6.1, one can show that $(8\epsilon - 7\epsilon^2)\mathcal{O}$ is a square in the ideal group of F_l. Here the critical primes are $23, 31, 47, 137$ but only $31, 47$ need a special treatment because the other two are excluded by assumption. Since $8\epsilon - 7\epsilon^2 \equiv \epsilon^2 \bmod 4$ we have $\sqrt{8\epsilon - 7\epsilon^2} \in \mathcal{H}^+(F_l)$, and (i) follows because $8\epsilon - 7\epsilon^2$ is not totally positive.

Put $\delta = (8 - 7\epsilon)(\epsilon - 1)/\epsilon$. To prove (ii) it is enough to show that $8 - 7\epsilon$ and δ are not squares in F_l for $l \geq l_0$. We have

$$\begin{aligned} \mathrm{Irr}(8 - 7\epsilon, \mathbb{Q}) &= x^3 - (7l + 17)x^2 + (63l + 80)x - 56l + 279, \\ \mathrm{Irr}(\delta, \mathbb{Q}) &= x^3 - (15l + 38)x^2 + (56l^2 + 281l + 361)x - 56l + 279, \end{aligned}$$

and as before we are led to two Diophantine equations $P(l,t) = 0$, $Q(l,t) = 0$ where

$$\begin{aligned} P(l,t) &= 2401l^2 - (126t^2 + 448t - 14084)l \\ &\quad + t^4 - 160t^2 + 2232t + 25372, \\ Q(l,t) &= 3136l^4 + 31472l^3 - (112t^2 - 116033)l^2 \\ &\quad - (562t^2 + 448t - 211110)l + t^4 - 722t^2 + 2232t + 172729. \end{aligned}$$

The substitution $343l = 2y + 9t^2 + 32t - 1006$ transforms $P(l,t) = 0$ into the hyperelliptic equation

$$y^2 = (t^2 + 2t - 342)(8t^2 + 128t + 169)$$

which does not have integer solutions for $l \geq l_0$, by a theorem of Baker. The curve $Q(l,t) = 0$ is rational. It has the parametric representation

$$\begin{aligned} 56l &= -561 + b(k)c(k)/a(k), \quad t = 56 - b(k)d(k)/a(k)\ , \\ a(k) &= 492276938161k^4 + 263721192000k^3 \\ &\quad + 52979866434k^2 + 4730349120k + 158382001\ , \\ b(k) &= 701623k^2 + 188160k + 12615\ , \\ c(k) &= 589365839k^2 + 157678304k + 10546271\ , \\ d(k) &= 2(6272k + 839)^2\ . \end{aligned}$$

As in the proof of Theorem 6.1 it follows that the Diophantine equation $Q(l,t)=0$ is impossible for $l \geq l_0$. □

Consider the exact sequence

$$1 \longrightarrow C(2) \longrightarrow H^+ \longrightarrow H \longrightarrow 1, \tag{6.3}$$

where the maps have obvious meanings. Plainly, (6.3) splits if and only if $\mathcal{H}^+(F_l) = \mathcal{H}(F_l)E$ for some quadratic extension E/F_l. Therefore, as noted by Leutbecher and Martinet [14], it follows from Proposition 6.1 that (6.3) splits for each $l \not\equiv 2 \bmod 4$. They further note that the sequence does not split for $l = 6, 10, 14$ and they ask about its behaviour for other values of l. From Theorem 6.3 we get infinitely many numbers $l \equiv 2 \bmod 4$ such that (6.3) splits. There are also infinitely many different fields F_l corresponding to these numbers l, as can be seen, e.g., by means of a method of Nagell [19] Th.5.

Let $l = 16m^2 - 2$. It follows from Theorem 6.2 that (6.3) does not split for $l = 14, 62, 142$, because Table 5.2 shows that in these cases H does not contain elements of order 4. It would be tempting to conjecture that (6.3) does not split for any $l = 16m^2 - 2$.

In Table 6.1 we determine the fields $\mathcal{H}_2(F_l)$, $\mathcal{H}_2^+(F_l)$ for $3 \leq l \leq 20$ and even h. In most cases we get the result directly from the above theorems in conjunction with Proposition 6.1. In some cases we obtain $\mathcal{H}_2(F_l)$ by adjoining to a subfield the square root of an element β given in the third column. As regards the use of Theorem 6.3 (ii), there are no integers t satisfying the equation $Q(l,t) = 0$ in the proof of the theorem if $l = 8$ or 20. On the other hand, note that, e.g., $Q(5,56) = Q(9,-86) = 0$.

Table 6.1

l	Theorem	β
6	6.1 (i)	
8	6.3 (ii)	
10	6.1 (i)	
11		$-1-13\epsilon+13\epsilon^2$
14	6.2	
15		$2-5\epsilon+3\epsilon^2$
18	6.3 (i)	
20	6.3 (ii)	$-2-43\epsilon+43\epsilon^2$

7 Arithmetic in $F_l(\rho)$

Let $\rho = \exp(2\pi i/3)$. In order to determine the field $\mathcal{H}_3(F_l)$ we must first investigate the CM-extension $F_l(\rho)/F_l$. We let $\mathcal{O}^*$, U^*, h^*, H^* denote the maximal order, unit group, class number and class group of the field $F_l(\rho)$. The bar will denote complex conjugation. $N_{6/3}$ is the norm from $F_l(\rho)$ to F_l, and N is the norm from F_l to $\mathbb{Q}$ as before. For any subset A of a group G, let $\langle A\rangle$ denote the subgroup generated by A.

In the next proposition we summarize the basic arithmetical facts, and in the subsequent theorem we determine the number of ambiguous classes in $F_l(\rho)/F_l$. The result shows that h^* is always even.

Proposition 7.1. **(i)** *$F_l(\rho)$ has relative integral basis 1, ρ over F_l.*

(ii) $U^* = \langle\rho\rangle U$.

(iii) *The natural mapping $H \to H^*$, induced by inclusion, is injective.*

Proof. Since 3 is not ramified in $F_l/\mathbb{Q}$, by Proposition 3.1, (i) follows by elementary reasoning. (ii) and (iii) are easily established by using the arguments in the proofs of [24] Th. 4.12, Cor. 4.13, Th. 4.14. □

Theorem 7.1. *For any $l \geq 3$, let a denote the number of ideal classes in H^* invariant under complex conjugation. Then $a = 2h$.*

Proof. Let ν denote the total number of primes ramified in $F_l(\rho)/F_l$. Since the infinite primes are ramified, we have $\nu = 3 + \nu'$ where ν' is the number of prime ideals of $\mathcal{O}$ dividing $3\mathcal{O}$. It is easily seen that 3 is inert in $F_l/\mathbb{Q}$ if and only if $3 \nmid l$, and never splits completely. Therefore, $\nu = 4$ if $3 \nmid l$, and $\nu = 5$ if $3 \mid l$.

Put $q = [U \cap N_{6/3}(F_l(\rho)) : U^2]$. From [9] Ia, Satz 13 we obtain $a = hq2^{\nu-4}$. Hence we have to show that $q = 2^{5-\nu}$. Of the coset representatives (3.8) only $\epsilon(\epsilon - 1)$ is totally positive, and is therefore the only one which can belong to $N_{6/3}(F_l(\rho))$. To prove the theorem it is now enough to show that $\epsilon(\epsilon - 1)$ is a norm from $F_l(\rho)$ if and only if $3 \nmid l$. By [9] II, §11, (13.), the assertion is equivalent to the following condition concerning the quadratic Hilbert symbol:

$$\left(\frac{\epsilon(\epsilon-1),\ -3}{\mathbf{p}}\right) = 1 \text{ for every prime } \mathbf{p} \text{ of } F_l \quad \Leftrightarrow \quad 3 \nmid l.$$

If $\mathbf{p}$ is a finite prime not dividing 3, then the Hilbert symbol is $= 1$, because -3 is $\mathbf{p}$-primary. The same is true if $\mathbf{p}$ is infinite, because $\epsilon(\epsilon - 1)$ is totally positive. For $3 \nmid l$ there is only one prime missing, and the result follows from the product formula. Suppose therefore that $3 \mid l$. Take $\mathbf{p} = 3\mathcal{O} + (\epsilon - 2)\mathcal{O}$. Since $N(\epsilon - 2) = -2l - 3 \equiv 0 \bmod 3$, $\mathbf{p}$ has norm 3, and

$$\left(\frac{\epsilon(\epsilon-1),\ -3}{\mathbf{p}}\right) = \left(\frac{-3,\ \epsilon(\epsilon-1)}{\mathbf{p}}\right) = \left(\frac{\epsilon(\epsilon-1)}{\mathbf{p}}\right) = \left(\frac{2}{\mathbf{p}}\right) = -1.$$

□

By analogy with Proposition 6.1, we have here the following result pertaining to the structure of $\mathcal{H}_3(F_l(\rho))$, and showing that the class number h^* is mostly divisible by 3.

Theorem 7.2. *If $l \geq 3$ and $l \equiv 0,1,5,6,7,8 \bmod 9$, there is exactly one cubic extension $F_l(\rho, \sqrt[3]{\lambda})/F_l(\rho)$ such that*

(i) *λ is a unit of $F_l(\rho)$;*

(ii) *$F_l(\rho, \sqrt[3]{\lambda})/F_l(\rho)$ is unramified.*

This extension is obtained by taking

$$\lambda = \begin{cases} \epsilon^2(\epsilon - 1), & \text{if } l \equiv 0,8 \bmod 9, \\ \epsilon(\epsilon - 1), & \text{if } l \equiv 1,5 \bmod 9, \\ \epsilon, & \text{if } l \equiv 6,7 \bmod 9. \end{cases}$$

Consequently, $h^* \equiv 0 \bmod 3$ for these values of l. If $l \equiv 2,3,4 \bmod 9$, there is no such extension.

Proof. For any nonzero ideal $\mathbf{A}$ of $\mathcal{O}^*$, let $G(\mathbf{A})$ denote the unit group of the residue class ring $\mathcal{O}^*/\mathbf{A}$.

Let $\lambda \in U^* \setminus U^{*3}$. It follows from [9] Ia, Satz 9 that (ii) is true if and only if there is a $\xi \in \mathcal{O}^*$ such that $\xi^3 \equiv \lambda \bmod (1-\rho)^3$.

Suppose first that $3 \nmid l$. Then 3 is inert in $F_l/\mathbb{Q}$ so that $\mathbf{P} = (1-\rho)\mathcal{O}^*$ is a prime ideal of the ring $\mathcal{O}^*$ having absolute inertial degree 3. We have $G(\mathbf{P}^3) = G_0 \times G_1$ where

$$\begin{aligned} G_0 &= \{\alpha + \mathbf{P}^3 \mid \alpha^{26} \equiv 1 \bmod \mathbf{P}^3\} && \text{is cyclic of order 26 ,} \\ G_1 &= \{\alpha + \mathbf{P}^3 \mid \alpha \equiv 1 \bmod \mathbf{P}\} && \text{is elementary abelian of order } 3^6 . \end{aligned}$$

Therefore, $G_0 = G(\mathbf{P}^3)^3$ so that (ii) holds if and only if $\lambda^{26} \equiv 1 \bmod \mathbf{P}^3$. Since $\lambda^{26} \equiv 1 \bmod \mathbf{P}^2$ for any $\lambda \in U$ and $\rho^{26a} \not\equiv 1 \bmod \mathbf{P}^2$ for $3 \nmid a$, we can suppose that $\lambda \in U$, by Proposition 7.1 (ii). It now follows from Theorem 4.1 that it is enough to take λ as a member of the set $\mathcal{S} = \{\epsilon,\ \epsilon - 1,\ \epsilon(\epsilon - 1),\ \epsilon^2(\epsilon - 1)\}$. The truth of the assertion can be established by means of the following congruences modulo 9, which in $\mathcal{O}$ gives the same result as $\mathbf{P}^3$:

CASE $l \equiv 1 \bmod 3$.

$$\begin{aligned} \epsilon^{26} &\equiv l + 3 + (2l+4)\epsilon && \equiv 1 \quad \Leftrightarrow \quad l \equiv 7 , \\ (\epsilon - 1)^{26} &\equiv 7 + (2l+1)\epsilon && \not\equiv 1 , \\ (\epsilon(\epsilon - 1))^{26} &\equiv l + (l-1)\epsilon && \equiv 1 \quad \Leftrightarrow \quad l \equiv 1 , \\ (\epsilon^2(\epsilon - 1))^{26} &\equiv 2l + 2 + 6\epsilon && \not\equiv 1 . \end{aligned}$$

CASE $l \equiv 2 \bmod 3$.

$$\begin{aligned} \epsilon^{26} &\equiv 2l + 6\epsilon + 3\epsilon^2 && \not\equiv 1 , \\ (\epsilon - 1)^{26} &\equiv 2l + (l-2)\epsilon + (2l-4)\epsilon^2 && \not\equiv 1 , \\ (\epsilon(\epsilon - 1))^{26} &\equiv l + 5 + (l+4)\epsilon + (2l-1)\epsilon^2 && \equiv 1 \quad \Leftrightarrow \quad l \equiv 5 , \\ (\epsilon^2(\epsilon - 1))^{26} &\equiv 1 + (l+1)\epsilon + (2l+2)\epsilon^2 && \equiv 1 \quad \Leftrightarrow \quad l \equiv 8 . \end{aligned}$$

Suppose now that $3 \mid l$. In this case $(1-\rho)\mathcal{O}^* = \mathbf{P}\mathbf{P}'$ where

$$\mathbf{P} = (\epsilon - 2)\mathcal{O}^* + (1-\rho)\mathcal{O}^*, \quad \mathbf{P}' = (2+\epsilon+\epsilon^2)\mathcal{O}^* + (1-\rho)\mathcal{O}^*$$

are prime ideals. We have

$$G(\mathbf{P}^3\mathbf{P}'^3) = G(\mathbf{P}^3) \times G(\mathbf{P}'^3) = G_0 \times G_1 \times G_0' \times G_1',$$

say, where $G_0 \cong C(2)$, $G_0' \cong C(8)$, and G_1, G_1' are elementary abelian of order 3^2, 3^4 respectively. It is again enough to take $\lambda \in \mathcal{S}$, and we find that (ii) holds if and only if $\lambda^8 \equiv 1 \bmod 9$. The rest of the theorem is a consequence of the following congruences modulo 9:

$$\begin{aligned}
\epsilon^8 &\equiv l+4+(l+3)\epsilon+(2l+6)\epsilon^2 && \equiv 1 \quad \Leftrightarrow \quad l \equiv 6\,,\\
(\epsilon-1)^8 &\equiv l+4+3\epsilon+6\epsilon^2 && \not\equiv 1\,,\\
(\epsilon(\epsilon-1))^8 &\equiv 2l+7+(l+6)\epsilon+(2l+3)\epsilon^2 && \not\equiv 1\,,\\
(\epsilon^2(\epsilon-1))^8 &\equiv 1+2l\epsilon+l\epsilon^2 && \equiv 1 \quad \Leftrightarrow \quad l \equiv 0.
\end{aligned}$$

□

The ensuing proposition is a simple but useful tool in the actual computation of cubic extensions cyclic over F_l.

Proposition 7.2. (i) *If E/F_l is a cyclic cubic extension, there is a number $\alpha \in F_l(\rho)$ such that*
(a) *α is not a third power in $F_l(\rho)$;*
(b) *$\alpha\overline{\alpha} = \beta^3$ where $\beta \in F_l$;*
(c) *$E = F_l(\vartheta + \beta/\vartheta)$ where $\vartheta^3 = \alpha$.*

(ii) *Conversely, suppose that $\alpha \in F_l(\rho)$ satisfies the conditions* (a) *and* (b). *If $\vartheta^3 = \alpha$, then $F_l(\vartheta+\beta/\vartheta)/F_l$ is a cyclic cubic extension which is independent of the choice of ϑ.*

(iii) *If (i) holds, then E/F_l is unramified if and only if $F_l(\rho,\vartheta)/F_l(\rho)$ is unramified.*

Proof. (i) If E/F_l is cyclic of degree 3, then $E(\rho)/F_l(\rho)$ is a Kummer extension so that $E(\rho) = F_l(\rho,\vartheta)$ for some ϑ with $\vartheta^3 = \alpha \in F_l(\rho)$. Put $\beta = \vartheta\overline{\vartheta}$ and $\mu = \vartheta + \overline{\vartheta} = \vartheta + \beta/\vartheta$. Then $\beta, \mu \in E$. But $\beta^3 = \alpha\overline{\alpha} \in F_l$ whence $\beta \in F_l$. Since ϑ is of degree at most 2 over $F_l(\mu)$, we must have $E = F_l(\mu)$.

(ii) Suppose that (a) and (b) are valid. Let ϑ be one of the values of $\sqrt[3]{\alpha}$ and put $\mu = \vartheta + \beta/\vartheta$. We have

$$\mathrm{Irr}(\mu, F_l) = x^3 - 3\beta x - \alpha - \overline{\alpha}. \tag{7.1}$$

Then $E = F_l(\mu)$ is a cyclic cubic extension of F_l because the discriminant $D(\mathrm{Irr}(\mu, F_l)) = -27(\alpha-\overline{\alpha})^2$ is a square in F_l. Since (7.1) is independent of the choice of ϑ, so is E.

(iii) The absolute ramification index of any finite prime of $F_l(\rho,\vartheta)$ is divisible by 3 if and only if the prime is ramified in either one of the two cubic extensions. □

8 Some 3-class fields

This final section is devoted to (mainly numerical) results about the structure of the fields $\mathcal{H}_3(F_l)$ and $\mathcal{H}_3(F_l(\rho))$. Theorem 7.2 already gives a piece of information about the latter. In many questions concerning $\mathcal{F}$ the prime 7 plays a peculiar special rôle, to some extent explained by the fact that $f_l(x)$ is never an Eisenstein polynomial mod p, if $p \neq 7$. An illustration is given by

Proposition 8.1. *Suppose that* $l \geq 3$, $l \equiv 2$ *mod* 7, $l \not\equiv 23$ *mod* 49. *Then the cyclic cubic extension* $F_l(\cos(2\pi/7))/F_l$ *is unramified so that* $h \equiv 0$ *mod* 3.

Proof. The result follows immediately from Proposition 3.1 and [11] Theorem 1. An alternative direct proof can be derived from the fact that the polynomial (cf. (3.4))

$$\operatorname{Irr}((\epsilon - 2)/\psi, \mathbb{Q}(\psi)) = \psi^{-3} f_l(\psi x + 2), \quad \text{where } \psi = 2\cos(2\pi/7) - 2,$$

has algebraic integer coefficients and discriminant prime to 7, so that every prime divisor of 7 is unramified in $F_l(\psi)/\mathbb{Q}(\psi)$ and therefore also in $F_l(\psi)/F_l$. □

We have computed $\mathcal{H}_3(F_l)$ for each l such that $3 \leq l \leq 18$ and $3 \mid h$. We exclude the cases $l = 9, 16$ which are covered by Proposition 8.1. If $\mathbf{A}$ is any nonzero ideal of $\mathcal{O}^*$, we let $\operatorname{Cl}(\mathbf{A})$ denote the ideal class in H^* containing $\mathbf{A}$. A^* denotes the subgroup of H^* consisting of elements invariant under complex conjugation. From Theorem 7.1 we know that A^* is of order $2h$.

We have determined a subgroup H_0^* of H^* and we conjecture that in each case in fact $H_0^* = H^*$. We have carried out the verification that H_0^* is really a subgroup, i.e., there are no further relations between the given generators. There is also very strong numerical evidence supporting the conjecture, because a rather comprehensive and systematic inspection of ideals with small norms has not revealed any further ideal classes, but a complete proof is missing in each of the cases investigated.

From Kummer theory we obtain a subextension $\widetilde{\mathcal{H}}_3^*$ of $\mathcal{H}_3(F_l(\rho))/F_l(\rho)$ and we contend correspondingly that $\widetilde{\mathcal{H}}_3^*$ is the maximal unramified 3-extension of $F_l(\rho)$ with elementary abelian Galois group over $F_l(\rho)$.

For the values of l in question, $\mathcal{H}_3(F_l)/F_l$ is a cubic extension, and we give a pair of numbers α, β satisfying the conditions of Proposition 7.2. The minimal polynomial for a primitive element μ is then obtained from (7.1). To see that $F_l(\rho, \sqrt[3]{\alpha})/F_l(\rho)$ is unramified it is enough to verify three things: (i) α is an integer prime to 3; (ii) $\alpha\mathcal{O}^*$ is a cube in the ideal group of $F_l(\rho)$; (iii) $\alpha \equiv \xi^3 \bmod (1-\rho)^3$ for some $\xi \in \mathcal{O}^*$. The results are as follows:

Case $l = 12$. Define

$$\begin{aligned}
\mathbf{P}_7 &= (7, \epsilon - 6, \rho - 2), & \mathbf{P}_{11} &= (11, \epsilon - 6),\\
\mathbf{P}_{67} &= (67, \epsilon - 18, \rho - 37), & \mathbf{P}_{211} &= (211, \epsilon - 53, \rho - 14);\\
K &= \operatorname{Cl}(\mathbf{P}_7), & L &= \operatorname{Cl}(\mathbf{P}_{67});\\
\alpha &= (\epsilon - 1)^2(21\epsilon + 2\epsilon^2 + 6\epsilon\rho), & \alpha_1 &= -1718 + 1461\epsilon + 133\epsilon^2.
\end{aligned}$$

Then

$$\begin{aligned}
K^{24} &= L^3 = 1;\\
H_0^* &= \langle K, L\rangle \cong C(8)\times C(3)\times C(3);\\
\overline{K} &= K^7, \quad \overline{L} = K^8L^2, \quad A^* = \langle K^4\rangle; \quad \beta = (\epsilon-1)(4\epsilon-1);\\
\alpha\mathcal{O}^* &= \mathbf{P}_{211}^3\ , \mathbf{P}_{211} \in K^8L, \ \alpha \equiv -\epsilon^6 \bmod (1-\rho)^3;\\
\alpha_1\mathcal{O}^* &= \mathbf{P}_{11}^3\ , \ \mathbf{P}_{11} \in K^8, \quad \alpha_1 \equiv \epsilon^3 \bmod 9;\\
\widetilde{\mathcal{H}}_3^* &= F_l\ (\rho,\ \sqrt[3]{\alpha},\ \sqrt[3]{\alpha_1}).
\end{aligned}$$

CASE $l = 13$. Define

$$\begin{aligned}
\mathbf{P}_5 &= (5, \epsilon-3), & \mathbf{P}_7 &= (7, \epsilon-4, \rho-2),\\
\mathbf{P}_{223} &= (223, \epsilon-56, \rho-39); & &\\
K &= \mathrm{Cl}(\mathbf{P}_7), & L &= \mathrm{Cl}(\mathbf{P}_{223});\\
\alpha &= 74-60\epsilon-5\epsilon^2+6\rho, & \alpha_1 &= (\epsilon^2-\epsilon)(-1-2\epsilon+3\epsilon^2).
\end{aligned}$$

Then

$$\begin{aligned}
K^{24} &= L^3 = 1;\\
H_0^* &= \langle K, L\rangle \cong C(8)\times C(3)\times C(3);\\
\overline{K} &= K^7, \quad \overline{L} = L^2; \quad A^* = \langle K^4\rangle; \quad \beta = 17-12\epsilon-\epsilon^2;\\
\alpha\mathcal{O}^* &= \mathbf{P}_{223}^3\ , \ \alpha \equiv (\epsilon+\epsilon^2)^3 \bmod (1-\rho)^3;\\
\alpha_1\mathcal{O}^* &= \mathbf{P}_5^3\ , \quad \mathbf{P}_5 \in K^{16}, \quad \alpha_1 \equiv \epsilon^6 \bmod 9;\\
\widetilde{\mathcal{H}}_3^* &= F_l\ (\rho,\ \sqrt[3]{\alpha},\ \sqrt[3]{\alpha_1}).
\end{aligned}$$

CASE $l = 15$. Define

$$\begin{aligned}
\mathbf{P}_3 &= (3, \epsilon-2, \rho-1), & \mathbf{P}_7 &= (7, \epsilon-5, \rho-2),\\
\mathbf{P}_{13} &= (13, \epsilon-10, \rho-3), & \mathbf{P}_{17} &= (17, \epsilon-6),\\
\mathbf{P}_{19} &= (19, \epsilon-5, \rho-11); & &\\
K &= \mathrm{Cl}(\mathbf{P}_7), & L &= \mathrm{Cl}(\mathbf{P}_{17}),\\
M &= \mathrm{Cl}(\mathbf{P}_3\mathbf{P}_{17}); & &\\
\alpha &= (\epsilon-1)^2(24\epsilon+\epsilon^2+6\epsilon\rho). & &
\end{aligned}$$

Then

$$\begin{aligned}
K^{18} &= L^6 = M^2 = 1;\\
H_0^* &= \langle K, L, M\rangle \cong C(2)\times C(2)\times C(2)\times C(3)\times C(9);\\
\overline{K} &= K^{17}L^5, \quad \overline{L} = L, \quad \overline{M} = M; \quad A^* = \langle L, M\rangle; \quad \beta = (\epsilon-1)(4\epsilon-1);\\
\alpha\mathcal{O}^* &= (\mathbf{P}_{13}\mathbf{P}_{19})^3, \ \mathbf{P}_{13} \in K^{17}LM, \ \mathbf{P}_{19} \in K^7L^5M, \ \alpha \equiv \epsilon^6 \bmod (1-\rho)^3;\\
\widetilde{\mathcal{H}}_3^* &= F_l\ (\rho,\ \sqrt[3]{\epsilon},\ \sqrt[3]{\alpha}).
\end{aligned}$$

CASE $l = 17$. Define

$$\begin{aligned}
\mathbf{P}_7 &= (7, \epsilon-3, \rho-2), & \mathbf{P}_{19} &= (19, \epsilon-7, \rho-7),\\
\mathbf{P}_{37} &= (37, \epsilon-23, \rho-26), & \mathbf{P}_{439} &= (439, \epsilon-5, \rho-171);\\
K &= \mathrm{Cl}(\mathbf{P}_7), & L &= \mathrm{Cl}(\mathbf{P}_7\overline{\mathbf{P}}_7),\\
M &= \mathrm{Cl}(\mathbf{P}_7^2\mathbf{P}_{37}), & N &= \mathrm{Cl}(\mathbf{P}_7^2\mathbf{P}_{19});\\
\alpha &= -10+\epsilon-12\rho+6\epsilon\rho, & \alpha_1 &= \epsilon^2(\epsilon+18).
\end{aligned}$$

Then

$$\begin{aligned}
K^3 &= L^3 = M^2 = N^9 = 1;\\
H_0^* &= \langle K, L, M, N\rangle \cong C(2)\times C(3)\times C(3)\times C(9);\\
\overline{K} &= K^2L,\ \overline{L} = L,\ \overline{M} = M,\ \overline{N} = LN^8;\ A^* = \langle L, M\rangle;\ \beta = 5-\epsilon;\\
\alpha\mathcal{O}^* &= \mathbf{P}_{439}^3\ ,\ \mathbf{P}_{439}\in KLN^3,\quad \alpha\equiv\epsilon^6 \bmod (1-\rho)^3;\\
\alpha_1\mathcal{O}^* &= (\mathbf{P}_7\overline{\mathbf{P}}_7)^3,\quad \alpha_1\equiv\epsilon^3 \bmod 9;\\
\widetilde{\mathcal{H}}_3^* &= F_l\ (\rho,\ \sqrt[3]{\epsilon^2(\epsilon-1)},\ \sqrt[3]{\alpha},\ \sqrt[3]{\alpha_1}).
\end{aligned}$$

Case $l = 18$. Define

$$\begin{aligned}
\mathbf{P}_3 &= (3,\epsilon-2,\rho-1), & \mathbf{P}_{13} &= (13,\epsilon-2,\rho-3),\\
\mathbf{P}_{67} &= (67,\epsilon-9,\rho-29), & \mathbf{P}_{607} &= (607,\epsilon-306,\rho-210);\\
K &= \mathrm{Cl}(\mathbf{P}_3), & L &= \mathrm{Cl}(\mathbf{P}_3\mathbf{P}_{13}),\\
M &= \mathrm{Cl}(\mathbf{P}_{67});\\
\alpha &= 12-14\epsilon+3\rho.
\end{aligned}$$

Then

$$\begin{aligned}
K^{12} &= L^{18} = M^2 = 1,\ L^9 = K^6;\\
H_0^* &= \langle K, L, M\rangle \cong C(2)\times C(4)\times C(3)\times C(9);\\
\overline{K} &= K,\ \overline{L} = L^{17},\ \overline{M} = MK^6;\ A^* = \langle K\rangle;\ \beta = 5-2\epsilon;\\
\alpha\mathcal{O}^* &= \mathbf{P}_{607}^3\ ,\ \mathbf{P}_{607}\in L^{12},\quad \alpha\equiv(1+\epsilon^2)^3 \bmod (1-\rho)^3;\\
\widetilde{\mathcal{H}}_3^* &= F_l\ (\rho,\ \sqrt[3]{\epsilon^2(\epsilon-1)},\ \sqrt[3]{\alpha}).
\end{aligned}$$

References

[1] A. Baker, "Transcendental Number Theory", Cambridge University Press, Cambridge, 1975.

[2] S. Chowla, *Proof of a conjecture of Julia Robinson*, Norske Vid. Selsk. Forh. (Trondheim) **34**, No. 20 (1961).

[3] H. Cohn, *A device for generating fields of even class number*, Proc. Amer. Math. Soc. **7** (1956), 595–598.

[4] B. N. Delone and D. K. Faddeev, "The Theory of Irrationalities of the Third Degree", Trudy Mat. Inst. Steklov, vol. 11, 1940; English transl., Transl. Math. Monographs, vol 10, Amer. Math. Soc., Providence, R. I., Second printing 1978.

[5] V. Ennola, *On a conjecture of H. J. Godwin on cubic units*, Ann. Acad. Sci. Fenn., Ser. A I **12**, No. 2 (1987), 319–328.

[6] V. Ennola and R. Turunen, *On totally real cubic fields*, Math. Comp. **44** (1985), 495–518.

[7] M.-N. Gras, *Sur les corps cubiques cycliques dont l'anneau des entiers est monogène*, C. R. Acad. Sci., Paris, Sér. A **278** (1974), 59–62.

[8] M.-N. Gras, *Lien entre le groupe des unités et la monogénéité des corps cubiques cycliques*, Publ. Math. Fac. Sci., Besancon 1975–76, Fasc. 1, 19p.

[9] H. Hasse, *Bericht über neuere Untersuchungen und Probleme aus der Theorie der algebraischen Zahlkörper I*, Jber. dt. MatVerein **35** (1926), 1–55; *Ia* **36** (1927), 233–311; *II*, Erg. Bd. **6** (1930), 1–204.

[10] H. Hasse, *Arithmetische Theorie der kubischen Zahlkörper auf klassenkörpertheoretischer Grundlage*, Math. Z. **31** (1930), 565–582.

[11] M. Ishida, *Class numbers of algebraic number fields of Eisenstein type*, J. Number Theory **2** (1970), 404–413.

[12] S. Lang, *Integral points on curves*, Publ. Math., Inst. Hautes Etud. Sci. **6** (1960), 27–43.

[13] E. Lehmer, *Connection between Gaussian periods and cyclic units*, Math. Comp. **50** (1988), 535–541.

[14] A. Leutbecher and J. Martinet, *Lenstra's constant and Euclidean number fields*, Journées arithmétiques, Metz 1981, Astérisque **94** (1982), 87-131.

[15] A. Leutbecher and J. Martinet, *Constante de Lenstra et corps de nombres euclidiens*, Sémin. Théor. Nombres, Univ. Bordeaux I 1981-1982, Exposé No. 4, 7p (1982).

[16] A. Leutbecher and G. Niklasch, *On cliques of exceptional units and Lenstra's construction of Euclidean fields*, Number Theory, Ulm 1987, Lect. Notes Math. **1380** (1989), 150–178.

[17] J. Martinet, *Sur la constante de Lenstra des corps de nombres*, Sémin. Théor. Nombres, Univ. Bordeaux I 1979-1980, Exposé No. 17, 21p (1980).

[18] J.-F. Mestre, *Corps euclidiens, unités exceptionnelles et courbes elliptiques*, J. Number Theory **13** (1981), 123–137.

[19] T. Nagell, *Quelques problèmes relatifs aux unités algébriques*, Ark. Mat. **8** (1969), 115–127.

[20] T. Nagell, *Sur un type particulier d'unités algébriques*, Ark. Mat. **8** (1969), 163–184.

[21] D. Shanks, *The simplest cubic fields*, Math. Comp. **28** (1974), 1137–1152.

[22] C.L. Siegel, *Approximation algebraischer Zahlen*, Math. Z. **10** (1921), 173–213.

[23] E. Thomas, *Fundamental units for orders in certain cubic number fields*, J. Reine Angew. Math. **310** (1979), 33–55.

[24] L.C. Washington, "Introduction to Cyclotomic Fields", Springer-Verlag, New York, 1982.

[25] L.C. Washington, *Class numbers of simplest cubic fields*, Math. Comp. **48** (1987), 371–384.

[26] M. Watabe, *On certain cubic fields I*, Proc. Japan Acad., Ser. A **59** (1983), 66–69; *II*, **59** (1983), 107–108; *III*, **59** (1983), 260–262; *IV*, **59** (1983), 387–389; *V*, **60** (1984), 302–305; *VI*, **60** (1984), 331–332.

Enumeration of totally complex quartic fields of small discriminant

D. Ford *

Abstract. All totally complex algebraic number fields with discriminant $< 10^6$ have been enumerated. Software derived from that used in the enumeration of totally real quartic fields of small discriminant [2], but substantially re-written, was used. The LLL algorithm was adapted to the problem of determining the isomorphism of two fields. The T_2 criterion of Pohst proved effective in reducing the number of redundant examples.

1 Introduction

In [2] all totally real quartic number fields of discriminant less than 10^6 were enumerated. We followed a method originating with Delone and Faddeev [3], taking advantage of number-geometric considerations that apply in the totally real case. For the totally complex case these considerations do not apply, and we use instead the more general methods of [4].

Procedures from the earlier computation that were used without substantial modification (e.g., testing for polynomial irreducibility, computing the signature and discriminant of a number field) are treated here in summary fashion; for details the reader is referred to [2].

2 Existence of small integers

A consequence of [4] Theorems 1-3 is

Proposition 2.1. *If F is an algebraic number field of degree 4 with discriminant d_F and ring of integers $\mathcal{O}_F$ then there exists $\rho \in \mathcal{O}_F \setminus \mathbb{Z}$ such that*

$$T_2(\rho) \;\leq\; 1 + \sqrt[3]{d_F/2}$$

*Research supported by the Natural Sciences and Engineering Research Council (Canada)

Computational Number Theory

where $\rho_1, \ldots, \rho_4$ are the conjugates of ρ, and

$$T_k(\rho) = \sum_{j=1}^{4} |\rho_j|^k.$$

We let ρ be an algebraic number satisfying the conditions of Proposition 2.1, with characteristic polynomial given by

$$f(x) = \prod_{j=1}^{4}(x - \rho_j) = x^4 - sx^3 + px^2 - qx + n.$$

Without loss of generality, ρ may be assumed to satisfy the following.

$$0 \le s \le 2, \tag{2.1}$$

$$\frac{s^2 - T_2(\rho)}{2} \le p \le \frac{s^2/2 + T_2(\rho)}{2}, \tag{2.2}$$

$$\frac{-s^3 + 3sp - T_3(\rho)}{3} \le q \le \frac{-s^3 + 3sp + T_3(\rho)}{3}, \tag{2.3}$$

$$0 \le n \le \frac{T_2^2(\rho)}{16}. \tag{2.4}$$

Relations (2.1), (2.2), and (2.4) come directly from [4] Theorem 3, while (2.3) is a consequence of $|s^3 - 3sp + 3q| = |\sum_{j=1}^{4} \rho_j^3| \le T_3(\rho)$. We take

$$\widehat{T}_{21}(d) = \lfloor 1 + \sqrt[3]{d/2} \rfloor, \qquad \widehat{T}_3(d) = \lfloor (1 + \sqrt[3]{d/2})^{3/2} \rfloor,$$
$$\widehat{T}_{22}(d) = \lfloor 2(1 + \sqrt[3]{d/2}) \rfloor, \qquad \widehat{T}_4(d) = \lfloor (1 + \sqrt[3]{d/2})^2 \rfloor.$$

and define $G_1(d)$ to be the set of all polynomials $f(x) = x^4 - sx^3 + px^2 - qx + n \in \mathbb{Z}[x]$ with coefficients satisfying

$$0 \le s \le 2, \tag{2.5}$$

$$\frac{s^2 - \widehat{T}_{21}(d)}{2} \le p \le \frac{s^2 + \widehat{T}_{22}(d)}{4}, \tag{2.6}$$

$$\frac{-s^3 + 3sp - \widehat{T}_3(d)}{3} \le q \le \frac{-s^3 + 3sp + \widehat{T}_3(d)}{3}, \tag{2.7}$$

$$0 \le n \le \frac{\widehat{T}_4(d)}{16}. \tag{2.8}$$

The integer quantities $\widehat{T}_{21}(d)$, $\widehat{T}_{22}(d)$, $\widehat{T}_3(d)$, $\widehat{T}_4(d)$ are easy to compute, so the question of whether a given polynomial belongs to $G_1(d)$ can be answered quickly.

Observing that

$$\lfloor T_2(\rho) \rfloor \;\leq\; \widehat{T}_{21}(d_F)\,, \quad \lfloor 2T_2(\rho) \rfloor \;\leq\; \widehat{T}_{22}(d_F)\,,$$

$$\lfloor T_3(\rho) \rfloor \;\leq\; \lfloor T_2(\rho)^{3/2} \rfloor \;\leq\; \widehat{T}_3(d_F)\,, \quad \lfloor T_2^2(\rho) \rfloor \;\leq\; \widehat{T}_4(d_F)$$

establishes

Proposition 2.2. *If F is a totally complex algebraic number field of degree 4 with discriminant d_F and ring of integers $\mathcal{O}_F$ then there exists $\rho \in \mathcal{O}_F \setminus \mathbb{Z}$ with characteristic polynomial belonging to $G_1(d_F)$.*

3 Fields with small quartic integers

For each polynomial $f(x) \in G_1(10^6)$ we perform the following tests (cf. [2]).

- If $f(x)$ is reducible in $\mathbb{Z}[x]$ we exclude f.
- If $f(x)$ is irreducible but $\mathbb{Q}(\rho)$ is not totally complex, we exclude f.
- We compute the discriminant d_F of the field $F = \mathbb{Q}(\rho)$.
- If $d_F \geq 10^6$ we exclude f.
- If $f \notin G_1(d_F)$ we exclude f.

Among the polynomials surviving these tests is a generating polynomial for every totally complex number field of discriminant less than 10^6, excepting those fields for which no choice of ρ in Proposition 2.1 gives a quartic integer. In these fields every such "small" ρ generates a quadratic subfield; we enumerate these fields separately.

4 Fields with quadratic subfields

We now assume that F is a totally complex algebraic number field of degree 4 with discriminant $d_F < 10^6$, and that F contains a quadratic subfield K with discriminant d_K.

Let $\omega, \omega' = \frac{1}{2}(\sigma \pm \sqrt{d_K})$, where $\sigma \in \{0, 1\}$, $\sigma \equiv d_K \bmod 4$, so that $K = \mathbb{Q}(\omega)$, and $\mathcal{O}_K = \mathbb{Z}[\omega]$ is the ring of integers of K.

We take $\rho \in \mathcal{O}_F \setminus \mathcal{O}_K$, and let

$$g(x) = x^2 - \alpha x + \beta \in \mathcal{O}_K[x]$$

be the minimal polynomial of ρ over K, with

$$\begin{aligned} \alpha &= a_1 + a_2\omega\,, & \beta &= b_1 + b_2\omega, \\ \alpha' &= a_1 + a_2\omega', & \beta' &= b_1 + b_2\omega'. \end{aligned}$$

Then the characteristic polynomial of ρ over $\mathbb{Q}$ is

$$\begin{aligned} f(x) &= (x^2 - (a_1 + a_2\omega)x + (b_1 + b_2\omega))(x^2 - (a_1 + a_2\omega')x + (b_1 + b_2\omega')) \\ &= (x^2 - \alpha x + \beta)(x^2 - \alpha' x + \beta') \qquad (4.1) \\ &= x^4 - sx^3 + px^2 - qx + n \quad \in \mathbb{Z}[x]. \end{aligned}$$

Following [1] we construct a $\mathbb{Z}$-module homomorphism $\lambda : \mathcal{O}_F \longrightarrow \mathbb{R}^4$, depending on d_K. We define

$$\lambda(\xi) = \begin{cases} (0, 0, \Im\xi_1, \Im\xi_2), & \text{if } d_K > 0 \\ \frac{1}{2}(\Re\xi_1 - \Re\xi_2, \Re\xi_2 - \Re\xi_1, \Im\xi_1 - \Im\xi_2, \Im\xi_2 - \Im\xi_1), & \text{if } d_K < 0 \end{cases}$$

where ξ_1, $\overline{\xi}_1$, ξ_2, $\overline{\xi}_2$ are the conjugates of ξ. Then $\lambda(\mathcal{O}_F)$ is a 2-dimensional lattice of determinant $\Delta = \frac{1}{4}\sqrt{d_F/|d_K|}$, and the kernel of λ is $\mathcal{O}_K$.

As in [2] we choose $\rho \in \mathcal{O}_F$ so that $\|\lambda(\rho)\|^2 \leq 2\Delta/\sqrt{3} = \sqrt{d_F/12|d_K|}$ and $\mathrm{Tr}_{F/K}(\rho) = \alpha \in \{0+\omega, 1+\omega, 0+2\omega, 1+2\omega\}$. (This choice for α ensures that ρ has degree 4 over $\mathbb{Q}$.)

There are two cases to consider, depending on the sign of d_K, corresponding to whether K is fixed by complex conjugation.

If $d_K > 0$ we have

$$\begin{aligned} \alpha &= \rho_1 + \overline{\rho}_1 = 2\Re\rho_1, & \beta &= \rho_1\overline{\rho}_1 = |\rho_1|^2, \\ \alpha' &= \rho_2 + \overline{\rho}_2 = 2\Re\rho_2, & \beta' &= \rho_2\overline{\rho}_2 = |\rho_2|^2. \end{aligned}$$

Therefore

$$\begin{aligned} \beta + \beta' &= \Re\rho_1{}^2 + \Re\rho_2{}^2 + \Im\rho_1{}^2 + \Im\rho_2{}^2 = \tfrac{1}{4}(\alpha^2 + \alpha'^2) + \|\lambda(\rho)\|^2, \\ \beta - \beta' &= \Re\rho_1{}^2 - \Re\rho_2{}^2 + \Im\rho_1{}^2 - \Im\rho_2{}^2 = \tfrac{1}{4}(\alpha^2 - \alpha'^2) + \Im\rho_1{}^2 - \Im\rho_2{}^2 \end{aligned}$$

so that

$$\begin{aligned} (\beta + \beta') - \tfrac{1}{4}(\alpha^2 + \alpha'^2) &= \|\lambda(\rho)\|^2, \\ |\,(\beta - \beta') - \tfrac{1}{4}(\alpha^2 - \alpha'^2)\,| &\leq \|\lambda(\rho)\|^2. \end{aligned}$$

In the case $d_K < 0$ we have

$$\begin{aligned} \alpha &= \rho_1 + \rho_2, & \beta &= \rho_1\rho_2, \\ \alpha' &= \overline{\rho}_1 + \overline{\rho}_2, & \beta' &= \overline{\rho}_1\overline{\rho}_2. \end{aligned}$$

We define $\delta = \rho_1 - \rho_2$, giving

$$\begin{aligned}\beta + \beta' &= \tfrac{1}{4}(\alpha^2 + \alpha'^2) - \tfrac{1}{4}(\delta^2 + \overline{\delta}^2),\\ \beta - \beta' &= \tfrac{1}{4}(\alpha^2 - \alpha'^2) - \tfrac{1}{4}(\delta^2 - \overline{\delta}^2)\end{aligned}$$

so that

$$\begin{aligned}|\,(\beta + \beta') - \tfrac{1}{4}(\alpha^2 + \alpha'^2)\,| &= \tfrac{1}{4}|\,\delta^2 + \overline{\delta}^2\,| \le \|\lambda(\rho)\|^2,\\ |\,(\beta - \beta') - \tfrac{1}{4}(\alpha^2 - \alpha'^2)\,| &= \tfrac{1}{4}|\,\delta^2 - \overline{\delta}^2\,| \le \|\lambda(\rho)\|^2.\end{aligned}$$

We define $A_1 = 4a_1^2 + 4a_1a_2\sigma + a_2^2(d_K + \sigma)$, $A_2 = a_2(2a_1 + a_2\sigma)$, so that

$$\begin{aligned}\alpha^2 + \alpha'^2 &= \tfrac{1}{2}A_1\,, & \beta + \beta' &= 2b_1 + b_2\sigma,\\ \alpha^2 - \alpha'^2 &= A_2\sqrt{d_K}\,, & \beta - \beta' &= b_2\sqrt{d_K}\,.\end{aligned}$$

It follows that

$$\frac{A_2 - M}{4} \le b_2 \le \frac{A_2 + M}{4} \tag{4.2}$$

$$\frac{A_1 - 8b_2\sigma - N_L}{16} \le b_1 \le \frac{A_1 - 8b_2\sigma + N_H}{16} \tag{4.3}$$

where

$$\begin{aligned}M &= \left\lfloor 2\sqrt{d_F/3|d_K|^2}\right\rfloor, \quad N_H = \left\lfloor 4\sqrt{d_F/3|d_K|}\right\rfloor,\\ N_L &= \begin{cases}0\,, & \text{if } d_K > 0,\\ N_H\,, & \text{if } d_K < 0.\end{cases}\end{aligned}$$

We define $G_2(d_F, d_K)$ to be the set of all polynomials $f(x) \in \mathbb{Z}[x]$ defined by (4.1), with $a_1 \in \{0,1\}$, $a_2 \in \{1,2\}$, and b_1, b_2 satisfying (4.2), (4.3).

Proposition 4.1. *If F is a totally complex algebraic number field of degree 4 with discriminant d_F and ring of integers $\mathcal{O}_F$, and if F contains a quadratic subfield K with discriminant d_K and ring of integers $\mathcal{O}_K$, then there exists $\rho \in \mathcal{O}_F \setminus \mathcal{O}_K$ with characteristic polynomial belonging to $G_2(d_F, d_K)$, such that $F = \mathbb{Q}(\rho)$.*

5 Fields with small quadratic integers

Let F be a totally complex quartic number field with discriminant $d_F < 10^6$. Suppose there exists $\rho \in \mathcal{O}_F \setminus \mathbb{Z}$ with characteristic polynomial in $G_1(d_F)$ such that $K = \mathbb{Q}(\rho)$ is a quadratic subfield of F. It follows from Proposition 2.2, using (2.5)–(2.8) with $d = 10^6$, that $|d_K| < 80$.

For each quadratic field discriminant d_K with $-80 < d_K < 80$, we generate the polynomials in $G_2(10^6, d_K)$ according to (4.1), taking $a_1 \in \{0, 1\}$, $a_2 \in \{1, 2\}$, and b_1, b_2 running through the values specified by (4.2) and (4.3). For each such polynomial $f(x)$ we perform the following tests.

- If $f(x)$ is reducible in $\mathbb{Z}[x]$ we exclude f.
- If $f(x)$ is irreducible but $\mathbb{Q}(\rho)$ is not totally complex, we exclude f.
- We compute the discriminant d_F of the field $F = \mathbb{Q}(\rho)$.
- If $d_F \geq 10^6$ we exclude f.
- If $f \notin G_2(d_F, d_K)$ we exclude f.

Among the polynomials surviving these tests is a generating polynomial for every totally complex quartic number field of discriminant less than 10^6, such that no $\rho \in \mathcal{O}_F \setminus \mathbb{Z}$ with characteristic polynomial in $G_1(d_F)$ is a quartic integer.

6 Eliminating redundant examples

If f_1 and f_2 give fields with the same discriminant, it must be decided whether the fields are isomorphic. We follow the same strategy as in [2], but with more efficient PASCAL and assembler software.

- Determine prime p, not dividing either polynomial discriminant, such that f_1 and f_2 each have at least one root modulo p. (If f_1 and f_2 have a different number of roots modulo p, their fields can not be isomorphic.)
- Approximate p-adic integer roots of f_1 and f_2 modulo $p^{2k} \leq (2^{31} - 1)^2$.
- Apply high-precision (128-bit) floating-point LLL to determine *suspected* relation (PASCAL + assembler). Confirm by computing *exact* characteristic polynomials (ALGEB).
- If no decision (roughly 1% of all cases), revert to rational integer LLL isomorphism test (ALGEB).

7 Execution time

To measure the efficiency of the new software, the enumeration of totally real quartic fields of [2] was repeated. The time shown, 5:30 CPU-hours, should be compared with the 140:41 CPU-hours reported in [2]. The computations were done on a Digital Equipment VAX 8550 computer at the Computer Centre of Concordia University.

Totally Complex:	104:12	CPU-hours,	81322	fields.
Totally Real:	5:30	CPU-hours,	13073	fields.

Table 1. Relative Distribution of Galois Groups

Group:	C_4	A_4	V_4	D_8	S_4
Fields:	54	90	818	36238	44122
Percent:	0.07	0.11	1.01	44.56	54.26

Table 2. Distribution of Galois Groups C_4, A_4, V_4 by Field Discriminant

	C_4	A_4	V_4
0:	8	9	78
25000:	2	6	42
50000:	5	3	45
75000:	2	5	36
100000:	2	4	29
125000:	3	1	23
150000:	1	2	22
175000:	0	1	33
200000:	1	2	27
225000:	4	5	14
250000:	2	6	14
275000:	0	2	16
300000:	2	0	29
325000:	1	3	14
350000:	1	0	14
375000:	1	2	26
400000:	1	1	15
425000:	0	2	27
450000:	2	5	10
475000:	0	2	15
500000:	0	1	14
525000:	0	5	19
550000:	0	0	14
575000:	3	0	15
600000:	4	2	17
625000:	1	2	14
650000:	1	1	18
675000:	0	1	7
700000:	1	1	26
725000:	2	1	11
750000:	0	2	16
775000:	0	2	17
800000:	0	3	10
825000:	0	1	22
850000:	0	0	17
875000:	0	2	10
900000:	1	1	14
925000:	0	2	5
950000:	2	1	14
975000:	1	1	9

Table 3. Distribution of Galois Group D_8 by Field Discriminant

0:	629
20000:	694
40000:	694
60000:	698
80000:	702
100000:	717
120000:	755
140000:	707
160000:	689
180000:	726
200000:	712
220000:	715
240000:	743
260000:	725
280000:	756
300000:	735
320000:	716
340000:	746
360000:	696
380000:	746
400000:	728
420000:	717
440000:	722
460000:	723
480000:	741
500000:	723
520000:	724
540000:	714
560000:	737
580000:	734
600000:	728
620000:	724
640000:	745
660000:	757
680000:	745
700000:	709
720000:	736
740000:	748
760000:	711
780000:	740
800000:	738
820000:	740
840000:	726
860000:	771
880000:	722
900000:	747
920000:	704
940000:	729
960000:	717
980000:	737

Table 4. Distribution of Galois Group S_4 by Field Discriminant

0:	501
20000:	670
40000:	688
60000:	739
80000:	776
100000:	754
120000:	805
140000:	819
160000:	834
180000:	810
200000:	851
220000:	823
240000:	885
260000:	849
280000:	867
300000:	863
320000:	882
340000:	879
360000:	913
380000:	880
400000:	897
420000:	899
440000:	904
460000:	909
480000:	890
500000:	929
520000:	865
540000:	962
560000:	956
580000:	926
600000:	912
620000:	899
640000:	953
660000:	945
680000:	864
700000:	999
720000:	935
740000:	930
760000:	980
780000:	954
800000:	972
820000:	979
840000:	973
860000:	882
880000:	921
900000:	959
920000:	955
940000:	963
960000:	974
980000:	948

References

[1] J. Buchmann, *private communication*, 1989.

[2] J. Buchmann and D. Ford, *On the Computation of Totally Real Quartic Fields of Small Discriminant*, Math. Comp. **52** (1989), 161–174.

[3] B.N. Delone and D.K. Faddeev, *The Theory of Irrationalities of the Third Degree*, "Transl. Math. Monographs", vol. 10, Amer. Math. Soc., Providence, R.I., 1964.

[4] M. Pohst, *On the Computation of Number Fields of Small Discriminants Including the Minimum Discriminants of Sixth Degree Fields*, J. Number Theory **14** (1982), 99–117.

Class number computation by cyclotomic or elliptic units

K. Nakamula[*]

Abstract. In [11] and [15], for an arbitrary subfield of a finite abelian extension of an imaginary quadratic field, the author developed a general theory for a method of computing its class number and fundamental units by using cyclotomic or elliptic units. In this paper, we study this *cyclo-elliptic* method from an algorithmic point of view. An outline of the process of computation is first described in general, and then explicit procedures are given for any sextic field composed of a quadratic subfield and a cubic not totally real subfield.

1 Cyclo - elliptic method

In the field $\mathbb{C}$ of complex numbers, let K be an arbitrary extension of degree finite over the field $\mathbb{Q}$ of rational numbers. We assume that K is contained in a certain abelian extension of an imaginary quadratic number field F. We first sketch a method to compute the class number h_K and the group E_K of units of K utilizing cyclotomic or elliptic units. This will be referred to as cyclo-elliptic method, and its general theory is fully described in [11] and [15].

Let W_K be the torsion part of E_K. As an application of the theory of cyclotomic fields or complex multiplication, we can derive a formula of the form

$$c_K h_K = \left(E_K : E^K\right) \prod_{i \in I} (E_i : \mathbb{E}_i), \qquad E^K = W_K \prod_{i \in I} E_i \tag{1.1}$$

from a well-known analytic class number formula. Here c_K is a simple natural number, I is a finite index set, E_i $(i \in I)$ are subgroups of E_K, and $\mathbb{E}_i$ $(i \in I)$ are groups of cyclotomic or elliptic units with explicit generators. Finding fundamental units of K from the generating cyclotomic or elliptic units, we can decide these group indices. This provides us a way to compute h_K and E_K at a time. The method consists of the following steps:

Step 1. We evaluate the generators of $\mathbb{E}_i$ $(i \in I)$ which are given by means of special values of the sine function or the elliptic functions with complex mul-

[*]On debugging programs, the author owes much to his colleague Professor Takafumi Sakurai.

tiplication. For that purpose, we first compute the ideal class group of $\mathbb{Q}$ or F corresponding to K as in [11], §3.2 and [15], §3.1

Step 2. Using the values of those generating units, we decide their minimal polynomials over $\mathbb{Q}$ or F, and, on the other hand, we compute numerical upper bounds of the indices in (1.1) beforehand.

Step 3. We then find a set of generators of E_K in terms of their minimal polynomials over $\mathbb{Q}$ or F in the sense in [11], §3.1, which together decide the indices in (1.1). This step is effectively bounded by the majorization above.

Thus, we obtain h_K from the formula (1.1) as c_K is always easily known.

This method has several significant features. First of all, we do not use any integral basis of K throughout the process. The fact that the ring of integers of $\mathbb{Q}$ or F is discrete makes it possible. The second important feature is that the process consists only of exact computation in $\mathbb{Q}$ or F and of approximate computation in $\mathbb{C}$. No geometric method is employed to find fundamental units of K. This comes from the explicitness of the classical theory of cyclotomic fields or complex multiplication, and it enables us to use the minimal polynomials. Another feature is that (1.1) gives a factorization of h_K and E_K. So we can compute each factor independently, and, in many cases, we can utilize the results for the subfields of K computed by the same method. We also note that some arithmetic of cyclotomic fields serves us a more efficient way to decide the indices $(E_i : \mathbb{E}_i)$ $(i \in I)$ in Step 3.

As we have started from the analytic formula, the structure of the ideal class group of K will not be decided by this method. And it gives no information on integral basis of K. G. Gras and M.-N. Gras first introduced such a method in [3] when K is abelian over $\mathbb{Q}$ and contained in the field $\mathbb{R}$ of real numbers. The author extended it in [11], [15] to a general precise algorithm, proving the effectiveness of the method.

Let $n = [K : \mathbb{Q}]$. When K is abelian over $\mathbb{Q}$ and contained in $\mathbb{R}$, valuable numerical tables have been given by this algorithm in [4], [5], [6] and [9] for $n = 3$, 4 and 6. In case K is a pure cubic field, the method was applied to make a table in [14]. But, when K is not abelian over $\mathbb{Q}$, there are only a few other numerical results as in [12], [8] and [13] for $n = 4$, 5 and 6 in case K does not contain F. In order to implement this method on an actual computer, there are two main difficulties in Step 2. One is to compute good approximate values of cyclotomic or elliptic units precise enough to decide their minimal polynomials, which requires multiple precision computation. Because the magnitude of those units increases very fast when the absolute value of the discriminant of K becomes large. The other is to prove a sharp estimation of the indices $(E_s : \mathbb{E}_s)$ $(s \in I)$ in (1.1) for each type of K we treat. General estimation is too rough to employ in actual computation. When K is not abelian over $\mathbb{Q}$, there is one more difficulty in Step 1 that programs to compute the ideal class group become a little bit complicated in general though their time complexity is usually not so large. We shall exhibit a good example in section 3.

When K is not abelian over $\mathbb{Q}$ and does not contain F, we have a class number relation in [15], Theorem 3 between K and other subfields of the composite

$L = FK$ which is abelian over F and galois over $\mathbb{Q}$. This settles a special important status of $L \cap \mathbb{R}$. In such a case, we first compute h_K and E_K by this algorithm when $K = L \cap \mathbb{R}$. Next, we compute those of other subfields of L from that class number relation. This way is much easier than to go directly on the steps above as is seen below.

From now on till the end of this paper, let $K = K_2 K_3$ be a non-galois sextic field composed of a real quadratic field K_2 with discriminant d_2 and a real cubic field K_3 with discriminant $-d_3 < 0$. Then K is contained in an abelian extension of $F = \mathbb{Q}\left(\sqrt{-d_3}\right)$. Moreover $L = FK = K\left(\sqrt{-d_3}\right)$ contains another non-galois sextic field $M = K_3\left(\sqrt{-d_3 d_2}\right)$, which is not conjugate to K and has an imaginary quadratic subfield $M_2 = \mathbb{Q}\left(\sqrt{-d_3 d_2}\right)$. As a special case, pure sextic fields $K = \mathbb{Q}\left(\sqrt[6]{m}\right)$ and $M = \mathbb{Q}\left(\sqrt[6]{-27m}\right)$ belong to this type. The essential part of the cyclo-elliptic method above will become clearer by a more detailed study of this typical example. Then (1.1) becomes

$$\begin{aligned} 6h_K &= \left(E_K : E^K\right)(E_2 : \mathbb{E}_2)(E_3 : \mathbb{E}_3)(E_6 : \mathbb{E}_6), \\ E^K &= \{\pm 1\} \times E_2 \times E_3 \times E_6. \end{aligned} \tag{1.2}$$

Here E_2 or E_3 respectively is the cyclic group of positive units of K_2 or K_3, E_6 is the cyclic group of positive relative units of K consisting of positive units with relative norms to K_2 and K_3 both equal to 1, $\mathbb{E}_2$ is the group of cyclotomic units of K_2 generated by the unit in (4.4) of [1], Chapter 5, and $\mathbb{E}_3$ or $\mathbb{E}_6$ respectively is the group of positive elliptic units of K_3 or K_6 defined in Example 3 or Example 4 of [15], §3.2. The cyclo-elliptic method applies to K_2 and K_3, too. For the class number h_{K_2} of K_2, nothing but the famous

$$h_{K_2} = (E_2 : \mathbb{E}_2) \tag{1.3}$$

is our formula (1.1). For the class number h_{K_3} of K_3, the simple

$$h_{K_3} = (E_3 : \mathbb{E}_3) \tag{1.4}$$

is our formula (1.1) as in Example 1.(i).(a) of [15], §3.2. Moreover, by [13], (14), the class numbers h_M, h_{M_2} of M, M_2 satisfy the relation

$$h_M h_{K_2} \gcd\left(3, \left(E_K : E^K\right)\right) = \left(E_{M/\mathbb{Q}} : W_M N_{L/M}\left(E_6'\right)\right) h_K h_{M_2}, \tag{1.5}$$

where W_M is the group of roots of unity in M, $E_{M/\mathbb{Q}}$ is the group of relative units of M consisting of units with relative norms to M_2 and K_3 both in W_M, and E_6' is an imaginary conjugate of E_6. Moreover, we have

$$\left(E_K : E^K\right) \mid 6 \quad \text{and} \quad \left(E_{M/\mathbb{Q}} : W_M N_{L/M}\left(E_6'\right)\right) \mid 3.$$

This is a special case of [15], Theorem 3 quoted above. The process of deciding $\left(E_K : E^K\right)$ or $\left(E_{M/\mathbb{Q}} : W_M N_{L/M}\left(E_6'\right)\right)$ is rather easy and not so much interesting in view of the cyclo-elliptic method. As it is investigated in [13], §5, §6 clearly, it will not been discussed in this paper. We study the process of deciding $(E_2 : \mathbb{E}_2)$, $(E_3 : \mathbb{E}_3)$ or $(E_6 : \mathbb{E}_6)$ in section 2, 3 or 4 respectively. We have good upper bounds for these group indices.

Computer programs in MACSYMA [1] will be displayed not to make a table but to reveal the algorithm of computing the class numbers and fundamental units of all fields with the same given discriminant. All MACSYMA functions not written plainly are predefined in UNIX [2] MACSYMA Release 309.2.

2 Quadratic case

In this section, we consider the real quadratic field K_2 with discriminant d_2 to explain a naive version of the cyclo-elliptic method.

In (1.3), Step 1 is almost trivial because the cyclotomic unit is explicitly expressed by the Kronecker symbol. Therefore the generator ξ of $\mathbb{E}_2$ is computed as follows.

Algorithm 2.1.

Input: *The discriminant $d > 0$ of a quadratic field.*
Output: *The cyclotomic unit of the field.*

```
cyclo2(d) := block(
[e : 1, n, q : quotient(d, 2), k],
if numer then (
  for n : 1 thru q do (
    k : kronecker(d, n),
    if k > 0 then
      e : e / sin(%pi * n / d)
    else if k < 0 then
      e : e * sin(%pi * n / d)  ),
  e )
else product
     (sin(%pi * n / d)^(-kronecker(d, n)), n, 1, q)  )  $
```

When cyclo2(d_2) is not evaluated numerically, it returns the definition of the cyclotomic unit ξ. The Kronecker symbol **kronecker**(d, n) is computed by the Jacobi symbol.

Algorithm 2.2.

Input: *n and a discriminant d.*
Output: *The kronecker symbol of n for d.*

```
kronecker(d, n) := block(
[m],
if evenp(d) then
```

[1]MACSYMA is a trademark of Symbolics Inc.
[2]UNIX is a trademark of AT & T Bell Laboratories.

```
  if oddp(n) then (
    m : abs(mod(d, 16)),
    mod((n - 1) * (4 * n + d + 4 - m) / (m + m) + 1, 4)
      * jacobi(n, abs(d / m))  )
  else 0
else jacobi(n, abs(d))  )  $
```

This completes Step 1.

As to Step 2, if $\varepsilon \in E_2$, $\varepsilon > 1$ and $\varepsilon \neq \frac{1}{2}\left(1+\sqrt{5}\right)$, then $\varepsilon > 2$ and

$$\mathrm{Tr}_{K_2/\mathbb{Q}}(\varepsilon) = [\varepsilon + 0.5] \quad \text{and} \quad N_{K_2/\mathbb{Q}}(\varepsilon) = \mathrm{sign}\left(\mathrm{Tr}_{K_2/\mathbb{Q}}(\varepsilon) - \varepsilon\right),$$

where $[\,\cdot\,]$ is the Gauß symbol. Therefore the coefficients of the minimal polynomial of ξ are computed as follows.

Algorithm 2.3.

Input: *A unit $u > 1$ of a real quadratic field.*
Output: *The minimal polynomial of u.*

```
minpoly2(u)  := block(
[m],
if u > 2 then
  m : fix(u + 1 / 2)
else
  m : 1,
[m, signum(m - u)]  )  $
```

A very sharp estimation of $(E_2 : \mathbb{E}_2)$ is given by

$$(E_2 : \mathbb{E}_2) \leq \frac{\log(\xi)}{\log\left(\left(\sqrt{d_2}+\sqrt{d_2-4}\right)/2\right)}$$

as is easily seen. Therefore its upper bound is computed by the following function.

Algorithm 2.4.

Input: *A unit $u > 1$ of a quadratic field with discriminant $d > 0$.*
Output: *An upper bound of $(< e >:< u >)$, where e is the fundamental unit of the field.*

```
bound2(u, d)  :=
fix(1 / 10 + log(u) / log((sqrt(d) + sqrt(d - 4)) / 2))  $
```

This completes Step 2.

The essential part of Step 3 is that, if the coefficients of the minimal polynomial of a unit $\varepsilon \in E_2$, $\varepsilon > 1$, are expressed by a recursive sequence, then ε should be a power of an $\eta \in E_2$, $\eta > 1$, and the coefficients of the minimal polynomial of the η are given by the initial term of the recursive sequence. This is realized as follows.

Algorithm 2.5.

Input: *A unit* $u>1$ *of a real quadratic field with minimal polynomial* c, *and* b.
Output: *The index* $(<e>:<u>)$ *not exceeding* b *and the minimal polynomial of the fundamental unit* $e>1$ *of the field.*

```
index2(u, b, c) := block(
[i : 1, e, p : 2, s, t],
local(s, t),
s : c,
unless p > b do (
  e : u^(1 / p), t : minpoly2(e),
  if sq2(p, t) = s then (
    i : i * p,  u : e, b : quotient(b, p), s : t  )
  else p : primen(p)  ),
[i, s]  )  $
```

Here the recursive sequence is Newton's formula for quadratic polynomials.

Algorithm 2.6.

Input: $q>1$ *and a minimal polynomial* c *of a quadratic integer.*
Output: *The minimal polynomial of the* q*-th power of the quadratic integer.*

```
sq2(q, c) :=
if q = 2 then
  [c[1] * c[1] - c[2] - c[2], 1]
else if q = 3 then
  [c[1] * (c[1] * c[1] - c[2] - c[2] - c[2]), c[2]]
else
  [c[1] * sq2(q - 1, c)[1] - c[2] * sq2(q - 2, c)[1],
    c[2] * sq2(q - 1, c)[2]]  $
```

The function `primen(p)` takes the prime next to p.

Algorithm 2.7.

Input: p.
Output: *The prime next to* p.

```
primen(p) :=
if p > 3 then (
  p : p + mod(p, 6) + 3,
  unless primep(p) do
    p : p + mod(p, 6) + 3,
  p  )
else if p = 3 then
  5
else
  3  $
```

This completes Step 3.

By (1.3), we are now able to decide h_{K_2} and E_2 as follows from a given discriminant d_2.

Algorithm 2.8.

Input: d.
Output: *The class number and the fundamental unit* > 1 *(resp. full representative ideals of the ideal class group) of a real (resp. imaginary) quadratic field with discriminant* d*, false for improper* d.

```
class2(d) := block(
[e],
if integerp(d) and discp2(d) then (
  if d > 0 then (
    e : ev(cyclo2(d), numer),
    index2(e, ev(bound2(e, d), numer), minpoly2(e))  )
  else
    rcl(d)  )  )  $
```

The returned value of $\texttt{class2}(d_2)$ is the list

$$[\, h_{K_2}, [\mathrm{Tr}_{K_2/\mathbb{Q}}(\eta)\,, N_{K_2/\mathbb{Q}}(\eta)]\,],$$

where η is the fundamental unit of K_2 greater than 1. The correctness of the input discriminant is checked in advance by the following function.

Algorithm 2.9.

Input: d.
Output: *True for a discriminant* d *of a quadratic field, else false.*

```
discp2(d) := block(
[m : mod(d, 16)],
member(m, [-4, 8, -8]) and sqfrp(d / 4)
  or member(m, [-7, -3, 1, 5]) and sqfrp(d) and d # 1  )  $
```

Here, to test square-freeness of a rational integer, we have applied the next routine.

Algorithm 2.10.

Input: n.
Output: *True for a non-zero square-free* n*, else false.*

```
sqfrp(n) :=
length(sqrt(abs(n))) = 1 or n = -1 or n = 1  $
```

The program $\texttt{class2}(d)$ above also contains a function $\texttt{rcl}(d)$ to compute the class group of an imaginary quadratic field, which will be explained in the next section.

3 Cubic case

In this section, we consider the real cubic field K_3 with negative discriminant $(-d_3)$ to explain the programming difficulty in Step 1 of the cyclo-elliptic method.

In (1.4), Step 1 is much more complicated since we should compute the ideal class groups of an imaginary quadratic field $F = \mathbb{Q}\left(\sqrt{-d_3}\right)$ which correspond to the cubic fields K_3 with discriminant $-d_3$. Of course, there may be many cubic fields with the same discriminant, and we compute all of them.

We first compute a canonical set of representatives of the ring class group $R(-d_3)$ with discriminant $-d_3$ in F by an ordinary method to list up reduced quadratic irrationals with discriminant $-d_3$.

Algorithm 3.1.

Input: *A quadratic discriminant* $d < 0$.
Output: *The ring class group with discriminant* d.

```
rcl(d) := block(
[a, b, e, f, g, h, c],
local(h, c),
h : [], g : fix(sqrt(-d / 3) + 1 / 10),
for b : mod(d, 2) thru g step 2 do (
  e : (b * b - d) / 4,
  f : isqrt(e),
  for a : b + 1 - signum(b) thru f do (
    c : divide(e, a),
    if c[2] = 0 and gcd(gcd(a, b), c[1]) = 1 then (
      if a # c[1] and a # b and b # 0 then
  h : cons([a, -b], h),
      h : cons([a, b], h)  )  )  ),
h )  $
```

This has already appeared in section 2 to get the class group of an arbitrary imaginary quadratic field. The returned value of rcl$(-d_3)$ is the list

$$[[a_1, b_1], \ldots, [a_k, b_k]] \quad \text{with} \quad k = \#R(-d_3), \tag{3.1}$$

where $[a_i, b_i]$ means a representative ring ideal of a class in $R(-d_3)$ generated by a_i and $\frac{1}{2}\left(-b_i + \sqrt{-d_3}\right)$. We next decide the 3-elementary factor group of $R(-d_3)$ by taking cubes of the representative ideals in $R(-d_3)$.

Algorithm 3.2.

Input: *The ring class group* g *with quadratic discriminant* $d < 0$.
Output: *The 3-elementary factor group of* g*, false if it is trivial.*

```
rcl3(g, d) := block(
[h : length(g), i, m, z, s, r, x, e, y],
```

```
local(z, s, r, x, e, y),
z : [], s : [], x : g,
if mod(h, 3) = 0 then (
  for i : 1 thru h do (
    e : mulrcl(x[i], mulrcl(x[i], x[i], d), d),
    if not member(e, s) then s : cons(e, s)  ),
  m : length(s),
  for i : 1 thru m do
    x : delete(s[i], x),
  r : s,
  while m + m + m < h do (
    y : last(x), z : cons(y, z),
    for i : 1 thru m do (
      e : mulrcl(r[i], y, d), r : cons(e, r), x : delete(e, x),
      e : invrcl(e, d), r : cons(e, r), x : delete(e, x)  ),
    m : m + m + m  ),
  [cons(last(x), z), s]  )  )  $
```

Here two more functions mulrcl(x, y, d) and invrcl(z, d) are necessary for the operation in $R(-d_3)$. They are defined below.

Algorithm 3.3.

Input: *The quadratic discriminant* $d < 0$, x *and* y.
Output: $x * y$ *in the ring class group with discriminant* d.

```
mulrcl(x, y, d) := block(
[m : 2 * x[1] * y[1], w],
local(w),
w : rowtrans2(
  matrix([x[1], -x[1] * y[2]], [y[1], -y[1] * x[2]])  ),
m : gcd(m, w[2, 2]),
w : rowtrans2(
  matrix(w[1], [(x[2] + y[2]) / 2, (-x[2] * y[2] - d) / 2])  ),
redrcl(
  [gcd(m, w[2, 2]) / 2 / abs(w[1, 1]),
 -w[1, 2] / w[1, 1]], d  )  )  $
  /* -- elementary row transformation of integral 2-2-matrix m */
  rowtrans2(m) := block(
  [w, q],
  local(w, q),
  w : m,
  while w[2, 1] # 0 do (
    q : divide(w[1, 1], w[2, 1]),
    w : matrix(w[2], [q[2], w[1, 2] - q[1] * w[2, 2]])  ),
  w  )  $
```

Algorithm 3.4.

Input: *The quadratic discriminant $d < 0$, and z.*
Output: *The inverse of z in the ring class group with discriminant d.*

```
invrcl(z, d) :=
if z[1] # z[2] and z[2] * z[2] - 4 * z[1] * z[1] # d then
  [z[1], -z[2]]
else
  z $
```

In the former mulrcl(x, y, d), usual reduction of imaginary quadratic numbers is applied. That is given by the routine below.

Algorithm 3.5.

Input: *The discriminant $d < 0$, and z.*
Output: *Reduced z in the ring class with discriminant d.*

```
redrcl(z, d) := block(
[m, w],
local(w),
w : z, w[2] : mod(w[2], w[1] + w[1]),
if w[2] = -w[1] then
  w[2] : -w[2],
m : (w[2] * w[2] - d) / 4 / w[1],
if w[1] > m or w[1] = m and w[2] < 0 then
  redrcl([m, -w[2]], d)
else
  w ) $
```

As we have decided the 3-elementary group $R(-d_3)/R(-d_3)^3$, we can list up all its elements of order 3. Hence, we can compute all the subgroups S_3 of $R(-d_3)$ of index 3 together with a generator $z_3 \in R(-d_3) \setminus S_3$ of the factor group by the following procedure.

Algorithm 3.6.

Input: *The 3-elementary factor group g of the ring class group with quadratic discriminant $d < 0$.*
Output: *The subgroups of index 3 for g.*

```
rclind3(g, d) := block(
[m : length(g[1]) - 1, k, t, s, z, y, x, r],
local(s, z, y, x, r),
s : [], z : g[1], y : g[2], x : makelist(0, k, 1, m),
for k : m thru 1 step -1 do (
  t : false,
```

```
  while not t do (
    r : y,
    for m : 1 thru k do
      r : genrcl3(z[m], r, d),
    s : cons([z[k+1], r], s), t : true,
    for m : 1 thru k while t do (
      if x[m] = 2 then (
z[m] : g[1][m], x[m] : 0  )
      else (
z[m] : mulrcl(z[m], z[k+1], d), x[m] : x[m] + 1,
t : false  )  )  ),
  y : genrcl3(z[k+1], y, d)  ),
cons([z[1], y], s)  )  $
  /* -- the subgroup <z, s> of the ring class group with quadratic
  discriminant d < 0 for a subgroup s and an element z of
  order 3 modulo s -- */
  genrcl3(z, s, d) := block(
  [n : length(s), m, g],
  local(g),
  g : s,
  for m : 1 thru n do (
    e : mulrcl(z, g[m], d), g : append(g, [e, invrcl(e, d)])  ),
  g  )  $
```

This very complicated process is necessary even though the 3-rank of $R(-d_3)$ is not so big in ordinary practical cases. But, the time complexity of this part is usually small.

Those subgroups S_3 of $R(-d_3)$ of index 3 do not necessarily fulfill the required property yet. We should further examine whether each S_3 is primitive or not. First, if $-d$ is the discriminant of the quadratic field F, then $-d_3 = -df_3^2$ with an f_3 in the set $\mathbb{N}$ of natural numbers, and prime factors of f_3 should satisfy a certain condition in [7]. We prepare the following function to check this.

Algorithm 3.7.

Input: d.
Output: *The prime factorization for a discriminant $d < 0$ of a cubic field, false if necessary conditions do not hold.*

```
factdisc3(d) := block(
[f, m : mod(d, 4), i, p, x],
local(x),
if m = 0 or m = 1 then (
  x : cons(d, factsqfr(-d)), x[2] : -x[2],
  if mod(x[2], 4) # 1 then
    x : [d, 4 * x[2], x[3] / 2],
```

```
  f : x[3],
  while mod(f, 3) = 0 do
    f : f / 3,
  if x[3] # f then
    x : endcons(x[3] / f, x),
  if (x[3] = f or x[4] = 3 and mod(x[2], 3) = 0 or
      x[4] = 9 and mod(x[2], 9) # 3) and sqfrp(f) then (
    if f > 1 then (
      f : factor(f), m : length(f),
      for i : 0 thru m unless x = false do (
if m = 0 then
  p : f
else
  p : inpart(f, i),
if p # "*" then
  if kronecker(x[2], p) = mod(p, 3) then
    x : endcons(p, x)
  else
    x : false  )  ),
    x  )  )  )  $

  /* -- square-free factorization of n > 0 -- */
  factsqfr(n) := block(
  [m],
  n : sqrt(n), m : length(n),
  if m = 2 then
    [inpart(n, 2, 1), inpart(n, 1)]
  else if m = 1 then
    [inpart(n, 1), 1]
  else
    [1, n]  )  $
```

Moreover, the subgroups S_3 of $R(-d_3)$ which can correspond to K_3 should be with conductor exactly f_3, or equivalently primitive. To assure it, we use the following algorithm.

Algorithm 3.8.

Input: *d and s.*
Output: *True if s is a primitive subgroup of the ring class group with factored imaginary quadratic discriminant d, else false.*

```
rprimp(s, d) := block(
[k : length(d), i, j, p, q, t : true, m],
local(m),
if k > 4 or k = 4 and not (
```

```
    member(d[2], [-3, -4, -7, -8, -11, -19, -43, -67, -163])
      and sqfrp(d[4]) ) then
  for i : 4 thru k while t do (
    if sqfrp(d[i]) then
      p : d[i]
    else
      p : inpart(factor(d[i]), 1),
    q : p * p, m : divide(1 - d[1] / q, 4), m[2] : 1 - m[2],
    for j : 0 thru p - 1 while t do
      if mod(j * (j - m[2]) + m[1], p) # 0 then
t : member(redrcl([q, p * (m[2] - j - j)], d[1]), s),
    t : not t  ),
t  )  $
```

To check the primitiveness, we have used in this function the fact that any subgroup S_3 of $R(-d_3)$ is with conductor exactly f_3 if and only if S_3 does not contain the kernel of the canonical onto map $R(-d_3) \to R\left(-d_3/p^2\right)$ for every prime divisor p of f_3. The image and the inverse image of the map is explicitly known as in [2] §8 or [1], Chapter 2 for example, and the validity of the algorithm is verified easily.

The last part of Step 1 is very simple as a procedure. Because we are now able to compute the generating elliptic unit of K_3 from z_3 and S_3 above by a finite product of the quotients of ring class invariants.

Algorithm 3.9.

Input: *A ring class subgroup s of quadratic discriminant $d < 0$ with an element z not in s.*
Output: *The elliptic unit of a cubic field for s.*

```
ellip3(z, s, d) := block(
[i, m : length(s), r],
local(r),
for i : 1 thru m do
  r[i] : mulrcl(s[i], z, d),
if numer then (
  product(rinvar_minor(r[i], d) / rinvar_minor(s[i], d), i, 1, m)
  * exp(%pi * sqrt(-d) *
      sum(1 / s[i][1] - 1 / r[i][1], i, 1, m) / 12)
  * sqrt(product(s[i][1] / r[i][1], i, 1, m))  )
else
  product(rinvar(r[i], d) /  rinvar(s[i], d), i, 1, m)  )  $
```

This also returns the definition of the elliptic unit if not evaluated numerically. Each ring class invariant essentially is a special value of the absolute value of the Dedekind eta function. Hence it is computed by an infinite product approximately.

Algorithm 3.10.

Input: *A ring class z of quadratic discriminant $d < 0$.*
Output: *The invariant of z.*

```
/*  -- its definition --  */
rinvar(z, d) :=
sqrt(sqrt(-d) / 2 / z[1]) *
     abseta((-z[2] + sqrt(d)) / 2 / z[1])^2  $
/*  -- its minor part --  */
rinvar_minor(z, d) := block(
[n, e : 0, u : 1],
for n : 1 unless e = u do (
  e : u,
  u : e +
    exp(-%pi * n * sqrt(-d) / z[1]) *
       (exp(-%pi * n * sqrt(-d) / z[1])
      - 2 * cos(-%pi * n * z[2] / z[1])) * e  ),
e )  $
```

A dummy function **abseta**(z) is used in the above to imply the absolute value of the Dedekind eta function $\eta(z)$. Thus, we complete Step 1. As is shown, the main algorithmic problem of this step stays mainly in computing the ideal class groups of the imaginary quadratic field F. But, our method applies to any other types of K we consider, since the structure of the ring class group of F can be decided by using mulrcl(x, y, d) and since the primitiveness of the subgroups is verified by **rprimp**(s, d) with a minor change. On the other hand, when $|-d_3|$ is large, there is another problem to make the precision of **ellip3**(z, s, d) large so that we can decide the minimal polynomial by **minpoly3**(u) below. This problem always occurs when we go on the cyclo-elliptic method including the function **cyclo2**(d) in the last section.

Steps 2 and 3 go mostly in a similar manner as the quadratic case. As it is described in [10] in detail, we only give the functions to be used. By [10], Theorem 2, we decide the minimal polynomial of the generating elliptic unit of $\mathbb{E}_3$ as follows.

Algorithm 3.11.

Input: *A unit $u > 1$ of a cubic field with negative discriminant.*
Output: *The minimal polynomial of u.*

```
minpoly3(u) := block(
[m, i, e, x, c],
local(x, c),
x : minpoly3_possible(u), m : length(x), c : x[1][1], e : x[1][2],
for i : 2 thru m do if x[i][2] < e then (
  c : x[i][1], e : x[i][2]  ),
c )  $
```

Here the subroutine `minpoly3_possible`(u) is used to restrict the possibility of the minimal polynomial in a finite number.

Algorithm 3.12.

Input: *A unit $u > 1$ of a cubic field with negative discriminant.*
Output: *The possible minimal polynomial of u.*

```
minpoly3_possible(u) := block(
[i, j, e, m : fix(u + 2 / sqrt(u) + 1 / 10)],
local(c),
c : [],
for i : fix(u + 9 / 10 - 2 / sqrt(u)) thru m do (
  e : u * (i - u) + 1 / u, j : fix(e + 1 / 2),
  c : cons([[i, j], abs(e - j)], c)  ),
c  )  $
```

On account of [10], Theorem 1, we estimate the index $(E_3 : \mathbb{E}_3)$ by the following function.

Algorithm 3.13.

Input: *A unit $u > 1$ of a cubic field with discriminant $d < 0$.*
Output: *An upper bound of $(< e >:< u >)$, where e is the fundamental unit of the field.*

```
bound3(u, d) :=
if d < -23 then
  fix(1 / 10 + 3 * log(u) / log(-d / 4 - 6))
else
  1.27 * 1.27  $
```

This completes Step 2.

The index $(E_3 : \mathbb{E}_3)$ is decided by [10] Theorem 2 again. An exact process of calculation is given as follows.

Algorithm 3.14.

Input: *A unit $u > 1$ of a cubic field with negative discriminant with minimal polynomial c, and b.*
Output: *The index $(< e >:< u >)$ not exceeding b and the minimal polynomial of the fundamental unit $e > 1$ of the field.*

```
index3(u, b, c) := block(
[i : 1, e, p : 2, x, s, t],
local(x, s, t),
s : c,
unless p > b do (
```

```
  e : u^(1 / p), x : minpoly3_possible(e),
  unless x = [] or x = p do (
    t : first(x)[1], x : rest(x),
    if sq3(p, t) = s then (
      i : i * p, u : e, s : t, x : p, b : quotient(b, p)  )  ),
  if x # p then p : primen(p)  ),
[i, s]  )  $
```

The recursive sequence utilized now is Newton's formula for cubic polynomials.

Algorithm 3.15.

Input: $q > 1$, *and the minimal polynomial* c *of a cubic unit with norm* 1.
Output: *The minimal polynomial of the* q*-th power of the cubic unit.*

```
sq3(q, c) :=
[sq3_(q, c[1], c[2]), sq3_(q, c[2], c[1])]  $
  /*  -- the recursive sequence --  */
  sq3_(q, i, j) :=
  if q = 2 then
    i * i - j - j
  else if q = 3 then
    i * (i * i - j - j - j) + 3
  else if q = 4 then
    i * (i * (i * i - 4 * j) + 4) + 2 * j * j
  else
    i * sq3_(q - 1, i, j) - j * sq3_(q - 2, i, j)
      + sq3_(q - 3, i, j)  $
```

This completes Step 3.

By (1.4), we are now able to decide h_{K_3} and E_3 as follows from a given discriminant $-d_3$.

Algorithm 3.16.

Input: d.
Output: *The class number and the minimal polynomial of the fundamental unit of each cubic field with discriminant* $d < 0$, *false for improper* d.

```
class3(d) := block(
[m, i, e, f, g, h],
local(f, g, h),
f : [], h : [],
if integerp(d) and d < 0 then (
  f : factdisc3(d),
  if f # false then (
    g : rcl3(rcl(d), d),
```

```
    if g # false then (
      g : rclind3(g, d), m : length(g),
      for i : 1 thru m do
if rprimp(g[i][2], f) then (
  e : ev(ellip3(g[i][1], g[i][2], d), numer),
  h : cons(cons(index3(e, ev(bound3(e, d), numer),
    minpoly3(e)), g[i]), h)  ),
      if h # [] then
cons(f, h)  )  )  )  )  $
```

The returned value of class3($-d_3$) is the list

$$\Big[[-d_3, -d, f_3, \text{ prime power factors of } f_3], \quad \Big[h_{K_3}, \left[\mathrm{Tr}_{K_3/\mathbb{Q}}(\eta), \mathrm{Tr}_{K_3/\mathbb{Q}}(\eta^{-1})\right], [z_3, S_3] \Big], \ldots \text{similar} \ldots \Big]$$

Here η is the fundamental unit of K_3 greater than 1. We mention that the corresponding subgroup S_3 of $R(-d_3)$ to K_3 is returned together with $z_3 \in R(-d_3)\backslash S_3$ by this function in the form of a list of representatives like in (3.1), which will be used in the next section.

4 Sextic case

In this section, we consider the relative units of the composite sextic field $K = K_2K_3$ to explain the merit to decompose h_K and E_K.

In (1.2), Step 1 is easier than cubic case. As we have mentioned in section 3, we may employ a general algorithm to compute representatives of the primitive subgroup S of $R(-df^2)$ corresponding to the sextic field $K = K_2K_3$, but we go on a simpler way. To describe it, we prepare a few more notations. We have already observed that the discriminant $-d_3$ of K_3 is factored as $-d_3 = -df_3^2$ with the discriminant $-d$ of $F = \mathbb{Q}(\sqrt{-d_3})$ and with $f_3 \in \mathbb{N}$. The discriminant d_2 of K_2 satisfies the relation $-d_2'd_2 = -df_2^2$ with the discriminant $-d_2'$ of another quadratic field $M_2 = \mathbb{Q}(\sqrt{-dd_2})$ and with $f_2 \in \mathbb{N}$. The discriminant D of K is factored as $D = dd_2d_3f^2$ with $f \in \mathbb{N}$, where f is decided as the least common multiple of f_2 and f_3. Let $\Phi : R(-df^2) \to R(-d_3)$ be the canonical surjection. Utilizing the subgroup S_3 of $R(-d_3)$ obtained above, we can then decide S by

$$S = \left\{ \mathbf{C} \in \Phi^{-1}(S_3) \,\middle|\, \left(\frac{d_2}{N(\mathbf{a}_\mathbf{C})}\right) = 1 \quad \text{or} \quad \left(\frac{-d_2'}{N(\mathbf{a}_\mathbf{C})}\right) = 1 \right\},$$

where $\left(\frac{\cdot}{\cdot}\right)$ is the Kronecker symbol, and an integral ideal $\mathbf{a}_\mathbf{C}$ of F prime to f is taken so that its restriction to the order of F with conductor f belongs to each class $\mathbf{C} \in R(-df^2)$. This is a nice consequence of the Verschiebungssatz of class field theory. Hence, all the necessary information on the cyclic sextic factor group

$R\left(-df^2\right)/S$ to compute the generating elliptic unit of $\mathbb{E}_6$ is obtained by the next procedure.

Algorithm 4.1.

Input: *A quadratic discriminant* $d = w * f * f < 0$*, discriminants* x, y *of quadratic fields and a subgroup* g *of index* 3 *of the ring class group with quadratic discriminant* $z < 0$.

Output: *The subgroup of index* 6 *of the ring class group with quadratic discriminant* d *for the discriminants* x, y *and for the subgroup* g.

```
rclind6(g, d, w, f, x, y, z) := block(
[l, i, p, s],
local(s),
d : d / z,
while evenp(d) do (
  d : d / 2, g : rclind6_invmodrcl(g, z, 2), z : z + z  ),
while mod(d, 3) = 0 do (
  d : d / 3, g : rclind6_invmodrcl(g, z, 3), z : z+z+z ),
if d >1 then (
  d : factor(d), l : length(d),
  for i : 0 thru l do (
    if l : 0 then
      p : d
    else
      p : inpart(d, i),
    if p # "*" then (
      g : rclind6_invmodrcl(g, z, p), z : z * p  )  )  ),
s : [], p : 0, l : length(g[2]),
for i : 1 thru l do
  if normres(g[2][i], z, w, f, x, y) then
    s : cons(g[2][i], s)
  else if p = 0 then
    p : i,
[g[2][p], mulrcl(g[1], g[1], z), s]  )  $
  /* -- the inverse image of a ring class subgroup g with
        quadratic discriminant d < 0 by the reduction
        modulo a prime p --      */
  rclind6_invmodrcl(g, d, p) := block(
  [i, j, m, s],
  local(s),
  s : invmodrcl(g[2][1], d, p), m : length(g[2]),
  for i : 2 thru m do (
    s : cons(invmodrcl(g[2][i], d, p), s)  ),
  [invmodrcl(g[1], d, p)[1], s]  )  $
```

Here two more general utility functions are defined to get the inverse image via Φ and to test the norm residue condition for those ring classes. The first one is as follows.

Algorithm 4.2.

Input: *A ring class z with quadratic discriminant $d < 0$, and a prime p.*
Output: *The inverse image of z by the reduction modulo p.*

```
invmodrcl(z, d, p) := block(
[a : p * p, b, i, j, y],
local(y),
y : [], b : a * d, i : (z[2] * z[2] - d) / 4 / z[1],
for j : 0 thru p - 1 do
  if mod((z[1] * j - z[2]) * j + i, p) # 0 then
    y : cons(redrcl([a * z[1],
                     p * (z[2] - 2 * z[1] * j)], b), y),
if mod(z[1], p) # 0 then
  cons(redrcl([z[1], p * z[2]], b), y)
else
  y ) $
```

The correctness of this algorithm is proved easily by [2] §8 or [1], Chapter 2 for example. The second one is prepared in order to find out prime ideals not dividing f.

Algorithm 4.3.

Input: *A ring class z with quadratic discriminant $w = d * f * f < 0$, and discriminants x, y.*
Output: *True if z belongs to the group corresponding to the bicyclic biquadratic field with discriminants x, y, else false.*

```
normres(z, w, d, f, x, y) := block(
[p, i, q, t],
local(q),
if gcd(z[1], f) # 1 then (
  q : divide(1 - d, 4), q[2] : 1 - q[2], p : 2, t : true,
  while t do (
    while (mod(f, p) = 0 or kronecker(d, p) # 1) do
      p : primen(p),
    i : 0,
    while mod((i + q[2]) * i + q[1], p) # 0 do
      i : i + 1,
    if redrcl([p, (i + i + q[2]) * f], w) = z then (
      z[1] : p, t : false  )
    else
      p : primen(p)  )  ),
kronecker(x, z[1]) = 1 or kronecker(y, z[1]) = 1  )  $
```

This method seems to be not so efficient, but the author could find no other better procedure. The generating elliptic unit ξ of $\mathbb{E}_6$ is now given by the next function similarly as in the cubic case.

Algorithm 4.4.

Input: *A ring class subgroup s of quadratic discriminant $d < 0$ with x, y of order 2, 3 modulo s.*
Output: *The elliptic unit of a sextic field for s.*

```
ellip6(x, y, s, d) := block(
[i, m : length(s), r],
local(r),
for i : 1 thru m do
  r[i] : mulrcl(s[i], x, d),
for i : 1 thru m do
  r[i+m] : mulrcl(s[i], y, d),
for i : 1 thru m do
  s[i+m] : mulrcl(r[i], y, d),
if numer then (
  product(rinvar_minor(r[i], d) /
                       rinvar_minor(s[i], d), i, 1, m) *
  exp(%pi * sqrt(-d) * sum(1 / s[i][1] - 1 /
                              r[i][1], i, 1, m) / 12) *
  sqrt(product(s[i][1] / r[i][1], i, 1, m))  )
else
  product(rinvar(r[i], d) /  rinvar(s[i], d), i, 1, m)  )  $
```

This completes Step 1.

Steps 2 and 3 goes mostly in a similar manner as before. As it is described in [13] in detail, we only give the functions to be used. By [13], Lemma 1, we decide the minimal polynomial of the generating elliptic unit of $\mathbb{E}_6$ as follows.

Algorithm 4.5.

Input: *A relative unit $u > 1$ of a sextic field with a real quadratic and a non-galois cubic subfields.*
Output: *The minimal polynomial of u.*

```
minpoly6(u) := block(
[m, i, e, x, c],
local(x, c),
x : minpoly6_possible(u), m : length(x), c : x[1][1],
                                          e : x[1][2],
for i : 2 thru m do if x[i][2] < e then (
  c : x[i][1], e : x[i][2]  ),
c  )  $
```

Here the subroutine minpoly6_possible(u) is used to restrict the possibility of the minimal polynomial in a finite number.

Algorithm 4.6.

Input: *A relative unit $u > 1$ of a sextic field with a real quadratic and a non-galois cubic subfields.*
Output: *The possible minimal polynomial of u.*

```
minpoly6_possible(u) := block(
[i, j, e, a : sqrt(u), m],
local(c),
a : a + 1 / a u : u + 1 / u, m : fix(u + a + a + 1 / 10)],
                                                   c : [],
for i : fix(u + 9 / 10 - a - a) thru m do (
  e : (i + 2) / a, e : u * (i + 2 - u) + e * e - i - i - 1,
  j : fix(e + 1 / 2), c : cons([[i, j], abs(e - j)], c)  ),
c  )  $
```

On account of [13], Theorem 1, we estimate the index $(E_6 : \mathbb{E}_6)$ by the following function.

Algorithm 4.7.

Input: *A relative unit $u > 1$ of a sextic field with discriminant d with a real quadratic and a non-galois cubic subfields.*
Output: *An upper bound of $(< e >:< u >)$, where e is the relative fundamental unit.*

```
bound6(u, d) :=
fix(1 / 10 + log(u) / log((sqrt(d) /
                              4 + 290) ^ (1 / 7) - 9 / 7))
 $
```

This completes Step 2.

The index $(E_6 : \mathbb{E}_6)$ is decided by [13], Theorem 2. An exact process of calculation is given as follows.

Algorithm 4.8.

Input: *A relative unit $u > 1$ of a sextic field with a real quadratic and a non-galois cubic subfields with minimal polynomial c, and b.*
Output: *The index $(< e >:< u >)$ not exceeding b and the minimal polynomial of the fundamental unit $e > 1$ of the field.*

```
index6(u, b, c) := block(
[i : 1, e, p : 2, x, s, t],
```

```
local(x, s, t),
s : c,
unless p > b do (
  e : u^(1 / p), x : minpoly6_possible(e),
  unless x = [] or x = p do (
    t : first(x)[1], x : rest(x),
    if sq6(p, t) = s then (
      i : i * p, u : e, s : t, x : p, b : quotient(b, p) )),
  if x # p then
    p : primen(p)  ),
[i, s]  )  $
```

The recursive sequence utilized here is defined as follows.

Algorithm 4.9.

Input: *$q>1$, the minimal polynomial c of a sextic relative unit.*
Output: *The minimal polynomial of the q-th power of the sextic relative unit.*

```
sq6(q, c) :=
[sq6_(q, c[1], c[2]),
  sq6_(q, c[2] - c[1] - 3, sq6_(3, c[1], c[2]) + c[2] - c[1])] $
  /*  -- the recursive sequence --  */
  sq6_(q, i, j) :=
  if q = 2 then
    i * i - j - j
  else if q = 3 then
    i * (i * (i + 3) + 6) - (3 * i + 6) * j + 6
  else if q = 4 then
    i * (i * (i * (i + 4) + 8) + 8)
    - j * ((4 * i + 8) * i - j - j + 4)
  else if q = 5 then
    i * (i * (i * (i * (i + 5) + 10) + 10) + 5)
    + j * ((5 * i + 10) * j - i * ((5 * i + 15) * i + 15)-10)
  else if q = 6 then
    i * (i * (i * (i * (i * (i + 6) + 15) + 24) + 30) + 24)
    + j * ((i * (9 * i + 24) - j - j + 18) * j
          - i * (i * ((6 * i + 24) * i + 42) + 48) - 24) + 6
  else if q = 7 then
    i * (i * (i * (i * (i * (i * (i + 7)
                                + 21) + 42) + 63) + 56) + 21)
    + j * (j * (i * (i * (14 * i + 49) + 56)
                                    - (7 * i + 14) * j  + 28)
          - i * (i * (i * ((7 * i + 35)
                                  * i + 77) + 105) + 77) - 14)
```

```
else
  i * sq6_(q - 1, i, j) - j * sq6_(q - 2, i, j)
  + (i * (i + 2)- j - j + 2) * sq6_(q - 3, i, j)
  - j * sq6_(q - 4, i, j)  + i * sq6_(q - 5, i, j)
  - sq6_(q - 6, i, j)  $
```

The returned value of index6(ξ,bound6 (ξ, D), minpoly6(ξ)) for the elliptic unit ξ and the discriminant D of K is the list

$$\left[(E_6 : \mathbb{E}_6), [s,t] \right],$$

where the relative fundamental unit of K, which generates the cyclic group E_6, is given by any one of the two real zeros of the polynomial

$$X^6 - sX^5 + tX^4 - \left(s^2 + 2s - 2t + 2\right) X^3 + tX^2 - sX + 1.$$

This completes Step 3.

To decide h_K, we should further compute $\left(E_K : E^K\right)$ in (1.2). The class number h_{M_2} of another quadratic field $M_2 = \mathbb{Q}\left(\sqrt{-d_3 d_2}\right)$ is computable by class2($-d_2'$) in section 2. To decide the class number h_M of $M = M_2 K_3$, we should also compute $\left(E_{M/\mathbb{Q}} : W_M N_{L/M}\left(E_6'\right)\right)$ in (1.5). But, these are omitted as is explained in section 1.

References

[1] Z.I. Borevich and I.R. Shafarevich, "Number Theory", Academic Press, New York-London 1966.

[2] M. Deuring, *Die Klassenkörper der komplexen Multiplikation*, Enzycl. der Math. Wissensch. I 2, Heft 10, Teil II **23** 1858.

[3] G. Gras and M.-N. Gras, *Calcul du nombre de classes et des unités des extensions abéliennes réelles de* $\mathbb{Q}$. Bull. Sci. Math. (2) **101**, 1977, 97 129.

[4] M.-N. Gras, *Méthodes et algorithmes pour le calcul numérique du nombre de classes et des unités des extensions cycliques cubiques do* $\mathbb{Q}$. J. Reine Angew. Math. **277**, 1975, 89 116.

[5] M.-N. Gras, *Table numérique du nombre de classes et des unités des extensions cycliques réelles de degré 4 de* $\mathbb{Q}$, Publ. Math. Univ. Besançon 1977/78, fasc 2, 1978.

[6] M.-N. Gras, *Classes et unités des extensions cycliques réelles degré 4 de* $\mathbb{Q}$, Ann. Inst. Fourier **29**, 1979, 107 124.

[7] H. Hasse, *Arithmetische Theorie der kubischen Zahlkörper auf klassenkörpertheoretischer Grundlage*, Math. Z. **31**, 1930, 565 582.

[8] H. Hayashi, *On elliptic units and the class numbers of a certain dihedral extension of degree 2ℓ*, Acta Arith. **44**, 1984, 43 53.

[9] S. Mäki, *The determination of units in real cyclic sextic fields*, Lecture Notes in Math., No. 797, Springer-Verlag, 1980.

[10] K. Nakamula, *Class number calculation of a cubic field from the elliptic unit*, J. Reine Angew. Math. **331**, 1982, 114–123.

[11] K. Nakamula, *Calculation of the class numbers and fundamental units of abelian extensions over imaginary quadratic fields from approximate values of elliptic units*, J. Math. Soc. Japan **37**, 1985, 245-273.

[12] K. Nakamula, *Class number calculation of a quartic field from the elliptic unit*, Acta Arith. **45**, 1985, 215–227.

[13] K. Nakamula, *Class number calculation of a sextic field from the elliptic unit*, Acta Arith. **45**, 1985, 229–247.

[14] K. Nakamula, *A table for pure cubic fields*, Advanced Studies in Pure Mathematics, Vol. 13. Kinokuniya, Tokyo, 1987, 461-477.

[15] K. Nakamula, *Elliptic units and the class numbers of non-galois fields*, J. Number Theory **31**, 1989, 142–166.

Computing fundamental units from independent units

B. Arenz

Abstract. We give some computational results of a new method for the determination of fundamental units in an algebraic number field of arbitrary degree. For a given set of independent units it is shown how it is possible to test whether it generates the whole unit group. Otherwise a set of fundamental units is computed.

1 Introduction

Let $f(t) \in \mathbb{Z}[t]$ be a monic, irreducible polynomial of degree $n = r_1 + 2r_2$. The field $\mathcal{F} := \mathbb{Q}(\rho)$ of discriminant $d_{\mathcal{F}}$ is generated by a zero ρ of f. An integral basis $o_{\mathcal{F}}$ is denoted by $\omega_1, \ldots, \omega_n$, its dual basis is given by $\omega_1^*, \ldots, \omega_n^*$. By the theorem of Dirichlet there are $r := r_1 + r_2 - 1$ (fundamental) units $\eta_1, \ldots, \eta_r$ which generate the unit group

$$U_{\mathcal{F}} = TU_{\mathcal{F}} \times \langle \eta_1 \rangle \times \ldots \times \langle \eta_r \rangle .$$

The torsion subgroup $TU_{\mathcal{F}}$ is generated by an element ζ of finite order.

Let $U_\epsilon := TU_{\mathcal{F}} \times \langle \epsilon_1 \rangle \times \ldots \times \langle \epsilon_r \rangle$ be a subgroup of $U_{\mathcal{F}}$ generated by r independent units ϵ_i $(1 \le i \le r)$. The group index I of U_ϵ in $U_{\mathcal{F}}$ is given by the quotient of the regulators, whereas the regulator of a unit group is defined as the mesh of the corresponding logarithmic space

$$I = [U_{\mathcal{F}} : U_\epsilon] = \frac{\mathrm{Reg}(U_\epsilon)}{\mathrm{Reg}(U_{\mathcal{F}})}.$$

By the lower estimation (Pohst/Zassenhaus) for the regulator of $U_{\mathcal{F}}$, we get an upper bound $q \in \mathbb{N}$ for the index I, e.g.

$$I \le \left\lfloor \frac{\mathrm{Reg}(U_\epsilon)}{\mathrm{Reg}_{est}} \right\rfloor =: q.$$

If $q = 1$, we know that $U_\epsilon = U_{\mathcal{F}}$ and $\eta_i = \epsilon_i$ $(1 \le i \le r)$. For $q \ge 2$, we need a criterion for whether $U_\epsilon = U_{\mathcal{F}}$ and if not, how the fundamental units η_i can be expressed in terms of powers of ϵ_j $(1 \le i, j \le r)$.

As a consequence of the fundamental theorem on finitely generated abelian groups we obtain the following Theorem.

Computational Number Theory

Theorem 1.1. *Let $\eta_1, \ldots, \eta_k$ $(0 \le k < r)$ be part of a system of fundamental units of $o_{\mathcal{F}}$. Then $\eta_{k+1} \in U_{\mathcal{F}}$ also belongs to that system, if and only if the equation*

$$\eta_{k+1} = \xi \eta_1^{m_1} \cdot \ldots \cdot \eta_k^{m_k} \omega^m$$

with $\xi \in TU_{\mathcal{F}}$, $m, m_i \in \mathbb{Z}$ $(1 \le i \le k)$, $\omega \in o_{\mathcal{F}}$ is unsolvable for $|m| \ge 2$.

This theorem leads to the following criterion.

Corollary 1.1. *Let $\eta_1, \ldots, \eta_r$ be independent units of $o_{\mathcal{F}}$. If for all primes p less than or equal to the bound q, all rational integers $0 \le m_i \le p-1$ $(1 \le i \le r)$ and all torsion units $\xi \in TU_{\mathcal{F}}$ the equation*

$$\omega^p = \xi \eta_1^{m_1} \cdot \ldots \cdot \eta_r^{m_r} \tag{1.1}$$

is unsolvable for $\omega \in o_{\mathcal{F}}$, then $\eta_1, \ldots, \eta_r$ are fundamental units.

Remark 1.1. This new method is available in every order $\mathcal{R}$ of the field $\mathcal{F}$.

2 The main idea

The main idea for this new method is to reduce the number of exponents $m_1, \ldots, m_r$ to be tested in (1.1) by excluding certain units η_i $(1 \le i \le r)$.

We start our computation by setting

$$\eta_i \leftarrow \epsilon_i \ (1 \le i \le r), \quad \mathrm{Reg}(U_{\mathcal{F}}) \leftarrow Reg(U_\epsilon),$$

$$q \leftarrow \left\lfloor \frac{\mathrm{Reg}(U_{\mathcal{F}})}{\mathrm{Reg}_{est}} \right\rfloor, \qquad I \leftarrow 1.$$

Let p be an arbitrary prime number below the index bound q. First we try to solve the equation

$$\omega^p = \xi \eta_1 \tag{2.1}$$

in $o_{\mathcal{F}}$ for a torsion unit $\xi \in TU_{\mathcal{F}}$. If there is a solution ω we know that η_1 is not a fundamental unit. We replace the first unit and the index bound improves

$$\eta_1 \leftarrow \omega, \qquad \mathrm{Reg}(U_{\mathcal{F}}) \leftarrow Reg(U_{\mathcal{F}})/p,$$

$$q \leftarrow \left\lfloor \frac{\mathrm{Reg}(U_{\mathcal{F}})}{\mathrm{Reg}_{est}} \right\rfloor, \qquad I \leftarrow I * p.$$

If $q = 1$ we are done. Else we repeat solving the equation (2.1).

If we find no solution we can reduce the number of the units η_i $(1 \le i \le r)$ to be tested by the following argument. Because of $\mathcal{F}(\sqrt[p]{\eta_1}) \supset \mathcal{F}$ the polynomial $t^p - \eta_1 \in o_{\mathcal{F}}[t]$ is irreducible. By Čebotarev [3] there exists a prime ideal $\wp \not\supseteq d_{\mathcal{F}}$ such that $t^p - \eta_1$ remains irreducible in $o_{\mathcal{F}}/\wp[t]$ and therefore $p | (N(\wp) - 1)$. (Namely, for $p \nmid (N(\wp) - 1)$ there is $q \in \mathbb{N}$ subject to $pq \equiv 1 \bmod (N(\wp) - 1)$ and $\mu := \eta_1^q$ satisfies $\mu^p = \eta_1^{pq} \equiv \eta_1 \bmod \wp$ in contradiction to the irreducibility of $t^p - \eta_1$ in $o_{\mathcal{F}}/\wp[t]$.)

Since $(o_{\mathcal{F}}/\wp)^\times$ is cyclic with $p|\#(o_{\mathcal{F}}/\wp)^\times$ and η_1 is not a pth power in $(o_{\mathcal{F}}/\wp)^\times$ the order of η_1 is divisible by p. Hence, for every unit η_i $(2 \le i \le r)$ there is precisely one rational integer ν_i $(0 \le \nu_i < p)$ such that $\eta_1^{\nu_i}\eta_i$ is congruent to a pth power modulo $\wp$. By defining $\tilde{\eta}_i := \eta_1^{\nu_i}\eta_i$ $(2 \le i \le r)$ we obtain

$$\begin{aligned} U_\epsilon &= TU_{\mathcal{F}} \times \langle \eta_1 \rangle \times \langle \eta_2 \rangle \times \ldots \times \langle \eta_r \rangle \\ &= TU_{\mathcal{F}} \times \langle \eta_1 \rangle \times \langle \tilde{\eta}_2 \rangle \times \ldots \times \langle \tilde{\eta}_r \rangle \end{aligned}$$

and every element of $\langle \tilde{\eta}_2 \rangle \times \ldots \times \langle \tilde{\eta}_r \rangle$ is a pth power modulo $\wp$.

As a consequence we have the following. If there is a pth power in U_ϵ, then there exist $\omega \in o_{\mathcal{F}}$, $\xi \in TU_{\mathcal{F}}$ and exponents $0 \le m_i < p$ $(1 \le i \le r)$ with

$$\omega^p = \xi\eta_1^{m_1} \cdot \tilde{\eta}_2^{m_2} \cdot \ldots \cdot \tilde{\eta}_r^{m_r}.$$

This equation holds modulo the ideal $\wp$ too. Since ω^p and $\tilde{\eta}_i^{m_i}$ $(2 \le i \le r)$ are congruent to pth powers modulo $\wp$, $\eta_1^{m_1}$ has to be congruent to a pth power modulo the ideal $\wp$ in contradiction to the existence of $\wp$. Hence, m_1 has to be zero and the requested pth power lies in $TU_{\mathcal{F}} \times \langle \tilde{\eta}_2 \rangle \times \ldots \times \langle \tilde{\eta}_r \rangle$.

Repeating this process we finally end up with one unit and we repeat the whole procedure for the next prime number p.

The search for a suitable ideal $\wp$ is as follows. We take the first prime number P which fulfills

$$P \nmid d_{\mathcal{F}} \qquad \text{and} \qquad p|(P-1).$$

We factorize the generating polynomial f modulo $P\mathbb{Z}[t]$, e.g.

$$f(t) \equiv f_1(t) \cdot \ldots \cdot f_l(t) \bmod P\mathbb{Z}[t].$$

The ideals $\wp_i := Po_{\mathcal{F}} + f_i(\rho)o_{\mathcal{F}}$ $(1 \le i \le l)$ then are candidates for the ideal $\wp$. Each of them is tested as to whether the polynomial $t^p - \eta_1$ is irreducible modulo $o_{\mathcal{F}}/\wp_i[t]$. Therefore we use the following criterion.

Corollary 2.1. *The polynomial $t^p - \eta_1$ is reducible modulo $o_{\mathcal{F}}/\wp_i[t]$*

$$\begin{aligned} &\Leftrightarrow \quad \eta_1 \text{ is a } p\text{th power modulo the ideal } \wp_i \\ &\Leftrightarrow \quad \eta_1^{\frac{N(\wp_i)-1}{p}} \equiv 1 \bmod \wp_i. \end{aligned}$$

Because of $N(\wp_i) = P^{\deg(f_i)}$ the sufficient condition $p|(N(\wp_i)-1)$ $(1 \le i \le l)$ is already fulfilled.

Remark 2.1. The handling of the torsion units can be done as well by applying the main idea to the generating element of the torsion subgroup. Let $\langle \zeta \rangle = TU_{\mathcal{F}}$. If ζ is not a pth power there exists a prime ideal $\wp$ such that ζ is not congruent to a pth power modulo $\wp$ and there exist exponents ν_i $(1 \le i \le r)$ such that $\zeta^{\nu_i}\eta_i$ is congruent to a pth power modulo $\wp$. We replace η_i by $\zeta^{\nu_i}\eta_i$ $(1 \le i \le r)$ and need not look at torsion units anymore. If ζ is a pth power we neglect it too.

The following diagram shows how the main idea is realized for all units η_i $(1 \le i \le r)$ and all primes $p \le q$.

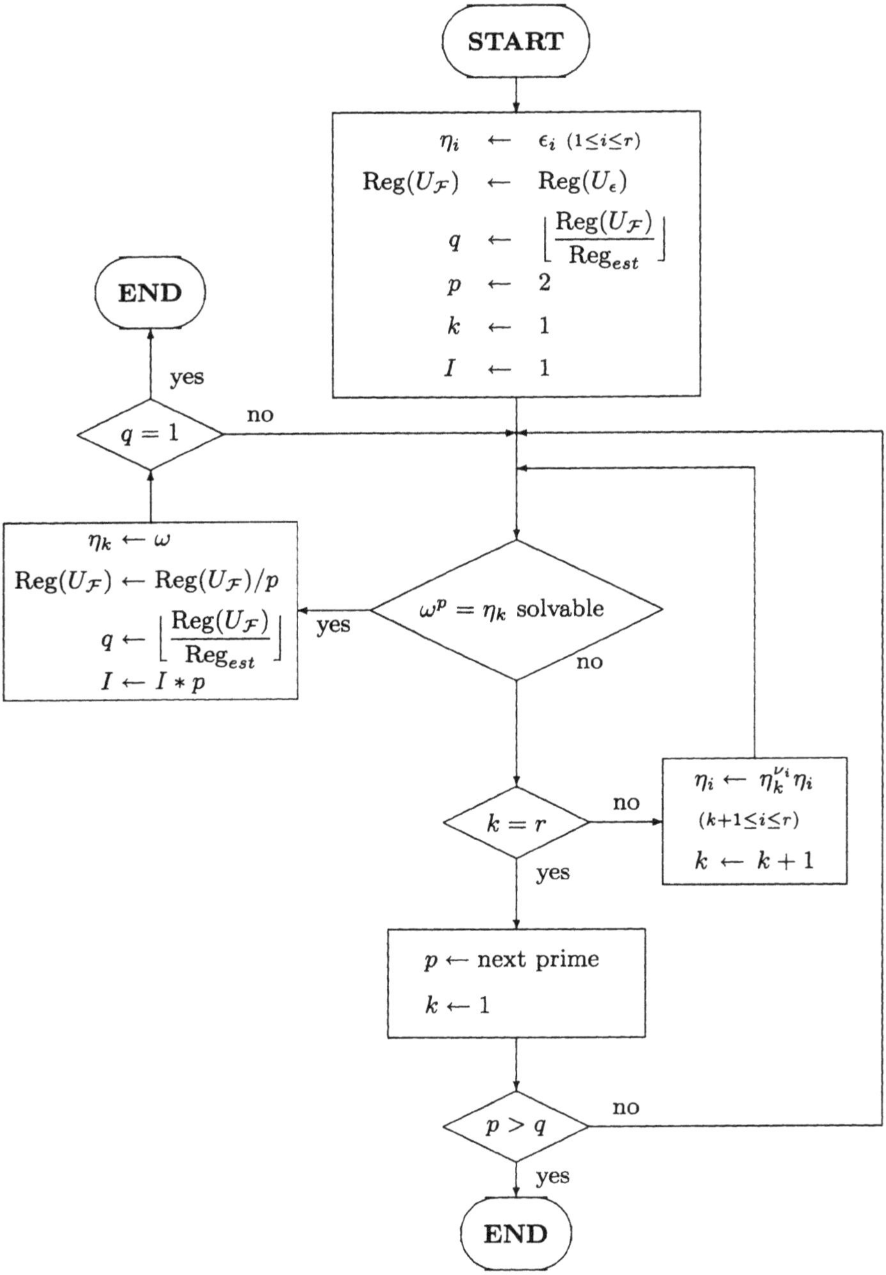

Figure 1. Diagram for the main algorithm

3 Solving the equation $\omega^p = \eta$

Another advantage of this method is that an equation like (1.1) only has to be solved for one unit. We describe three different techniques for this task.

3.1 Inverting the equation modulo a suitable ideal

If $\omega \in o_{\mathcal{F}}$ is a solution then we estimate the coefficients e_i $(1 \le i \le r)$ of ω in the integral basis representation

$$\begin{aligned} |e_i| = |Tr(\omega\omega_i^*)| &= \left|\sum_{j=1}^{n} \omega^{(j)}\omega_i^{*(j)}\right| \\ &\le \sum_{j=1}^{n} |\omega^{(j)}|\,|\omega_i^{*(j)}| \\ &= \sum_{j=1}^{n} |\eta^{(j)}|^{1/p}|\omega_i^{*(j)}| =: T_i \ (1 \le i \le n). \end{aligned}$$

We take the first prime P which fulfills the three conditions

(i) $P > 2\max\{T_i \mid 1 \le i \le n\}$,
(ii) $P \not| \, d_{\mathcal{F}}$,
(iii) $p \not| \, h_P := \#(o_{\mathcal{F}}/Po_{\mathcal{F}})^\times$.

Hence, there exists $q \in \mathbb{N}$ such that $pq \equiv 1 \bmod h_P$ and $\eta^q = \omega^{pq} \equiv \omega \bmod Po_{\mathcal{F}}$. We determine a possible solution ω by computing η^q with coefficients e_i $(1 \le i \le r)$ lying in the interval $]\frac{-P}{2}, \frac{P}{2}[$. If there is a solution then we will find it by this construction. If not, the equation $\omega^p = \eta$ does not hold for this ω, hence it has no solution at all.

The advantage of this method is that we need only test one candidate ω as a solution. The disadvantage is that the existence of the prime P is guarented only for primes $p > n+1$ and the search can take a long time.

For determing the order h_P we factorize the polynomial f modulo $P\mathbb{Z}[t]$, e.g.

$$f(t) \equiv f_1(t) \cdot \ldots \cdot f_l(t) \bmod P\mathbb{Z}[t].$$

The ideals $\wp_i := Po_{\mathcal{F}} + f_i(\rho)o_{\mathcal{F}}$ $(1 \le i \le l)$ fulfill $N(\wp_i) = P^{\deg(f_i)}$ $(1 \le i \le l)$ and we get

$$h_P = \#(o_{\mathcal{F}}/Po_{\mathcal{F}})^\times = \prod_{i=1}^{l} \#(o_{\mathcal{F}}/\wp_i)^\times = \prod_{i=1}^{l}(P^{\deg(f_i)} - 1).$$

3.2 Taking complex roots

If ω is a solution, its conjugates satisfy $\omega^{(j)^p} = \eta^{(j)}$ $(1 \le j \le n)$. If we take complex roots we get

$$\begin{aligned} \omega^{(j)} &= \eta^{(j)1/p} & (1 \le j \le r_1) \\ \omega_k^{(j)} &= |\eta^{(j)}|^{1/p}\exp(\tfrac{2\pi i}{p}(\phi_j + x_j)) & (r_1+1 \le j \le r_1+r_2) \\ \omega_k^{(j+r_2)} &= |\eta^{(j)}|^{1/p}\exp(-\tfrac{2\pi i}{p}(\phi_j + x_j)) & (r_1+1 \le j \le r_1+r_2) \end{aligned}$$

for $\phi_j = \arctan\left(\frac{\Im(\eta^{(j)})}{\Re(\eta^{(j)})}\right)$ and $0 \le x_j \le p-1$ $(r_1+1 \le j \le r_1+r_2)$. The number of possible candidates for ω is p^{r_2}. Hence, this method is only useful for small p and r_2, for example in the totally real case.

3.3 Enumerating an ellipsoid

We have just seen that we have bounds for the absolute values of the conjugates $|\omega^{(j)}| = |\eta^{(j)}|^{1/p}$ $(1 \le j \le n)$. Since every solution ω is a unit we can use the fast method of Fincke/Pohst [4, 8] for solving the norm equation $|N(\omega)| = 1$. This method of solving norm equations consists in enumerating integer points in suitable ellipsoids. The advantage of this technique is that the number of ellipsoids to be enumerated is bounded by 2^r for every prime p.

We can improve this idea of enumerating ellipsoids by looking at a different quadratic form. Let ω be a solution of $\omega^p = \eta$. Then $\eta^{-1/p}\omega = 1$ is a torsion unit and $\omega = \sum_{i=1}^n x_i\omega_i$ is a solution of the following quadratic form

$$Q(x_1, \dots, x_n) := \sum_{j=1}^n |\eta^{(j)}|^{-2/p} \cdot \left|\sum_{i=1}^n x_i\omega^{(j)}\right|^2$$

with right side n. The advantage of this method is that we need only enumerate the integer points within one ellipsoid. This idea is developed in detail in [6].

4 Implementation

The program has been implemented in ANSI-Fortran-77 on Apollo workstations DN 3000 and DN 4000. The number of all subroutines used is about 240, where the whole program is a part of the subroutine library package KANT ([9]).

For example, we use

- various methods for computing independent units ([1, 2, 8])
- Pohst/Zassenhaus methods for lower regulator bounds ([5, 8]) and for determining torsion subgroups ([8])
- the Fincke/Pohst algorithm for solving norm equations based on interval arithmetic ([4, 8])
- the Berlekamp algorithm for factoring polynomials modulo a prime.

5 Examples

The program has been tested for algebraic number fields of degrees 4–19, especially for large numbers of
- totally real quartic fields with discriminant $d_{\mathcal{F}} < 1,000,000$
- totally real quintic fields with discriminant $d_{\mathcal{F}} < 2,000,000$
- totally complex sextic fields with absolute discriminant $< 23,100$
- totally real octic fields.

The next three polynomials generate the three octic fields of smallest discriminant, the fourth example is one of degree 11. The independent units were computed by the method of Buchmann [1]. All fundamental units are LLL-reduced in the logarithmic space.

Example 5.1. Let $f(x) = x^8 + 2x^7 - 7x^6 - 8x^5 + 15x^4 + 8x^3 - 9x^2 - 2x + 1$; $r = 7$; $o_{\mathcal{F}} = \mathbb{Z}[\rho]$. Coefficients of fundamental units (referring to the power basis):

η_1	η_2	η_3	η_4	η_5	η_6	η_7
−2	−1	−1	0	3	−2	0
0	6	2	−4	−5	3	0
19	0	3	9	−15	11	−2
2	−12	−8	6	5	−3	−9
−31	5	−3	−11	15	−14	6
−7	6	3	−4	1	−1	6
10	−2	1	3	−4	4	−2
3	−1	0	1	−1	1	−1

Discriminant: 282,300,416,
Regulator of fundamental units: 22.45,
Lower regulator bound: 4.59,
Index: $696 = 2^3 * 3 * 29$.

Example 5.2. Let $f(x) = x^8 + 3x^7 - 5x^6 - 21x^5 - 3x^4 + 35x^3 + 28x^2 + 4x - 1$; $r = 7$; $o_{\mathcal{F}} = \mathbb{Z}[\rho]$. Coefficients of fundamental units (referring to the power basis):

η_1	η_2	η_3	η_4	η_5	η_6	η_7
2	−3	3	−1	1	−2	1
−16	4	−7	−1	−8	15	−9
−40	25	−29	0	−24	40	−37
−4	11	−10	0	−10	4	−21
22	−14	15	0	14	−22	21
6	−7	7	0	7	−6	14
−3	2	−2	0	−2	3	−3
−1	1	−1	0	−1	1	−2

Discriminant: 309,593,125,
Regulator of fundamental units: 23.70,
Lower regulator bound: 4.93,
Index: $378 = 2 * 3^2 * 7$.

Example 5.3. Let $f(x) = x^8 - 8x^6 + 14x^4 - 7x^2 - 1$; $r = 7$; $o_{\mathcal{F}} = \mathbb{Z}[\rho]$.
Coefficients of fundamental units (referring to the power basis):

η_1	η_2	η_3	η_4	η_5	η_6	η_7
5	0	0	0	4	4	1
0	0	−1	0	6	−1	−6
−21	0	0	−7	−13	−13	0
0	7	0	−7	−14	0	14
15	0	0	7	8	8	0
0	−7	0	7	8	0	−8
−2	0	0	−1	−1	−1	0
0	1	0	−1	−1	0	1

Discriminant: 324,000,000,
Regulator of fundamental units: 24.39,
Lower regulator bound: 5.02,
Index: $360 = 2^3 * 3^2 * 5$.

Example 5.4. Let $f(x) = x^{11} + 11x + 11$; $r = 5$; $o_{\mathcal{F}} = \mathbb{Z}[\rho]$.
Coefficients of fundamental units (referring to the power basis):

η_1	η_2	η_3	η_4	η_5
1	11188	21583	−27301	−1059719
1	264	132767	−46373	347422
0	485	109242	−33360	31931
0	−1200	−67464	−19619	−266222
0	167	−70707	−10227	338006
0	−834	32812	−3923	−411640
0	900	44716	1117	358757
0	−540	−14961	2350	−252674
0	931	−27721	4197	160712
0	−886	6114	3398	−43272
0	820	16879	3413	−46114

Discriminant: 10,253,366,650,368,160,101,211,
Regulator of fundamental units: 4,537,331.46,
Lower regulator bound: 13,480.45,
Index: 7.

References

[1] J. Buchmann, *A generalization of Voroni's unit algorithm I,II*, J. Number Theory **20** (1985), 177-209.

[2] J. Buchmann and A. Pethö, *Computation of independent units in number fields by Dirichlet's method*, Math. Comp. **52** (1989), 149-159.

[3] N. Čebotarev, *Die Bestimmung der Dichtigkeit einer Menge von Primzahlen, welche zu einer gegebenen Substitutionsklasse gehören*, Math. Ann. **95** (1926), 191-228.

[4] U. Fincke, *Ein Ellipsoidverfahren zur Lösung von Normgleichungen*, Dissertation, Düsseldorf 1984.

[5] M. Pohst, *Regulatorabschätzungen für total reelle algebraische Zahlkörper*, J. Number Theory **9** (1977), 459-492.

[6] M. Pohst, *On extracting roots of algebraic integers*, in preparation.

[7] M. Pohst and H. Zassenhaus, *An Effective Number Geometric Method of Computing the Fundamental Units of an Algebraic Number Field*, Math. Comp. **31** (1977), 754-770.

[8] M. Pohst and H. Zassenhaus, "Algorithmic algebraic number theory", Cambridge University Press 1989.

[9] J. Graf v. Schmettow, *KANT - a tool for computations in algebraic number fields*, submitted to Proceedings of Colloquium on Computational Number Theory.

A note on index divisors

M. Pohst

Abstract. Let $F = \mathbb{Q}(\rho)$ be an algebraic number field which is generated by a root ρ of a given monic irreducible polynomial $f(t) \in \mathbb{Z}[t]$. In this paper we develop methods for computing the ideal decomposition in F of those prime numbers p which divide the quotient of the discriminant $d(f)$ of the polynomial $f(t)$ and of the field discriminant d_F. They base on the use of reduced resultants and divisor cascading.

1 Introduction

Throughout this paper F denotes an algebraic number field of degree n over the rational numbers $\mathbb{Q}$. We assume that it is generated by a root ρ of a monic irreducible polynomial

$$f(t) = t^n + a_1 t^{n-1} + \ldots + a_n \in \mathbb{Z}[t]. \tag{1.1}$$

Over the complex numbers $\mathbb{C}$ the polynomial $f(t)$ splits into a product of linear factors

$$f(t) = \prod_{j=1}^{n} (t - \rho^{(j)}), \tag{1.2}$$

where the *conjugates* $\rho = \rho^{(1)}, \ldots, \rho^{(n)}$ are ordered as usual, i.e. $\rho^{(1)}, \ldots, \rho^{(r_1)} \in \mathbb{R}$ and $\rho^{(r_1+1)}, \ldots, \rho^{(n)} \in \mathbb{C} \setminus \mathbb{R}$ subject to $\rho^{(r_1+j)} = \overline{\rho^{(r_1+r_2+j)}}$ $(1 \le j \le r_2)$. Especially, we have

$$n = r_1 + 2r_2 . \tag{1.3}$$

The *maximal order* of F, i.e. the ring of algebraic integers contained in F, is denoted by o_F. We stipulate the knowledge of an *integral basis* $\omega_1, \ldots, \omega_n$, and thus of the *discriminant* d_F of F:

$$o_F = \mathbb{Z}\omega_1 + \ldots + \mathbb{Z}\omega_n , \tag{1.4}$$

$$d_F = (det((\omega_i^{(j)})_{1 \le i,j \le n}))^2. \tag{1.5}$$

Computational Number Theory

Let $\mathbb{P}$ denote the set of rational prime numbers. For $p \in \mathbb{P}$ we obtain a factorization of $f(t)$ modulo p (*congruence factorization*) in the form

$$f(t) \equiv \prod_{i=1}^{g} f_i(t)^{e_i} \bmod p\mathbb{Z}[t] \tag{1.6}$$

with monic polynomials $f_i(t) \in \mathbb{Z}[t]$ which are irreducible and coprime in $\mathbb{Z}/p\mathbb{Z}[t]$.

In case the prime number p does not divide the quotient of the *polynomial discriminant* $d(f)$ and the field discriminant d_F the prime ideal decomposition of po_F in F is simply

$$po_F = \mathbf{p}_1^{e_1} \cdots \mathbf{p}_g^{e_g} \tag{1.7}$$

with coprime prime ideals

$$\mathbf{p}_i = po_F + f_i(\rho)o_F \quad (1 \leq i \leq g). \tag{1.8}$$

This result is well-known and is proved in many textbooks on algebraic number theory. Hence it remains to consider the prime numbers p which indeed divide the quotient $d(f)/d_F$ (*index divisors*). To obtain their prime ideal decomposition in F is much more difficult.

The following result still holds for all prime numbers p.

Proposition 1.1. *Let $p \in \mathbb{P}$ and $f_i(t) \in \mathbb{Z}[t]$ $(1 \leq i \leq g)$ satisfying (1.6). Then any prime ideal $\mathbf{p}$ of F lying above po_F contains $f_j(\rho)$ for at most one index $j \in \{1, \ldots, g\}$.*

Proof. For distinct indices $i, j \in \{1, \ldots, g\}$ the premises on $f_i(t), f_j(t)$ guarantee the existence of $u_i(t), v_j(t) \in \mathbb{Z}[t]$ subject to

$$u_i(t)f_i(t) + v_j(t)f_j(t) \equiv 1 \bmod p\,\mathbb{Z}[t]$$

and the specialization $t \mapsto \rho$ provides a contradiction to $f_i(\rho)$, $f_j(\rho)$ both being contained in $\mathbf{p}$. □

Hence, to different factors $f_i(t)$ of a congruence factorization of $f(t)$ modulo $p\mathbb{Z}[t]$ there belong different prime ideals $\mathbf{p}$ lying above po_F. In case p does not divide the quotient $d(f)/d_F$ there is a bijection between the factors $f_i(t)$ and the prime ideals $\mathbf{p}$ lying above po_F via (1.6) – (1.8). This does not need to be true for prime numbers q dividing $d(f)/d_F$, however.

Example 1.1. For $f(t) = t^3 - t^2 - 2t - 8 \in \mathbb{Z}[t]$ the prime number $p = 2$ is an index divisor. Obviously, we obtain

$$g = 2\,,\ f_1(t) = t\,,\ f_2(t) = t + 1\,,\ e_1 = 2\,,\ e_2 = 1$$

in (1.6). But there are three prime ideals $\mathbf{p}_1$, $\mathbf{p}_2$, $\mathbf{p}_3$ lying above $2o_F$. Later we shall see that

$$\rho o_F = \mathbf{p}_1^2\mathbf{p}_2\,,\ (\rho + 1)o_F = \mathbf{p}_3^3\,.$$

For index divisors p we must therefore develop methods for separating the prime ideals $\mathbf{p}$ which belong to the same factor $f_i(t)$ of the congruence factorization (1.6).

If we know that they occur in distinct powers in $f_i(\rho)o_F$ the separation can be easily done via the p–radical and the method of divisor cascading which will be outlined in section 3.

In the next section we shall modify the congruence factorization (1.6) modulo higher powers of p to a congruence factorization (2.1) such that there is again a bijection between the factors $h_i(\rho)$ of the right hand side and the prime ideals above po_F. For this we improve a result of Ore from [2] (see Theorem 2.1). This guarantees that for each prime ideal $\mathbf{p}$ lying above po_F there is a maximal exponent μ and a unique index i such that $h_i(\rho)$ is contained in $\mathbf{p}^{\mu}$.

Hence, the intention of this paper is twofold. It improves a theoretical result of Ore and it also provides the means to compute the ideal decomposition of index divisors explicitly. The necessary calculations essentially involve arithmetic with polynomials in $\mathbb{Z}/p\mathbb{Z}[t]$ and with (algebraic) integers. We note that an integral basis of F is used for the computation of the p–radical in section 3. This is due to our philosophy that an algebraic number field F is in general given by a generating equation and that an integral basis of F should be known before ideal decompositions are considered. In practice one would only need a $\mathbb{Z}$–basis of a p–maximal overorder of $\mathbb{Z}[\rho]$ in F which can be computed by Zassenhaus' round two algorithm (see [1], for example).

Of course, there are other methods for obtaining the ideal decomposition of index divisors. For example, J. Buchmann and H. W. Lenstra Jr. [1] use primitive element constructions of $\mathbb{Z}/p\mathbb{Z}$–algebras for this purpose. But this requires to carry out computations over various rings instead of over $\mathbb{Z}$ and $\mathbb{Z}/p\mathbb{Z}$ only.

The author is indebted to the referee for his careful reading of the manuscript and several useful hints.

2 The idea of Ore

Throughout this section we fix a prime number p dividing $d(f)/d_F$ and denote the corresponding normalized p–adic exponential valuation of $\mathbb{Z}$ by ν_p. For $a \in \mathbb{Z}^{>0}$ the value $\nu_p(a)$ is the exponent of the maximal power of p dividing a. Analogously we write $\nu_{\mathbf{p}}$ for non–zero prime ideals $\mathbf{p}$ of o_F and the corresponding normalized $\mathbf{p}$–adic exponential valuation of o_F.

In a paper of 1927 [2] O. Ore proved the following theorem.

Theorem 2.1. *Let $\lambda \geq \nu_p(d(f))+1$. Then there exists a congruence factorization*

$$f(t) \equiv h_1(t) \cdots h_r(t) \ mod\ p^{\lambda}\mathbb{Z}[t] \tag{2.1}$$

of $f(t)$ into monic polynomials $h_i(t) \in \mathbb{Z}[t]$ which are distinct and irreducible in $\mathbb{Z}/p^{\lambda}\mathbb{Z}[t]$. There are r distinct prime ideals $\mathbf{p}_1, \dots, \mathbf{p}_r$ lying above po_F, and for each such prime ideal $\mathbf{p}_i$ there is exactly one index $j = j(i)$ for which $\nu_{\mathbf{p}_i}(h_j(\rho))$ is maximal.

The proof essentially uses the properties of polynomial resultants. We denote the resultant of two polynomials $g(t)$ and $h(t)$ by $\mathrm{res}\,(g(t),\, h(t))$ in the sequel. Comparing the values of $\nu_p\,(\,d(f)\,)$ on one side and of $\nu_p\,(\,d(h_i)\,)$ as well of $\nu_p\,(\,\mathrm{res}\,(\,h_i(t),\, h_j(t)\,))$ on the other hand Ore shows that the congruence factorization can be lifted to all larger powers of p and establishes a bijection between the factors $h_i(t)$ and the prime ideals lying above $p o_F$. Namely, he proves that for each such prime ideal $\mathbf{p}$ there is exactly one factor $h_j(t)$ $(j = j(i))$ such that $\nu_{\mathbf{p}}(h_j(\rho))$ is maximal. A congruence factorization (2.1) can therefore be used to calculate all prime ideals explicitly.

Studying Ore's proof one easily sees that he mainly uses the properties of reduced resultants. Hence, in numerical calculations we need to lift (1.6) to a congruence factorization modulo $p^\lambda \mathbb{Z}[t]$ with a much smaller exponent λ than in (2.1) in general. This will be carried out in the remainder of this section.

By Hensel's method (see [3], for example) we can easily calculate monic polynomials $f_1^{(k)}(t), \ldots, f_g^{(k)}(t) \in \mathbb{Z}[t]$ $(k \in \mathbb{N})$ subject to

$$f(t) \;\equiv\; \prod_{i=1}^{g} f_i^{(k)}(t) \bmod p^k \mathbb{Z}[t]\,, \tag{2.2}$$

$$f_i^{(k)}(t) \;\equiv\; f_i^{e_i}(t) \bmod p\mathbb{Z}[t] \;\; (1 \le i \le g), \tag{2.3}$$

and for distinct indices $i, j \in \{1, \ldots, g\}$ there exist polynomials $u_i^{(k)}(t), v_j^{(k)}(t) \in \mathbb{Z}[t]$ with

$$u_i^{(k)}(t) f_i^{(k)}(t) + v_j^{(k)}(t) f_j^{(k)}(t) \;\equiv\; 1 \bmod p^k \mathbb{Z}[t]. \tag{2.4}$$

Hence, it remains to factorize each $f_i^{(k)}(t)$ $(1 \le i \le g)$ in $\mathbb{Z}/p^k\mathbb{Z}[t]$ if it is not irreducible. Because of (2.3) for any two monic non-constant polynomials $h_{i1}^{(k)}(t)\,,\, h_{i2}^{(k)}(t) \in \mathbb{Z}[t]$ satisfying

$$f_i^{(k)}(t) \;\equiv\; h_{i1}^{(k)}(t) h_{i2}^{(k)}(t) \bmod p^k \mathbb{Z}[t] \tag{2.5}$$

there is an exponent $\lambda_{ik} < e_i$ for which

$$h_{i1}^{(k)}(t) \;\equiv\; f_i(t)^{\lambda_{ik}} \bmod p\mathbb{Z}[t], \tag{2.6}$$

$$h_{i2}^{(k)}(t) \;\equiv\; f_i(t)^{e_i - \lambda_{ik}} \bmod p\mathbb{Z}[t]. \tag{2.7}$$

Without loss of generality we can assume $e_i \ge 2$ and $1 \le \lambda_{ik} \le e_i/2$ in the sequel.

The number of possibilities for λ_{ik} is rather limited. From (2.3), (2.6), (2.7) we know that there exist polynomials $d_j^{(k)}(t)\,,\, g_{i\nu,j}^{(k)}(t) \in \mathbb{Z}[t]$ $(\nu = 1, 2\,;\, 1 \le j \le k)$ whose coefficients (except for $d_k^{(k)}(t)$) are either 0 or not divisible by p subject to

$$f_i^{(k)}(t) = f_i(t)^{e_i} + \sum_{j=1}^{k} p^j d_j^{(k)}(t), \tag{2.8}$$

$$h_{i1}^{(k)}(t) = f_i(t)^{\lambda_{ik}} + \sum_{j=1}^{k-1} p^j g_{i1,j}^{(k)}(t), \tag{2.9}$$

$$h_{i2}^{(k)}(t) = f_i(t)^{e_i-\lambda_{ik}} + \sum_{j=1}^{k-1} p^j g_{i2,j}^{(k)}(t). \tag{2.10}$$

We note that there are the degree restrictions deg $(d_j^{(k)}(t)) <$ deg $(f_i^{(k)}(t))$, deg $(g_{i1,j}^{(k)}(t)) <$ deg $(f_i(t)^{\lambda_{ik}})$, deg $(g_{i2,j}^{(k)}(t)) <$ deg $(f_i(t)^{e_i-\lambda_{ik}})$. Hence, (2.5) is solvable if and only if there is a solution of

$$\begin{aligned}\sum_{j=1}^{k-1} p^j d_j^{(k)}(t) \equiv \\ \sum_{j=1}^{k-1} p^j \left(f_i(t)^{\lambda_{ik}} g_{i2,j}^{(k)}(t) + f_i(t)^{e_i-\lambda ik} g_{i1,j}^{(k)}(t) + \sum_{\nu=1}^{j-1} g_{i2,\nu}^{(k)}(t) g_{i1,j-\nu}^{(k)}(t) \right) \\ \bmod p^k \mathbb{Z}[t].\end{aligned} \tag{2.11}$$

To solve (2.11) we proceed inductively. If we consider that congruence modulo $p^2\mathbb{Z}[t]$ we obtain

$$d_1^{(k)}(t) \equiv f_i(t)^{\lambda_{ik}} (g_{i2,1}^{(k)}(t) + f_i(t)^{e_i-2\lambda_{ik}} g_{i1,1}^{(k)}(t)) \bmod p\mathbb{Z}[t]. \tag{2.12}$$

The last congruence is solvable if and only if $f_i(t)^{\lambda_{ik}}$ divides $d_1^{(k)}(t)$ in $\mathbb{Z}/p\mathbb{Z}[t]$. In case of solvability let $g_{i\nu,1}^{(k)}(t)$ $(\nu = 1, 2)$ be a fixed solution. Then all solutions of the congruence (2.12) are of the form

$$\widetilde{g}_{i1,1}^{(k)}(t) = g_{i1,1}^{(k)}(t) + r_{i,1}^{(k)}(t), \tag{2.13}$$

$$\widetilde{g}_{i2,1}^{(k)}(t) = g_{i2,1}^{(k)}(t) - r_{i,1}^{(k)}(t) f_i(t)^{e_i-2\lambda_{ik}}, \tag{2.14}$$

and the polynomial $r_{i,1}^{(k)}(t) \in \mathbb{Z}[t]$ is uniquely determined modulo $p\mathbb{Z}[t]$ and modulo $f_i(t)^{\lambda_{ik}}\mathbb{Z}[t]$. We set

$$s_1^{(k)}(t) = \left(d_1^{(k)}(t) - f_i(t)^{\lambda_{ik}} g_{i2,1}^{(k)}(t) - f_i(t)^{e_i-\lambda_{ik}} g_{i1,1}^{(k)}(t)\right) / p \tag{2.15}$$

and note that $s_1^{(k)}(t)$ remains invariant if we replace $g_{i\nu,1}^{(k)}(t)$ by $\widetilde{g}_{i\nu,1}^{(k)}(t)$ $(\nu = 1, 2)$.

Next we consider the congruence of (2.11) modulo $p^3\mathbb{Z}[t]$ and obtain

$$s_1^{(k)}(t) + d_2^{(k)}(t) \equiv f_i(t)^{\lambda_{ik}} g_{i2,2}^{(k)}(t) + f_i(t)^{e_i - \lambda_{ik}} g_{i1,2}^{(k)}(t) + g_{i2,1}^{(k)}(t) g_{i1,1}^{(k)}(t) \bmod p\mathbb{Z}[t],$$

respectively,

$$\begin{aligned} s_1^{(k)}(t) + d_2^{(k)}(t) - g_{i2,1}^{(k)}(t) g_{i1,1}^{(k)}(t) &\equiv \\ & f_i(t)^{\lambda_{ik}} \left(g_{i2,2}^{(k)}(t) + f_i(t)^{e_i - 2\lambda_{ik}} g_{i1,2}^{(k)}(t) \right) \bmod p\mathbb{Z}[t]. \end{aligned} \tag{2.16}$$

Replacing $g_{i\nu,1}^{(k)}(t)$ by $\widetilde{g}_{i\nu,1}^{(k)}(t)$ $(\nu = 1, 2)$ the condition that $f_i(t)^{\lambda_{ik}}$ divides the left hand side of (2.16) yields a quadratic congruence for $r_{i,1}^{(k)}(t)$. Again this can be solved recursively by studying it modulo increasing powers of $f_i(t)$. Modulo $f_i(t)$ this requires to solve a quadratic equation in a finite field and any solution of it (if it has one) can then be lifted to one or more solutions modulo $f_i(t)^{\mu}$ $(\mu = 2, \dots, \lambda_{ik})$. For any such solution $r_{i,1}^{(k)}(t)$ we analogously calculate all solutions of (2.16):

$$\widetilde{g}_{i1,2}^{(k)}(t) = g_{i1,2}^{(k)}(t) + r_{i,2}^{(k)}(t), \tag{2.17}$$

$$\widetilde{g}_{i2,2}^{(k)}(t) = g_{i2,2}^{(k)}(t) - r_{i,2}^{(k)}(t) f_i(t)^{e_i - 2\lambda_{ik}}. \tag{2.18}$$

We note that from now on the consideration of (2.11) modulo the next higher power of p only yields a linear congruence for the polynomials $r_{i,\mu}^{(k)}(t)$ $(\mu = 2, 3, \dots, k-2)$. In this way we either obtain a solution $h_{i\nu}^{(k)}(t)$ $(\nu = 1, 2)$ of (2.5) or conclude that $f_i^{(k)}(t)$ is irreducible modulo $p^k\mathbb{Z}[t]$.

Example 2.1. Let $f(t) = t^3 - t^2 - 2t - 8$ again. According to Theorem 2.1 we calculate $d(f) = -4 \cdot 503$ and see that a congruence factorization of $f(t)$ modulo $8\mathbb{Z}[t]$ will be sufficient. From $f_1(t) = t$, $e_1 = 2$, $f_2(t) = t + 1$ we obtain by an application of Hensel's method

$$f_1^{(3)}(t) = t^2 - 2t, \quad r^{(3)}(t) = -t, \quad f_2^{(3)}(t) = f_2(t).$$

Because of $e_1 = 2$ the only possibility for λ_{13} is 1. The equations (2.8)–(2.10) yield

$$-2t \equiv 2\left(t(g_{11}^{(3)}(t) + g_{12}^{(3)}(t)) + 2g_{11}^{(3)}(t) g_{12}^{(3)}(t) \right) \bmod 8\mathbb{Z}[t]$$

with the obvious solution $g_{11}^{(3)}(t) = 1$, $g_{12}^{(3)}(t) = 0$.

Once we have obtained a congruence factorization of $f_i^{(k)}(t)$ into a product of distinct irreducible polynomials the lifting procedure becomes much easier in general. It is clear that it suffices to show how to lift congruences for a product of two polynomials. Hence, for $h(t) \in \mathbb{Z}[t]$ we assume

$$h(t) \equiv h_1^{(\alpha)}(t) h_2^{(\alpha)}(t) \bmod p^{\alpha}\mathbb{Z}[t] \tag{2.19}$$

for all $\alpha = 1, \dots \delta$, $\delta \geq 1$ sufficiently large, subject to

$$h_i^{(\beta)}(t) \equiv h_i^{(\alpha)}(t) \bmod p^\alpha \mathbb{Z}[t] \quad (1 \leq \alpha \leq \beta \leq \delta\,,\; i = 1, 2). \tag{2.20}$$

For defining a sufficient size of δ we introduce the following definition.

Definition. *For polynomials $g(t), h(t) \in \mathbb{Z}[t]$ the reduced resultant $res_r(g(t), h(t))$ is defined as the positive generator of the ideal $\big(\mathbb{Z}[t]g(t) + \mathbb{Z}[t]h(t)\big) \cap \mathbb{Z}$.*

We note that the reduced resultant is in general much smaller than the ordinary resultant. For example, we have $\mathrm{res}_r\,(t^2+1, t^2+3) = 2 < 4 = \mathrm{res}\,(t^2+1, t^2+3)$.

Lemma 2.1. *The congruence factorization (2.1) can be lifted to all powers p^α for $\alpha \geq \delta$ provided that $\nu_p\big(res_r\,(h_1^{(\delta)}(t), h_2^{(\delta)}(t))\big) < \delta/2$.*

Proof. The proof is by induction on α. The initial value is $\alpha = \delta$. The induction hypothesis becomes

$$\begin{aligned} f(t) &= h_1^{(\alpha)}(t)\, h_2^{(\alpha)}(t) + p^\alpha r^{(\alpha)}(t), \\ \nu_p\big(\mathrm{res}_r(h_1^{(\alpha)}(t), h_2^{(\alpha)}(t))\big) &= \kappa < \delta/2, \\ h_i^{(\alpha)}(t) &= h_i^{(\delta)} + p^{\delta-\kappa} g_i^{(\alpha)}(t) \ (i = 1, 2), \end{aligned}$$

with initial polynomials $g_i^{(\delta)}(t) = 0$. For the lifting procedure we set

$$h_i^{(\alpha+1)}(t) = h_i^{(\alpha)}(t) + p^{\alpha-\kappa} g_i^{(\alpha+1)}(t) \quad (i = 1, 2)$$

and need to specify the polynomials $g_i^{(\alpha+1)}(t) \in \mathbb{Z}[t]$ appropriately. Let $X(t)$, $Y(t) \in \mathbb{Z}[t]$ subject to $X(t)h_1^{(\alpha+1)}(t) + Y(t)h_2^{(\alpha+1)}(t) \in \mathbb{Z}$. Then we obtain

$$\nu_p\big(Xh_1^{(\alpha+1)} + Yh_2^{(\alpha+1)}\big) = \nu_p\big(Xh_1^{(\alpha)} + Yh_2^{(\alpha)} + p^{\alpha-\kappa}(Xg_1^{(\alpha+1)} + Yg_2^{(\alpha+1)})\big) = \kappa$$

because of $\alpha - \kappa > \kappa$. Finally we have

$$\begin{aligned} &f(t) - h_1^{(\alpha+1)}(t)h_2^{(\alpha+1)}(t) = \\ &p^\alpha r^{(\alpha)}(t) - p^{\alpha-\kappa}\big(h_1^{(\alpha)}(t)g_2^{(\alpha)}(t) + h_2^{(\alpha)}(t)g_1^{(\alpha+1)}(t) + p^{\alpha-\kappa}g_1^{(\alpha+1)}(t)g_2^{(\alpha+1)}(t)\,\big) \end{aligned}$$

and we must determine $g_i^{(\alpha+1)}(t)$ $(i = 1, 2)$ according to

$$p^\kappa r^{(\alpha)}(t) \equiv h_1^{(\alpha)}(t)g_2^{(\alpha+1)}(t) + h_2^{(\alpha)}(t)g_1^{(\alpha+1)}(t) \bmod p^{\kappa+1}.$$

But that can be easily done in the course of computing the reduced resultant of $h_1^{(\alpha)}(t), h_2^{(\alpha)}(t)$ because of our premises on α, δ. □

Example 2.2. Let $f(t) = t^3 - t^2 - 2t - 8$ again. Here we see that the premises of Lemma 2.1 are not satisfied for $h(t) = f_1^{(3)}(t)$ and $\delta = 2$. However, the congruence factorization of $f_1^{(3)}(t)$ modulo $4\mathbb{Z}[t]$ coincides with that modulo $8\mathbb{Z}[t]$. Hence, we can as well assume $\delta = 3$ and then Lemma 2.1 already gives the result of Ore.

We remark that for linear polynomials the reduced resultant coincides with the resultant so that we can not expect an advantage in that case.

As in the example above also in other cases a congruence factorization of $f(t)$ modulo $p\mathbb{Z}[t]$ is sufficient for determining all prime ideals lying above po_F by the methods of section 3. However, the number of those prime ideals (and thus a criterion for finishing the procedure) can in general only be obtained from congruences modulo higher powers of p. Hence, for the subsequent section we assume that we have already computed a congruence factorization

$$f(t) \equiv \prod_{i=1}^{g} \prod_{j=1}^{g_i} h_{ij}(t) \bmod p^{\lambda}\mathbb{Z}[t] \tag{2.21}$$

subject to

$$\prod_{j=1}^{g_i} h_{ij}(t) \equiv f_i(t)^{e_i} \bmod p\mathbb{Z}[t] \tag{2.22}$$

for a sufficiently large exponent λ.

3 p–radical and divisor cascading

Definition. *Let Λ be an order of F and p be a prime number. Then*

$$J_p(\Lambda) := \{ \alpha \in \Lambda \mid \exists m \in \mathbb{Z}^{>0} : \alpha^m \in p\Lambda \}$$

is called the p-radical of Λ.

Lemma 3.1. *$J_p(\Lambda)$ is the intersection of all maximal ideals of Λ containing p.*

Proof. Clearly, $J_p(\Lambda)$ is contained in each maximal ideal $\mathbf{m}$ of Λ satisfying $p \in \mathbf{m}$.

On the other hand, let α be an element of the intersection of all maximal ideals of Λ which contain p. Then α is also contained in their product, say $\mathbf{a}$. But there is an exponent $m \in \mathbb{Z}^{>0}$ satisfying $p\Lambda \supseteq \mathbf{a}^m$. We note that we can stipulate $m \le n$ in the sequel. □

Lemma 3.2. *Let $\varphi : \Lambda/p\Lambda \to \Lambda/p\Lambda\ ;\ x \mapsto x^{p^k}$, where $k \in \mathbb{Z}^{>0}$ is chosen subject to $p^{k-1} < n \le p^k$. Then the kernel of φ is just $J_p(\Lambda)/p\Lambda$.*

Proof. Because of $ker\ \varphi = \{ \alpha + p\Lambda \mid \alpha^{p^k} \in p\Lambda \}$ the lemma is an easy consequence of the remark at the end of the proof of the preceding lemma. □

The last lemma makes it easy to compute $J_p(\Lambda)$. Let $\omega_1, \dots, \omega_n$ be a $\mathbb{Z}$–basis of Λ. (We note that we want to apply this method only in case $\Lambda = o_F$ so this notation is compatible to that in section 1.) Then $J_p(\Lambda)/p\Lambda$ can be represented by the residue classes $\alpha + p\Lambda$ with $\alpha = a_1\omega_1 + \ldots + a_n\omega_n$ $(0 \le a_i < p\ ,\ 1 \le i \le n)$. Since the map φ is linear it suffices to compute elements

$$\widetilde{\omega}_i = \sum_{j=1}^{n} m_{ij}\omega_j \quad (m_{ij} \in \mathbb{Z},\ 0 \le m_{ij} < p)$$

subject to

$$\omega_i^{p^k} \equiv \widetilde{\omega}_i \bmod p\Lambda \ \ (1 \le i \le n).$$

Then we obtain

$$\varphi(\alpha) \equiv \sum_{i=1}^{n} a_i \sum_{j=1}^{n} m_{ij}\omega_j \bmod p\Lambda,$$

and the elements α are calculated as the solutions of a homogeneous system of linear equations over the field of p elements. The computation of a $\mathbb{Z}$-basis for $J_p(\Lambda)$ from the representatives α is straightforward. Usually we will choose a basis $\tau_i = \sum_{j=1}^{n} t_{ij}\omega_j$ $(t_{ij} \in \mathbb{Z})$ such that the transition matrix $T = (t_{ij})$ is in Hermite normal form.

The p–radical $J_p(o_F)$ can now be used for determining all prime ideals $\mathbf{p}$ lying above po_F. From the previous sections we assume knowledge of congruence factorizations (1.6) and usually also (2.21),(2.22) (but compare Example 3.1). We already know from Proposition 1.1 that prime ideals belonging to elements $f_i(\rho)$ for different indices i are distinct. Hence, the following procedure needs to be carried out for each index $i = 1, \dots, g$.

Let $\mathcal{M}_i$ be the set consisting of the ideals

$$J_p(o_F) \ \text{ and } \ h_{ij}(\rho)o_F + p^{\nu_p(N(h_{ij}(\rho))}o_F \ \ (1 \le j \le g_i),$$

where $N(\ )$ denotes the norm of an element of the field F. An application of divisor cascading (see [3]) to $\mathcal{M}_i$ amounts to the computation of a multiplicative basis for $\mathcal{M}_i$. For any two ideals $\mathbf{a}, \mathbf{b} \in \mathcal{M}_i$ we calculate $\mathbf{c} = \gcd(\mathbf{a}, \mathbf{b})$. For $\mathbf{c} \ne o_F$ we remove $\mathbf{a}, \mathbf{b}$ from $\mathcal{M}_i$ and add $\mathbf{c}$ to $\mathcal{M}_i$. In that case we compute maximal exponents $k, l \in \mathbb{N}$ such that $\widetilde{\mathbf{a}} = \mathbf{a}\mathbf{c}^{-k}$, $\widetilde{\mathbf{b}} = \mathbf{b}\mathbf{c}^{-l}$ are still ideals of o_F. For $\widetilde{\mathbf{a}} \ne o_F$ (respectively $\widetilde{\mathbf{b}} \ne o_F$) we add $\widetilde{\mathbf{a}}$ (respectively $\widetilde{\mathbf{b}}$) to $\mathcal{M}_i$. This procedure stops if any two elements of $\mathcal{M}_i$ are coprime. From Theorem 2.1 we conclude that the elements of $\mathcal{M}_i$ are then prime ideals. Hence, we have proved the following proposition.

Proposition 3.1. *Let $\mathcal{M}$ be the union of the sets $\mathcal{M}_i$ $(1 \le i \le g)$ obtained after applying divisor cascading to the original sets. Then $\mathcal{M}$ consists exactly of the prime ideals lying above po_F.*

For explicit calculations with ideals of an algebraic number field we refer the reader to [3],[4].

We conclude with two examples to illustrate our methods. The first one is our standard example which already occured several times throughout the text.

Example 3.1. Let $f(t) = t^3 - t^2 - 2t - 8$ again. We easily calculate $J_2(o_F) = 2o_F$ and then $(1+\rho)o_F = \mathfrak{p}_3^3$ for $\mathfrak{p}_3 = 2o_F + (1+\rho)o_F$ as well as $\rho o_F = \mathfrak{p}_1^2\mathfrak{p}_2$ for $\mathfrak{p}_1 = 2o_F + 2^{-1}(\rho^2 - \rho)o_F$, $\mathfrak{p}_2 = 2o_F + 2^{-1}(\rho^2 - 3\rho + 1)o_F$.

The second example shows that higher congruences are indeed necessary and also that the exponent in Ore's theorem is too large in general.

Example 3.2. Let $f(t) = t^4 + 4t^2 - 13$, then the polynomial discriminant is $d(f) = -961792$ and the field discriminant $d_F = -15028 = d(f)/64$. An integral basis of $F = \mathbb{Q}(\rho)$ for a root ρ of $f(x) = 0$ is

$$\omega_1 = 1,\ \omega_2 = \rho,\ \omega_3 = (\rho^2+1)/2,\ \omega_4 = (\rho^3 + \rho^2 + \rho + 1)/4.$$

We obtain the congruence factorizations

$$\begin{aligned} f(t) &\equiv (t+1)^4 \bmod 2\mathbb{Z}[t], \\ f(t) &\equiv (t^2+1)(t^2-1) \bmod 4\mathbb{Z}[t], \\ f(t) &\equiv (t^2+1)(t^2+3) \bmod 8\mathbb{Z}[t], \end{aligned}$$

and the latter can be lifted to any 2–power according to Lemma 2.1. Hence, we have 2 prime ideals $\mathfrak{p}_1$ (of norm 2) and $\mathfrak{p}_2$ (of norm 4) lying above $2o_F$. Clearly, $(\rho+1)o_F = \mathfrak{p}_1\mathfrak{p}_2 = J_2(o_F)$ and a separation of the two prime ideals cannot be obtained from a congruence factorization modulo $2\mathbb{Z}[t]$. However, the congruence factorization modulo $8\mathbb{Z}[t]$ yields

$$\mathfrak{p}_1 = 2o_F + (\omega_2 + \omega_3)o_F\ , \qquad \mathfrak{p}_2 = 2o_F = (\omega_1 + \omega_2 + \omega_3 + \omega_4)o_F\ .$$

We note that the theorem of Ore would require a congruence factorization modulo $2^9\mathbb{Z}[t]$ instead.

References

[1] J. Buchmann and H.W. Lenstra Jr., *Computing maximal orders and factoring over* $\mathbb{Z}_p$, to appear.

[2] O. Ore, *Über den Zusammenhang zwischen den definierenden Gleichungen und der Idealtheorie in algebraischen Körpern*, Math. Ann. **96** (1927), 313–352.

[3] M. Pohst and H. Zassenhaus, "Algorithmic Algebraic Number Theory", Cambridge University Press 1989.

[4] H. Zassenhaus, *Ein Algorithmus zur Berechnung einer Minimalbasis über gegebener Ordnung*, in Funktionalanalysis, Birkhäuser Verlag 1967, 90–103.

Computation of independent units in number fields by a combination of the methods of Buchmann/Pethö and Pohst/Zassenhaus

U. Schröter

Abstract. The algorithm of Pohst and Zassenhaus [9] was one of the first for computing a full set of independent units in arbitrary algebraic number fields. One step of the algorithm is based on the choice of a suitable integer ω of the algebraic number field F which determines a transformation of the fundamental parallelotope for a fixed integral basis $\alpha_1, \ldots, \alpha_n$ of F. This transformation is crucial for the whole algorithm and in practice there are nearly always enough "good" ω to obtain a full set of independent units. However, until today there was no theoretical "good" choice of ω in order to make the algorithm efficient. In this paper such a choice of ω is given by means of an algorithm of Buchmann and Pethö [1] which itself also determines independent units.

1 Introduction

Let F be an algebraic number field of degree $n \geq 2$ over $\mathbb{Q}$. Let R be an order of F and $1 \leq r = s + t - 1$ be the unit rank of R. Let $\alpha_1, \ldots, \alpha_n$ be a $\mathbb{Z}$–basis of R with first element $\alpha_1 = 1$ and let $\alpha^{(i)}$ denote the i–th conjugate and $N(\alpha)$ the norm of an element $\alpha \in R$.

Pohst and Zassenhaus [9] gave an algorithm which produces many elements in R of bounded norm and thus many units which are then tested for independency. In practice this algorithm usually produces a full set of independent units but there is one step in which one has to choose suitable elements $\omega \in R$ which precede each new cycle of the algorithm. Up to now the choice of these elements was not determined in a satisfactory way.

In this paper we suggest a better choice of ω which is closely related to Dirichlets approximation theorem: Let $\gamma_1, \ldots, \gamma_n \in R$, $n \geq 2$. Then there exist integers $x_1, \ldots, x_n$ not all zero such that for any $Q \in \mathbb{R}$, $Q > 1$,

$$| x_k | \leq Q, \quad k = 2, \ldots, n, \qquad | \sum \gamma_k x_k | \leq | \gamma_1 | Q^{-(n-1)}. \tag{1.1}$$

Computational Number Theory

On this basis we develop an algorithm which produces independent units by solving an approximation problem similar to the Dirichlet approximation theorem.

2 The algorithm of Pohst/Zassenhaus

The algorithm of Pohst and Zassenhaus for the computation of independent units proceeds as follows.

Consider the bijective mapping

$$\varphi : \quad R \to \mathbb{Z}^n; \quad m_1\alpha_1 + \ldots + m_n\alpha_n \mapsto (m_1, \ldots, m_n)^t \tag{2.1}$$

for a fixed $\mathbb{Z}$-basis $\alpha_1, \ldots, \alpha_n$ of R. This defines an n–dimensional lattice of $\mathbb{Z}^n$ and the fundamental parallelotope

$$\Pi := \left\{ x = (x_1, \ldots, x_n)^t \in \mathbb{R}^n \mid -1 \leq x_j \leq 1, \ 1 \leq j \leq n \right\} \tag{2.2}$$

obviously contains non–trivial lattice points and satisfies the premises of Minkowski's Convex Body Theorem.

$$B := \max\Big\{ \prod_{i=1}^{n} \mid y_1{\alpha_1}^{(i)} + \ldots + y_n{\alpha_n}^{(i)} \mid \text{ for } y \in \Pi \Big\} \tag{2.3}$$

is an upper bound for $\mid N(\varphi^{-1}(x)) \mid$ for all lattice points $x \in \mathbb{Z}^n \cap \Pi$.

Lemma 2.1. *Let $\omega \in R$ and $M_\omega \in \mathbb{Z}^{n \times n}$ its right representation matrix defined by*

$$(\alpha_1, \ldots, \alpha_n)\omega = (\alpha_1, \ldots, \alpha_n)M_\omega \tag{2.4}$$

Then $\pi_\omega : x \mapsto \mid N(\omega) \mid^{-\frac{1}{n}} M_\omega x$ transforms Π such that

(i) $\Pi(\omega) := \pi_\omega(\Pi) = \Big\{ \mid N(\omega) \mid^{-\frac{1}{n}} M_\omega x \in \mathbb{R}^n \Big| -1 \leq x_j \leq 1, \ 1 \leq j \leq n \Big\}$
still satisfies the premises of Minkowski's Convex Body Theorem.

(ii) *The norms of elements of $\varphi^{-1}(\Pi(\omega) \cap \mathbb{Z}^n)$ stay bounded by B.*

Remark 2.1. The elements of $\Pi(\omega) \cap \mathbb{Z}^n$ are easy to compute. Pohst and Zassenhaus [10] give an efficient algorithm to which we will refer as algorithm PZ.

As the elements of $\varphi^{-1}(\Pi(\omega) \cap \mathbb{Z}^n)$ stay of bounded norm (it is well known that there are only finitely many non–associate such elements in R) one can find units directly or by dividing associate elements of small norm, which will be found by an application of algorithm PZ for several ω.

Let $(\omega_k)_{k \in \mathbb{N}}$ be the sequence of these elements. It is easy to see that it is useless to choose $\omega_k \in \mathbb{Z}$ because of $\Pi(\omega_k) = \Pi$. Practical experience [8] showed that it is useful to choose ω_k of small absolute norm, but there is no good algorithm for finding them. This was also mentioned by Pohst and Zassenhaus [10].

Weiler selected elements ω_k of small norm which came out of his enumeration of parallelotopes. In consecutive steps he took small powers of these elements. This choice provided reasonably good results.

It is obvious that there are choices of ω_k which will produce a full set of independent units after a finite number of steps. Such a choice will be called *a valid choice* in the sequel.

3 A theoretical choice of ω

Lemma 3.1. *If $\omega' = k\omega$ with $k \in \mathbb{Z}$ then $\Pi(\omega') = \Pi(\omega)$ for the transformation of Lemma 2.1.*

Lemma 3.2. *If $|N(\omega)| = 1$ then $\omega \in \Pi(\omega)$ if the first basis element of R is chosen as 1.*

Remark 3.1. It is also possible to show, that if the first basis element $\alpha_1 \neq 1$ then for each unit ε in R there is an $\omega \in R$ with $\varepsilon \in \Pi(\omega)$.

These Lemmata show that for our sequence ω_k it is only useful to choose new elements ω_k in such a way that they are not $\mathbb{Q}$–collinear to ω_j $(j < k)$ in order to find new lattice points in $\Pi(\omega_k)$.

We are sure to find a unit if we can construct such a sequence of $(\omega_k)_{k\in\mathbb{N}}$, which are not $\mathbb{Q}$–collinear to previous ones and end up in some ω_{k_0} with $|N(\omega_{k_0})| = 1$ for an index $k_0 \in \mathbb{N}$.

Then either we will find the unit ω_{k_0} in $\Pi(\omega_{k_0})$ or other units by enumerating the parallelotopes $\Pi(\omega_j)$ for $1 \leq j < k_0$.

Now it is interesting to recall how Dirichlet proved his famous unit theorem by means of his diophantine approximation theorem. The proof can be found in Dedekind's classical book [2] and essentially works as follows.

- Construct a sequence $(\eta_k)_{k\in\mathbb{N}}$ of integer elements $\eta_k \in F$ with

  $$|\eta_{k+1}{}^{(i)}| < |\eta_k{}^{(i)}|, \ |\eta_{k+1}{}^{(j)}| > |\eta_k{}^{(j)}| \text{ for } j \in \{1, \ldots, s+t\},\ j \neq i \tag{3.1}$$

  for $k \geq 2$ and fixed conjugate direction $i \in \{1, \ldots, s+t\}$.

- Show that among the quotients of these η_k there is a unit ε with the property

  $$|\varepsilon^{(i)}| < 1, \quad |\varepsilon^{(j)}| > 1 \text{ for } j \in \{1, \ldots, s+t\},\ j \neq i. \tag{3.2}$$

- Show that there is a full set of independent units contained in the set of units found by applying step 1 for all $i \in \{1, \ldots, s+t\}$.

Using these results we can prove the following theorem.

Theorem 3.1. *Let η_k be the approximations as used in the proof above for a fixed conjugate direction i. Let*

$$\begin{array}{llllll} Q_1 & := & \{ \eta_1 \}, & A_1 & := & \{\eta_1\} \\ Q_k & := & \{ q_{k,1}, \ldots, q_{k,k} \} & & := & \left\{ \frac{\eta_1}{\eta_k}, \ldots, \frac{\eta_{k-1}}{\eta_k}, \eta_k \right\} \quad (k \geq 2) \\ A_k & := & \{ a_{k,1}, \ldots, a_{k,k} \} & & := & \{ d_{k,1} \cdot q_{k,1}, \ldots, d_{k,k} \cdot q_{k,k} \} \end{array}$$

where $d_{k,j}$ is the common denominator of the coefficients of $q_{k,j}$, $j \in \{1, \ldots, k\}$.

Let N_k the smallest absolute norm of the elements of A_k and $\omega_k := a_{k,j_0}$ the element of A_k with $| N(a_{k,j_0}) | = N_k$ and smallest index j_0. Then $(\omega_k)_{k \in \mathbb{N}}$ is a valid choice for algorithm PZ.

Proof. For a fixed conjugate direction i the elements of A_k are algebraic integers and hence candidates for a choice of ω. They are multiples of quotients of approximations of (1.1) This means that there are units among them according to the proof of Dirichlet's unit theorem. These units have property (3.2). They are also contained in the transformed parallelotope and thus are found by algorithm PZ. If i runs from 1 to $s + t$ algorithm PZ therefore terminates with a full set of independent units. □

Remark 3.2. Unfortunately there exists no general practical method for solving the approximation problem (1.1) for $n > 2$. Only for $n = 2$ we know that the convergents of the continued fraction expansion of α_1/α_2 solve (1.1).

Up to now we presented a theoretical way of choosing ω_i for the algorithm of Pohst and Zassenhaus. The next section also provides a practical solution.

4 The algorithm of Buchmann / Pethö

Buchmann and Pethö [1] give an algorithm for the construction of independent units which uses the fact that the basis reduction algorithm of Lenstra, Lenstra and Lovàsz [5] – in the following LLL–algorithm – solves very fast the approximation problem

$$| x_k | \leq 2^{\frac{n}{4}} \cdot Q, \quad k = 2, \ldots, n, \qquad | \sum x_k \alpha_k | \leq | \alpha_1 | Q^{-(n-1)} \tag{4.1}$$

which is slightly weaker than (1.1).

Lemma 4.1. *There is a choice of algebraic numbers $(\beta_k)_{k \in \mathbb{N}} \in F$ with property (3.2) for a fixed conjugate direction i such that after a finite number of steps k_2*

there is an index $k_1 < k_2$ *such that*

$$\varepsilon_i = \prod_{l=k_1}^{k_2-1} \beta_l \tag{4.2}$$

is a unit which also satisfies property (3.2).

The algorithm of Buchmann and Pethö solves this problem. Buchmann and Pethö produce a sequence $(\beta_k)_{k\in\mathbb{N}}$ of algebraic numbers in F with property (3.2) for which the conjugates are also absolutely bounded from above and finally a nontrivial unit ε_i for any fixed conjugate direction i. We will refer to the algorithm as algorithm BP.

5 The new algorithm

Algorithm 5.1.

Input: *A* $\mathbb{Z}$*-basis* $\alpha_1, \ldots, \alpha_n$ *of the order* R *with* $\alpha_1 = 1$.
Output: *A system of independent units* $\varepsilon_1, \ldots, \varepsilon_{s+t-1}$ *in* R.

(1) *Initialization: Apply LLL–Reduction to the conjugates of the basis elements* $\alpha_1, \ldots, \alpha_n$. *Apply algorithm PZ to the fundamental parallelotope* $\Pi(1)$.

$$U \leftarrow \Big\{ \varepsilon \in \varphi^{-1}(\Pi(1) \cap \mathbb{Z}^n) \;\Big|\; |\, N(\varepsilon)\,| = 1 \Big\}.$$

(2) *For every conjugate direction* $i \in \{1, \ldots, s+t-1\}$ *repeat:*
 a) *Initialization:* $\beta_1 \leftarrow 1,\ k \leftarrow 2$.
 b) *Compute* β_k *by algorithm BP.*
 c) *Let* $P_k := \{ \prod_{j=i}^{k} \beta_j,\ i \in \{1, \ldots, k\} \}$ *and* ω_k *be definded as in Theorem 3.1 with* P_k *instead of* Q_k.
 d) *Compute the transformed fundamental parallelotope* $\Pi(\omega_k)$ *and enumerate it by algorithm PZ.*

$$U \leftarrow U \cup \Big\{ \varepsilon \in \varphi^{-1}(\Pi(\omega_k) \cap \mathbb{Z}^n) \;\Big|\; |\, N(\varepsilon)\,| = 1 \Big\}.$$

 e) *If* $|\, N(\omega_k)\,| \neq 1$, $k \leftarrow k+1$, *go to b).*
 f) *Next* i.

(3) *A full system of independent units is found by computing a generating set* $\{\varepsilon_1, \ldots, \varepsilon_{s+t-1}\}$ *for the set* U.

As in the theoretical considerations the algorithm will stop for any fixed conjugate direction i when algorithm BP will stop with a unit ε_i that has property (3.2) with $\varepsilon_i^{(i)} < 1$.

It is well known that a set of units of cardinality $s+t-1$ of this kind is a maximal system of independent units in R.

So the algorithm will certainly stop and produce a full set of independent units. The set U can be expected to contain more than $s+t-1$ units because of the enumeration process.

6 Numerical experiences

The algorithm has been implemented on an APOLLO Workstation using the number theoretical subroutine package KANT at the Heinrich-Heine-University of Düsseldorf. The first results show that there are different effects for different kinds of number fields:

In some totally complex examples that have been computed the first unit found in each direction was the unit produced by the algorithm of Buchmann and Pethö. The enumeration of parallelotopes did not yield any new units and the norms of the computed algebraic numbers were pairwise different. So there was no chance of finding more units by division of associate elements (The classical way of Pohst/Zassenhaus). Refined methods of treating those numbers of bounded norm are being developed.

In some totally real examples there were lots of units in the transformed parallelotopes. In many cases a full set of independent units was already found after considering only a few parallelotopes. In the following examples the given units were found already in the first transformed parallelotope.

Example 6.1. Coefficients of the generating polynomial (leading first):

$$1,\ 2,\ -7,\ -8,\ 15,\ 8,\ -9,\ -2,\ 1.$$

Coefficients of LLL–basis with respect to the power basis:

$$\begin{array}{llrrrrrrrr}
\alpha_1 & = & 1 & 0 & 0 & 0 & 0 & 0 & 0 & 0 \\
\alpha_2 & = & 0 & 1 & 0 & 0 & 0 & 0 & 0 & 0 \\
\alpha_3 & = & 1 & 4 & -5 & -11 & 0 & 4 & 1 & 0 \\
\alpha_4 & = & 1 & -2 & -3 & 8 & 3 & -3 & -3 & 0 \\
\alpha_5 & = & 0 & -6 & 0 & 12 & -5 & -6 & 2 & 1 \\
\alpha_6 & = & -1 & -1 & -4 & 5 & 11 & 0 & -4 & -1 \\
\alpha_7 & = & 4 & -1 & -18 & 1 & 23 & 4 & -7 & -2 \\
\alpha_8 & = & 1 & 2 & -14 & -7 & 20 & 7 & -6 & -2
\end{array}$$

Coefficients of independent units with respect to $\alpha_1, \ldots, \alpha_8$:

$$\begin{array}{llrrrrrrrr}
\varepsilon_1 & = & 1 & -1 & -1 & 0 & 0 & -1 & 0 & 0 \\
\varepsilon_2 & = & -3 & 0 & 0 & 2 & -3 & -1 & -2 & 2 \\
\varepsilon_3 & = & -2 & -1 & -1 & 1 & -1 & -1 & 0 & 1 \\
\varepsilon_4 & = & -1 & 0 & 0 & 0 & -1 & -1 & 0 & 0 \\
\varepsilon_5 & = & 0 & 0 & 0 & -2 & 1 & 1 & 2 & -1 \\
\varepsilon_6 & = & 0 & 0 & -1 & -1 & 0 & 0 & 0 & 1 \\
\varepsilon_7 & = & -2 & 1 & 1 & 0 & -2 & 2 & 0 & 2
\end{array}$$

Regulator of these units: 2873.1993 ,
Regulator of fundamental units: 22.4469 .

Example 6.2. Coefficients of the generating polynomial (leading first):

$$1,\ 3,\ -5,\ -21,\ -3,\ 35,\ 28,\ 4,\ -1.$$

Coefficients of LLL–basis with respect to the power basis:

α_1	=	1	0	0	0	0	0	0	0
α_2	=	−1	16	40	4	−22	−6	3	1
α_3	=	−3	4	25	11	−14	−7	2	1
α_4	=	5	−19	−65	−15	36	13	−5	−2
α_5	=	0	9	12	−6	−7	1	1	0
α_6	=	0	−4	−24	−11	14	7	−2	−1
α_7	=	−3	0	25	12	−14	−7	2	1
α_8	=	−1	19	49	15	−28	−13	4	2

Coefficients of independent units with respect to $\alpha_1, \ldots, \alpha_8$:

ε_1	=	1	−1	0	0	0	0	0
ε_2	=	0	0	−1	0	0	0	0
ε_3	=	1	0	0	1	1	0	−1
ε_4	=	0	1	−1	0	1	1	0
ε_5	=	−1	1	1	0	0	0	0
ε_6	=	−1	1	0	0	0	1	0
ε_7	=	−1	0	0	−1	0	0	1

Regulator of these units: 47.3935 ,
Regulator of fundamental units: 23.6968 .

7 Remarks

The algorithm provides a full set of independent units by the method of Pohst/Zassenhaus even if the elements of norm $\neq 1$ which are found in the transformed parallelotopes are not used to produce new units by dividing elements of small norm as in [8] (Pohst/Weiler/Zassenhaus).

On the one hand the algorithm gives a valid choice of the ω_k for the algorithm of Pohst/Zassenhaus which was not known until now (see [10]). On the other hand it will produce more units than the algorithm of Buchmann/Pethö so that the index of the generated subgroup in the whole unit group can be expected to be smaller than by an application of Buchmann/Pethö alone which only produces $s+t$ units.

Step 3 in the new algorithm can be carried out by using a modification of the LLL–algorithm by Pohst [6] which produces a basis for a lattice from a system of generators.

In order to obtain more than $s+t$ units from algorithm BP it is advisable to construct elements $\eta \in F$ with the following property rather than with property (3.2):

$$\begin{aligned} | \eta^{(i)} | &< 1, \quad (i \in S_1) \\ | \eta^{(j)} | &> 1, \quad (j \in S_2) \end{aligned}$$

where $S_1 \cup S_2 = \{1, \ldots, s+t\}$ and $S_1 \cap S_2 = \{\}$. This was done by B.v.Geul [4].

References

[1] J. Buchmann and A. Pethö, *Computation of independent units in number fields by Dirichlet's method*, Math. Comp. **52** (1989), 149–159.

[2] R. Dedekind, "Über die Theorie der ganzen algebraischen Zahlen", Vieweg, 1964.

[3] G. Lejeune Dirichlet, *Zur Theorie der complexen Einheiten*, Bericht über die Verhandlungen der Königl. Preuss. Akademie der Wissenschaften, 1846, pp. 103–107.

[4] B. v. Geul, *Einheitenberechnung nach einer Methode von Buchmann und Pethö*, Diplomarbeit, Heinrich-Heine-Universität Düsseldorf (1990).

[5] A.K. Lenstra, H.W. Lenstra Jr. and L. Lovasz, *Factoring polynomials with rational coefficients*, Math. Ann. **261** (1982), 515–534.

[6] M. Pohst, *A modification of the LLL reduction algorithm*, J. Symbolic Computation **4** (1987), 123–127.

[7] M. Pohst, *Some aspects of computational algebraic number theory*, to appear in the proceedings of the Malcev conference on algebra, Novosibirsk, 1987.

[8] M. Pohst, P. Weiler and H. Zassenhaus, *On effective computation of fundamental units II*, Math. Comp. **38** (1982), 293–329.

[9] M. Pohst and H. Zassenhaus, *On effective computation of fundamental units I*, Math. Comp. **38** (1982), 275–291.

[10] M. Pohst and H. Zassenhaus, "Algorithmic Algebraic Number Theory", Cambridge University Press 1989.

Hecke actions on classes of ternary quadratic forms

B.J. Birch

For a fixed discriminant D, let us write $\mathcal{C}(D)$ for the set of classes of positive definite ternary quadratic forms with integral coefficients and discriminant D, and $\mathcal{M}(D)$ for the free module with basis the elements of $\mathcal{C}(D)$. In this note, I will define the action of a Hecke algebra on the module $\mathcal{M}(D)$; this Hecke action is very easy to compute and is related to other Hecke actions which are already familiar in the literature.

The construction is very similar to the well-known construction of Mestre and Oesterlé [13]: they consider the free module on the set $\mathcal{S}(N)$ of supersingular elliptic curves of characteristic N, where N has to be prime, and we simply replace $\mathcal{S}$ by $\mathcal{C}$. The Hecke module of Mestre-Oesterlé is essentially the same as the Hecke action on the space $S_2(\Gamma_0(N))$ of cusp forms of weight 2 on $\Gamma_0(N)$, but is enormously easier to calculate, and maybe it is a more informative model. Unfortunately, though their method for describing the Hecke actions on $S_2(\Gamma_0(N))$ works in theory for all levels pN_1 with p prime and N_1 an integer prime to p, it is a rapid practical method of computation only when N_1 is small and fixed (typically 1, 2 or 3, certainly no more than 10); my hope was to find a model for the Hecke action which would work as well for all levels N.

It turns out that the Hecke action on $\mathcal{M}(D)$ does indeed mimic an action on cusp forms of level D, at least when D is squarefree; unhappily, it only gives about half the Hecke action on $S_2(\Gamma_0(D))$, essentially only the $-$-part. (The mapping $z \mapsto -1/Nz$ induces a standard involution on S_2; we say a form f is in S_2^- if the differential $f(z)dz$ is *not* fixed by the involution, in this case the associated L-function has a $+$ sign in its functional equation.) This deficiency seems to be inevitable for at least two reasons: the naïve reason is that a ternary quadratic form simply does not contain enough information, and a more detailed reason is that our action of T_p on $\mathcal{M}(D)$ leads most naturally to an action of T_{p^2} on forms of weight $\frac{3}{2}$, and these only lift to $S_2^-(\Gamma_0(D))$. So this attempt to find a computational method as satisfying as that of Mestre-Oesterlé that works for general level N, has so far failed in two ways : first, it usually gives only half the information needed, and, second, when the level is not square-free it gives even less information. At least the program is very fast!

Computational Number Theory

The layout of the paper is as follows:
Section 1. *Definition of the action.* Notations are set up, the classical theory of definite integral ternary quadratic forms is recalled, and the Hecke action on $\mathcal{M}(D)$ is defined. The crucial steps are Lemma 1.4, which is utterly elementary, and Lemma 1.5.
Section 2. *Computation.* A program for calculating the Hecke action is described.
Section 3. *Relation with modular forms.* In this section, I relate the action defined in section 1 to other work, particularly work relating to Hecke actions on modular forms; proofs are not given.
Section 4. *Calculations.* Here, I describe the sort of information the program gives, and continue to give applications to the arithmetic of elliptic curves and of modular forms. In particular, I describe some interesting things that happen when there are elliptic curves with Mordell-Weil rank 2; the relation (4.2.8) seems particularly significant.
In section 5, I indicate some possible extensions of the methods, and in section 6 I give a short table.

This Hecke action provided the material for talks I gave in Luminy and Oberwolfach in 1988, and again in Luminy in 1989; I thank my hosts at those conferences for their hospitality, and I also thank the organisers of the most enjoyable meeting at Debrecen (where in fact I talked about something else). I must also thank many friends, particularly Dick Gross and Don Zagier (but I guess I must apologise to Don for paying insufficient attention to Jacobi forms) for conversations that helped to form the 'background' sections 3 and 4.

1 Definition of the action

1.1 Standard preliminaries

Everything in this section is 'standard', but we need to fix notations and conventions; also, the standard results which we will quote include several with which I was not familiar. In what follows,

$$F(x,y,z) = ax^2 + hxy + by^2 + fyz + cz^2 + gzx \tag{1.1.1}$$

will always be a positive definite ternary quadratic form with integer coefficients a, b, c, f, g, h; it has matrix

$$\begin{pmatrix} 2a & h & g \\ h & 2b & f \\ g & f & 2c \end{pmatrix},$$

and discriminant

$$D(F) = 4abc - af^2 - bg^2 - ch^2 + fgh > 0. \tag{1.1.2}$$

I will always write $\mathcal{T}(D)$ for the set of positive definite integral ternary quadratic forms of discriminant D.

When I need to quote results, I will usually refer to Cassels' admirable book [4]; note however that my 'integral forms' are his 'integer valued forms', his 'classically integral forms' have f, g, h even and his determinant d is one quarter of my $D(F)$.

Definition. *If R is a ring, two quadratic forms F, G are* equivalent over R *if there is a unimodular transformation defined over R taking one to the other. In particular, F, G are* in the same class *if they are equivalent over $\mathbb{Z}$.*

If F and G are in the same class then they certainly have the same discriminant, and they take the same values with the same multiplicities. We may associate to any form $F \in \mathcal{T}(D)$ a θ-function

$$\theta_F(\tau) = \sum_{x,y,z \in \mathbb{Z}} exp(2\pi i \tau F(x,y,z)) \; ; \tag{1.1.3}$$

if F and G are in the same class then $\theta_F = \theta_G$. By a well-known theorem antedating Minkowski (Cassels p128), for each D there are only finitely many classes of forms in $\mathcal{T}(D)$; I will write $\mathcal{C}(D)$ for this set of classes.

In this definition, I have been superclassical. Nowadays, it is perhaps more fashionable to talk about lattices in a quadratic space. We regard $\mathbb{R}^3$ with the ordinary Euclidean distance as a quadratic space; if Λ is a lattice in $\mathbb{R}^3$ with basis e_1, e_2, e_3 then $|\sum x_i e_i|^2$ is a quadratic form. That this form should be integral-valued is a restriction on the lattice Λ; we call such lattices *integral lattices*; the set of integral lattices should not be confused with the set of sublattices of $\mathbb{Z}^n$! If Λ is such a lattice, then different choices of bases for Λ give quadratic forms in the same class; on the other hand, rotating the lattice does not alter the quadratic form. Accordingly, there is a 1-1 correspondence between classes of integral valued positive definite ternary quadratic forms and classes of integral lattices in $\mathbb{R}^3$, where we put rotationally equivalent lattices in the same class. Roughly speaking, it is usually more convenient to compute with quadratic forms, but to prove theorems for lattices, particularly when automorphisms are involved.

Definition. *Two forms $F, G \in \mathcal{T}(D)$ are* in the same genus *if they are equivalent over the p-adic integers $\mathbb{Z}_p$ for every p.*

Two forms in the same class are certainly in the same genus, so the set $\mathcal{C}(D)$ is a union of genera. It is a standard procedure to enumerate the $\mathbb{Z}_p$-equivalence classes of integral ternary forms with discriminant D, see Cassels p115 for odd p and pp117-122 for p=2. We will not usually consider the most general case. In the spirit of H.J.S. Smith [19], we say F has *type* $(D, 1)$ if it has discriminant D and for no p is its reduction modulo p a square; this is clearly a property shared by all forms of a genus. (So F is of type $(D, 1)$ when it has matrix with determinant $2D$ and coprime 2×2 minors.) I will write $\mathcal{T}^*(D)$ for the subset of $\mathcal{T}(D)$ consisting of forms of type $(D, 1)$, and $\mathcal{C}^*(D)$ for the corresponding classes; clearly, $\mathcal{T}^*(D)$ is the whole of $\mathcal{T}(D)$ if D is squarefree.

For $F \in \mathcal{T}^*(D)$, we define $c_p(F) = +1$ if $F(X) \equiv 0 (\text{mod } p)$, regarded as a projective locus over the field with p elements, has ordinary (non-singular) points,

$c_p(F) = -1$ if not. In particular, $c_p(F) = +1$ if p does not divide $D(F)$, and $c_q(F)$ is the Hasse-Minkowski invariant if q divides $D(F)$ to an odd power. It follows from the classification in Cassels that

Lemma 1.1. *Forms F_1, $F_2 \in \mathcal{T}^*(D)$ are $\mathbb{Z}_p$-equivalent if and only if $c_p(F_1) = c_p(F_2)$.*

The relation $D(F_1) = D(F_2)$ is of course implicit in the notation. As a consequence of quadratic reciprocity, we have the product formula (Lemma 1.1 on Cassels p76)

$$\prod c_q(F) = -1 \tag{1.1.4}$$

where the product is over all primes that divide $D(F)$ to an odd power. (The product formula for Minkowski-Hasse symbols, taken over 'all primes including ∞', has $+1$ on the right hand side. This is the same formula, since $c_\infty(F) = -1$ and the Minkowski-Hasse symbol is trivial for primes that divide D to an even power.) This is the only restriction on the invariants of F (c.f Cassels Theorem 1.3 on p77 and Theorem 1.2 on p129); we deduce

Lemma 1.2. *If D is not a perfect square, write r for the number of distinct primes that divide D; then $\mathcal{T}^*(D)$ is the union of just 2^{r-1} genera. If D is a square then $\mathcal{T}^*(D)$ is empty.*

In detail, the genera of $\mathcal{T}^*(D)$ are characterised by their invariants c_q for $q|D$; for q dividing D to an even power the c_q are ± 1 independently, for the primes q dividing D to an odd power there is the single restriction $\prod c_q = -1$. If D is a discriminant and $\{e\}$ is a suitable sequence of signs, one for each prime divisor of D, we may denote by $G(D, \{e\})$ the genus $G \subseteq \mathcal{T}^*(D)$ with $c_q(G) = e_q$ for q dividing D.

Later on, we will need

Lemma 1.3. *In every genus of $\mathcal{T}^*(D)$ there is a form that represents 1 integrally.*

For every form of type $(D, 1)$ represents 1 over $\mathbb{Z}_p$ for every p, and the result follows from Theorem 1.3 on p129 of Cassels.

If F is a form in $\mathcal{T}(D)$ whose reduction modulo p is a perfect square, then F is equivalent over $\mathbb{Z}$ to a form F' all of whose coefficients except a are divisible by p; and now $p^{-1}F'(px, y, z)$ is a form with determinant $p^{-1}D$ which is 'almost equivalent' to F. (Later on, in Lemma 3.1, we will refer to this process as 'flipping'.) In view of this, we will usually restrict ourselves to forms in $\mathcal{T}^*(D)$; life becomes much simpler! It might be reasonable to name the forms in $\mathcal{T}^*(D)$ 'semistable', they seem to have the similar nice properties to semistable elliptic curves over $\mathbb{Q}$.

1.2 The Hecke graph

Definition. *If p is a prime, and $F, G \in \mathcal{T}(D)$ are related by $F(x, y, pz) = G(px, y, z)$, then F, G are p-neighbours.*

If the forms F, G are p-neighbours then their classes are p-neighbours.

That is to say, classes C_1, C_2 are p-neighbours if we can find $F_1 \in C_1$, $F_2 \in C_2$ so that F_1, F_2 are p-neighbours.

Equivalently, we could define lattices Λ_1, Λ_2 to be neighbours if there is a basis e_1, e_2, e_3 for Λ_1 such that $\frac{1}{p}e_1, pe_2, e_3$ is a basis for Λ_2.

Lemma 1.4. *For a prime p not dividing D there is an algorithm that, for each class $C \in \mathcal{C}(D)$, determines precisely $p+1$ p-neighbours of C. (These neighbours are not necessarily distinct from each other or from C.)*

Let F be a form in the class C; since p does not divide D, the projective locus $S_p(F) : F(x, y, z) \equiv 0 \pmod{p}$ is a non-singular conic, so it has precisely $p+1$ points.

We will describe an algorithm that leads from a point of the conic $S_p(F)$ to a well-determined class C_1 neighbouring C, but first we will check that every class C_1 that is a p-neighbour of C 'comes from' a point of $S_p(F)$. Indeed, if C_1 is a p-neighbour of C then there are integral forms $G \in C_1$, $F' \in C$ such that $G(px, y, z) = F'(x, y, pz)$; it follows that $F'(1, 0, 0) \equiv 0\ (p)$, and since F' is equivalent to F the point $(1, 0, 0)$ of $S_p(F')$ is the same as a point of $S_p(F)$.

Now suppose that P is a point of $S_p(F)$; lift P to a primitive point with integer coordinates, then there is a unimodular integral transformation that takes P to $(1, 0, 0)$ and takes F to F' say. Then F' has its coefficient a divisible by p; since p does not divide D, it does not divide both of h, g , so (after a further transformation switching y and z if necessary) we may suppose that p does not divide g; by a further unimodular integral transformation we may force $p^2 | a$ and $p | h$, and now the form F' has an integral neighbour G given by $G(x, y, z) = F_1(p^{-1}x, y, pz)$. The p-neighbour of C corresponding to the point P of $S_p(F)$ is determined as the class of G. There are a number of choices in the algorithm, so it is necessary to check that the class of G depends only on P, F and p; but this presents no difficulty.

I have deliberately given the proof as an algorithm, since that is how I use it. It is arguable that it is clearer to talk about lattices. Suppose we start with an integral lattice Λ and a primitive point $e_1 \in \Lambda$ with $|e_1|^2$ divisible by p; extend to a basis $\{e_1, e_2, e_3\}$ of Λ, and adjust e_1 by multiples of pe_2, pe_3 so that $|e_1|^2$ is divisible by p^2. Then $\{e_2, e_3\}$ span a sublattice $\Lambda_1 \subseteq \Lambda$, and Λ_1 has an index p sublattice

$$\Lambda_1' = \{Q \ : \ p \text{ divides } |Q + e_1|^2 - |Q|^2 - |e_1|^2\} \ ;$$

$p^{-1}e_1$ and Λ_1' span an integral lattice Λ' with the same mesh as Λ; Λ' depends only on Λ and on $e_1 \bmod p\Lambda$, it is the required p-neighbour of Λ.

Given Lemma 1.4, we may define our Hecke action. As in the introduction, $\mathcal{M}(D)$ is the free module with basis $\mathcal{C}(D)$; it has an obvious submodule $\mathcal{M}^*$

spanned by $\mathcal{C}^*$. For each prime p not dividing D we define

$$T_p(C) = \sum \{p\text{--neighbours of } C\} \tag{1.2.1}$$

for each class C, and then extend T_p to $\mathcal{M}(D)$ by linearity. It is immediate from the Chinese Remainder Theorem that $T_p \circ T_q = T_q \circ T_p$ whenever p, q are distinct primes, so the algebra is commutative. (Those familiar with Eichler's book [6] will know that this is all a particular case of his theory of Anzahlmatrizen; but we are more down-to-earth.)

It is natural to describe the situation in terms of a family of *Hecke graphs*

$$\mathcal{G}(p; D) \supseteq \mathcal{G}^*(p; D) .$$

For fixed D and each prime p, $\mathcal{G}(p; D)$ is the graph whose vertices are the classes in $\mathcal{C}(D)$, with $p+1$ edges from each vertex to its p-neighbours, constructed by the algorithm of Lemma 1.2; we will usually consider only the subgraph $\mathcal{G}^*$ spanned by $\mathcal{C}^*$. Note that $\mathcal{G}$ has to be considered as a *directed* graph, since, though it is true that G is a neighbour of F if and only if F is a neighbour of G, the multiplicities need not be the same. To describe the Hecke action T_p it is enough to exhibit the graph; and we will describe a program to do this in section 2.

First, however, we will determine the connected components of $\mathcal{G}^*(p; D)$, and then we will discuss briefly what the Hecke operator should be in case $p|D$.

For each connected component $\mathcal{G}'$ there is an obvious 'trivial' eigenvector, the sum of the vertices of $\mathcal{G}'$, which corresponds to the eigenvalue $p+1$ of T_p for each p; in homage to Mazur, we will refer to this eigenvector as **Eis**, short for Eisenstein.

Lemma 1.5. *The connected components of $\mathcal{G}^*(p; D)$, for p not dividing D, correspond to the genera of $\mathcal{T}^*(D)$.*

[We will denote the subgraph of $\mathcal{G}^*(p; D)$ corresponding to the genus $G(D, \{e\})$ by $\mathcal{G}^*(p; D, \{e\})$.]

If F_1, F_2 are forms in $\mathcal{T}^*(D)$ that are p-neighbours, for some prime p, then F_1, F_2 are $\mathbb{Z}_q$-equivalent for all $q \neq p$; it follows by Lemma 1.1 and the remark above it that they are $\mathbb{Z}_q$-equivalent for all q, so they are in the same genus. [Note that we have not needed the hypothesis that p doesn't divide D for this part of the argument. We *do* need the restriction to $\mathcal{T}^*(D)$, for example $x^2 + xy + 3y^2 + 4z^2$ has $x^2 + y^2 + 11z^2$ as a 2-neighbour not in $\mathcal{T}^*$.]

To prove the converse, that forms in the same genus of $\mathcal{T}^*$ are connected by a chain of p-neighbours for each p not dividing D, we need the theory of spinor genera. Fortunately, we can quote everything we need from Cassels; the argument is even correct in the lugubrious case $p = 2$. In the first place, every $F \in \mathcal{T}(D)$ is isotropic over (has a non-trivial zero in) $\mathbb{Q}_p$ for all primes p not dividing D; so by Theorems 8.3 and 8.4 on pp 232-3 of Cassels, if $F_1, F_2 \in \mathcal{T}(D)$ are in the same spinor genus then, for each prime p not dividing D, F_1 is connected to F_2 by a chain of forms with each successive pair of forms of the chain a pair of p-neighbours. Using the arguments on pp 213-5 of Cassels, we may complete the proof by showing that there is precisely one spinor genus in each genus of $\mathcal{T}^*(D)$; we omit the details, as indeed does Cassels.

To finish off the section , we consider how to define the Hecke operator T_q when $q|D$. To cut the cackle, if F is a representative of the class C, we simply take

$$T_q(C) = c_q(F)C.$$

This operator is fairly useless, but is probably the right choice; in view of Lemma 1.5, the algebra remains commutative.

[It might seem natural to try to define a graph $\mathcal{G}^*(q, D)$, mimicking the process of Lemma 1.4, even when $q|D$. Let us consider what this would lead to. If $F \in \mathcal{T}^*(D)$ and $q|D$ then $S_q(F)$ is a pair of lines, meeting in a double point X. Since X is a singular point of the locus $F \equiv 0(\text{mod } q)$, it does not usually lift to a q-adic point of $F = 0$, so does not lead to a neighbour. As for the ordinary points of $S_q(F)$, there are two possibilities. If the two lines are defined over the finite field then between them they have $2q$ ordinary points, and F has accordingly $2q$ neighbours; so there is a well defined operator, which we will denote by $T_q^!$. This operator seems mysterious, but it is not necessarily useless, since it has the 'correct' eigenvectors and it may, for instance, be easier to calculate the eigenvectors of $T_2^!$ than of T_7 when D is divisible by 30. If the two lines are not defined over the finite field then they have no ordinary points, and we have given a complicated definition of the zero operator.]

1.3 The size of the genera

We would like to know the size of the connected subgraphs of $\mathcal{G}^*(D)$, which we have shown correspond to the genera contained in $\mathcal{T}^*(D)$. The number of vertices of $\mathcal{G}^*(D)$ is the number of classes in the genus, but this class number is not the natural measure of the size of a genus, and is decidedly unpleasant to calculate. There is a more natural quantity associated with a genus, namely its *weight*. For the weight there is a very beautiful formula which was stated by Eisenstein and proved by H.J.S Smith, but which nowadays is usually associated with the name of Siegel.

If F is a quadratic form, we write $A(F)$ for its group of proper automorphisms, that is to say the group of integral transformations of determinant +1 that take F to itself, and write $|A(F)|$ for its order; this is the same as the subgroup of the rotation group that takes the corresponding lattice to itself. The *weight* $w(F)$ is defined as the inverse of $|A(F)|$. Equivalent forms have isomorphic automorphism groups, so if C is a class of forms the weight $w(C)$ is well-defined. If G is a genus, we define its weight to be

$$w(G) = \sum_{C \subseteq G} w(C) . \tag{1.3.1}$$

Lemma 1.6. *If $G(D,\{e\}) \subseteq \mathcal{T}^*(D)$ is the genus with discriminant D and invariants e_q for q dividing D then*

$$w(G) = \frac{D}{12}\prod_{q|D}\frac{q+e_q(G)}{2q}\ , \tag{1.3.2}$$

where the product is taken over the distinct primes dividing D.

This formula can be retrieved from Smith's classic paper on ternary quadratic forms [19]; it is a particular case of the formula on p376 of Cassels. It may help to give a few examples; for $D = 2;3;5;6;7;8$ the possible values of $w(G)$ are $\frac{1}{24};\frac{1}{12};\frac{1}{6};\frac{1}{12},\frac{1}{8};\frac{1}{4};\frac{1}{6}$; the corresponding genera all happen to have class number 1, the relevant classes being represented by forms with $(a,b,c,h,g,f) = (1,1,1,1,0,1)$; $(1,1,1,1,0,0)$; $(1,1,2,1,0,1)$; $(1,1,2,1,0,0)$; $(1,1,2,0,1,1)$; $(1,1,2,0,0,1)$; $(1,1,3,1,0,1)$.

The weight of the vertices is an essential part of the structure of the graphs $\mathcal{G}^*(p;D)$; indeed, we have

Lemma 1.7. *Suppose that C, C' are two classes of $\mathcal{T}^*(D)$ that are p-neighbours, for a prime p not dividing D. Suppose that there are $e(C,C')$ edges of $\mathcal{G}^*$ from C to C', and $e(C',C)$ from C' to C. Then*

$$w(C)e(C,C') = w(C')e(C',C). \tag{1.3.3}$$

This is an easy exercise, so long as one thinks about lattices rather than quadratic forms!

It is clear from Lemma 1.7 that the total weight of $\mathcal{G}^*(p;D,\{e\})$ is at least as significant an invariant of the graph as the cardinality of its vertex set. None the less, it may happen that one wants the class number rather than the weight. In principle, the calculation is easy enough: one uses $w(G)$ as a first approximation — it would be correct if all forms in the genus had trivial proper automorphism group — and then adjusts to take account of forms with extra automorphs. In practice, the calculation is perfectly horrible, and we do not describe it.

However, it is certainly useful to know the forms in $\mathcal{T}^*(D)$ with extra automorphisms. To find about these, we may again look in Cassels; the appropriate place is Example 15 on pp165-6, but this time he is less helpful. It is more illuminating to think geometrically, about the possible subgroups of the rotation group that may leave an integral lattice fixed. Such subgroups are necessarily finite, and we know all the finite subgroups of the rotation group, so it is really no more than an exercise to work out the possibilities; in checking the following Lemma (I hope I have got it right!), one should think about how a lattice is constrained when it is fixed by a rotation of order 2,3 or 4 about a given axis.

Lemma 1.8. *Suppose that F is a positive definite ternary integral quadratic form, let Λ be a corresponding integral lattice in R^3, and let the automorphism group be $A(F)$. Suppose that $F \in \mathcal{T}^*(D)$ for some D. Then either*

(i) $A(F)$ is the octahedral group and F is equivalent to $x^2+y^2+xz+yz+z^2$ with $D(F)=2$; or
(ii) $A(F)$ is $C(2)\times C(6)$ and F is equivalent to a form $x^2+xy+y^2+cz^2$ with $D(F)=3c$; or
(iii) $A(F)$ is $C(2)\times C(4)$ and F is equivalent to a form $x^2+y^2+xz+yz+cz^2$ with $c\geq 2$, so that $D(F)=4c-2$; or
(iv) $A(F)$ is dihedral of order 6 and F is equivalent to a form $x^2+xy+y^2+yz+cz^2$ with $c\geq 2$, so that $D(F)=3c-1$; or
(v) $A(F)$ is $C(2)\times C(2)$ and F is equivalent to a form of shape $a(x^2+xz)+b(y^2+yz)+cz^2$ or $ax^2+b(y^2+yz)+cz^2$ or $a(2x^2+2xy+2xz+yz)+by^2+cz^2$; or
(vi) $A(F)$ is $C(2)$ and F is equivalent to a form of shape $ax^2+by^2+fyz+cz^2$ or $a(x^2+xy)+by^2+fyz+cz^2$.

In cases (i) to (iv), the forms listed are all normalised; in cases (v) and (vi), they need not be. One may count equivalence classes of the forms in cases (v) and (vi) by counting classes of definite binary forms of discriminant $4D$.

Remark. The condition that $F\in\mathcal{T}^*(D)$ is not essential for the Lemma, but saves listing a few extra cases. For instance, $x^2+y^2+z^2$ is another form with octahedral automorphism group.

Problem. This is probably a silly question to ask, but an answer would be extremely welcome: Are there natural maps from $\mathcal{M}^*(N)$ to $\mathcal{M}^*(MN)$ that preserve the Hecke action?

2 Implementation of the algorithm

We need to train a computer to give a description of the graph $\mathcal{G}^*(p,D)$, for given discriminant D and prime p not dividing D.

The vertices of $\mathcal{G}^*$ are the classes in $\mathcal{C}^*(D)$; and a class C is conveniently described by a *normalised* quadratic form F belonging to it (we will give a suitable method of normalisation in a moment). So a proper description of $\mathcal{G}^*$ is a list of normalised quadratic forms, one for each $C\in\mathcal{C}(D)$, labelled as vertices $v_1,\ldots,v_{h^*}$ where h^* is the cardinal of $\mathcal{C}^*$; together with a table listing for each v_i its $(p+1)$ p-neighbours.

Given such data, we will no doubt wish to use it in some way; in particular, we will probably form the adjacency matrix of $\mathcal{G}^*$ and find its eigenvalues and eigenvectors — at least its rational ones, which should conjecturally correspond to elliptic curves. In this section, we are not concerned with this further processing; we are simply concerned with calculating a list of normalised forms and their neighbourly relations. We will usually refer to a normalised form as a *vertex*.

There are four principal subroutines. The graph $\mathcal{G}^*$ is a union of connected subgraphs, corresponding (according to Lemma 1.5) to the genera in $\mathcal{C}^*$. The routine *Startlist* gives an initial list of vertices (normalised forms!), with at least one in each connected subgraph (genus!). Next, given a 'root' $v(0)$ provided by *Startlist*, the routine *Growtree* forms the maximum connected subgraph containing $v(0)$; an essential subroutine of *Growtree* is *Neighbours* which finds the p-neighbours of a given form; and all three routines use *Normalise*, which calculates a normalised form equivalent to a given form F. I will describe these routines in reverse order.

First, *Normalise*. We need a definition of normalisation so that in each class there is precisely one normalised form; and given such a definition we need an algorithm that goes from a form F to the normalised form equivalent to it. A suitable definition, essentially the one we will use, was given long ago by Seeber and by Eisenstein (see for instance [2] p188). The idea is to define a region $\mathcal{F}$ in (a, b, c, f, g, h)-space, described by linear inequalities, so that every form is equivalent to exactly one form in $\mathcal{F}$; using jargon, $\overline{\mathcal{F}}$ with its faces appropriately identified is a moduli space for equivalence classes of definite ternary quadratic forms. It is relatively easy to define $\mathcal{F}$ so that no two forms in the interior of $\mathcal{F}$ are equivalent; the difficulty is to ensure that the equivalence classes of forms on the *faces* of $\overline{\mathcal{F}}$ are included in $\mathcal{F}$ once and only once. Because of this, there are a large number of particular cases and the program is quite long to write; on the other hand, these particular cases hardly affect the average running time, as it is very rare for a form to correspond to a vertex or edge-point of $\mathcal{F}$ and comparatively unusual for it to be represented by a point on one of the faces of $\mathcal{F}$.

The polyhedron $\overline{\mathcal{F}}$ is defined by the inequalities

$$a \le b \le c\ ;\ 0 \le h \le a\ ; 0 \le f \le b\ ;\ -a \le g \le a\ ;\ h + f - g \le a + b\ .$$

It is not hard to write a program that, given a form F, finds a form $G \in \overline{\mathcal{F}}$ equivalent to F. Starting with a form F, a Step begins by reordering the variables so that $a \le b \le c$, adjusting the signs of y, z so that $h \ge 0$ and $f \ge 0$, and clearing a flag; then we adjust h, g, f: for instance, if $h > a$ we take t as the integer part $\left[\frac{h+a}{2a}\right]$, replace b, h, f by $b - ht + at^2, h - 2ta, f - gt$, and set the flag; if $g > a$, $g < -a$, $f > b$ or $h + f - g > a + b$ we adjust c, g, f appropriately and set the flag. If the flag is set, we return to the beginning, and we repeat the Step until we can exit with the flag clear. The process certainly terminates, since a or b or c is decreased every time we set the flag, and the process terminates with a form equivalent to F in $\overline{\mathcal{F}}$. The number of Steps needed will be bounded by $\log(D(F))$.

Tidying up the faces of $\overline{\mathcal{F}}$ so as to get a region $\mathcal{F}$ with exactly one representative of each class is a self-explanatory bit of C programming containing 35 more 'if's. (The program may be available via e-mail.)

Second, *Neighbours*. We may be brief, since the algorithm has been described in Lemma 1.4. Given a form F, we find the $(p+1)$ points of $S_p(F)$ by search: this takes time proportional to $(p+1)^2$. Then for each point of $S_p(F)$ we find a neighbouring form by the algorithm of Lemma 1.4, and normalise it. The time taken by this routine for each F is roughly $c_1(p+1)^2 + c_2(p+1)$ where c_1 is a small constant (a comparison) and c_2 is a big one (*Normalisation*); so the second term is dominant.

Third, *Growtree*. Again, we may be brief. We are concerned with the problem of growing a maximal connected graph, given a 'root', a recipe for finding the neighbours of a given vertex, and a method of recognising when two vertices are the same. This is a standard problem, and it is actually a very pleasant programming exercise. In the present instance, the 'root' is provided by *Startlist* and we have a *Neighbours* routine. When we compare vertices it is important that we think of them as being normalised forms -- it is easy to tell when two forms are the same -- rather than equivalence classes; in fact the prime purpose of the *Normalisation* routine is to make this comparison simple. Testing whether two forms are equivalent is really no easier than normalising them both; normalisation is quite an expensive routine, so it is important to normalise each form once and for all, rather than renormalising each time we compare two vertices.

If we grow a subgraph with $(g+1)$ vertices, we will need to call *Neighbours* approximately g times; we also need to compare vertices a number of times. My program was inefficient in this respect, as the number of vertex comparisons increased as g^2; however, it seems that the computing time is dominated by the approximately $g(p+1)$ calls of *Normalisation* that are needed.

Finally, *Startlist.* By Lemma 1.5, the connected subgraphs of $\mathcal{G}^*$ correspond to genera, and by Lemma 1.3 each of these genera contains a form that represents 1. So it is enough to make a list of all normalised forms in $\mathcal{T}^*$ that represent 1. Such a form has $a = 1$ and $(h,g) = (0,0), (1,0), (0,1)$ or $(1,1)$, so $(4b-h)(4c-g) = 4D + (4f - hg)^2$ with of course $0 \leq f \leq b \leq c$. It is easy to calculate a list of all forms satisifying these inequalities; a crude method is to search through positive values of b, f less than $1 + \left(\frac{4D}{3}\right)^{\frac{1}{2}}$. We check that the resulting forms are normalised, and so obtain a 'Startlist' of normalised forms, at least one in each genus. When we use this to grow the graph $\mathcal{G}^*$, we need to avoid growing the same subgraph of $\mathcal{G}^*$ more than once by including a few extra orders in *Growtree* to delete vertices from our 'Startlist' as soon as they are reached as vertices of the subgraph which is being grown.

It is worth noting that the algorithm consists entirely of manipulations of small integers; if one is reasonably careful, one need never use any number larger than about pD, so even hand computation (for D up to 100 or so) is quite fast. One may implement the algorithm on any inexpensive PC; my own program was written in Turbo-C for an IBM XT-compatible. The program appears to be 'fast enough' in a rather strong sense: if all we want is the description of $\mathcal{G}^*$, the program is output dominated for $p = 2$, even when D is 20000 or so; on the other hand, any program for using our description of $\mathcal{G}^*$ (by for instance checking an eigenvalue of the adjacency matrix) tends to be enormously slower.

3 Relation with modular forms

As in the introduction, $S_2(\Gamma_0(N))$ is the vector space of cusp forms of weight 2 invariant by $\Gamma_0(N)$; by restricting ourselves to the cusp forms with rational Fourier coefficients, we may conveniently regard $S_2(\Gamma_0(N))$ as a vector space over $\mathbb{Q}$. There is a Hecke algebra acting on it, and it is an important standard problem to sort out the action of this Hecke algebra. Suppose for a moment that N is prime; then following Mestre and Oesterlé we may consider the free module $\mathcal{M}'(N)$ on the set $\mathcal{S}(N)$ of supersingular elliptic curves of characteristic N. For each prime p not N they grow a graph of valency $(p+1)$, taking p-isogeny as their definition of p-neighbourliness; they thus define a Hecke action on $\mathcal{M}'(N)$, and they show that

$$\mathcal{M}'(N) \otimes \mathbb{Q} \cong \mathbb{Q} \oplus S_2(\Gamma_0(N)) \tag{3.1}$$

as Hecke modules, where the $\mathbb{Q}$ on the right hand side corresponds to the obvious Hecke eigenvector $\sum_{\mathcal{E}\in\mathcal{S}(N)} \mathcal{E}$, with eigenvalue $p+1$ for each T_p with $p \neq N$.

Now drop the assumption that N is prime. If $\{e\}$ is a sequence of signs, one for each prime dividing N, we denote by $S_2(\Gamma_0(N), \{e\})$, or $S_2(N, \{e\})$ for short, the subspace of S_2 which is in the e_q-eigenspace of W_q for each q dividing N, where of course the W_q are the Atkin-Lehner involutions. We have been imitating the construction of Mestre-Oesterlé, in the hope of finding a similar method that works for more general conductors. So we are hoping that the Hecke module $\mathcal{M}^*(N)$ will mimic $S_2(\Gamma_0(N))$, or at least $S_2^-(\Gamma_0(N))$. Recollect that for each prime p not dividing N and for each sequence $\{e\}$ of permissible invariants $c_q(G)$ for q dividing N, $\mathcal{G}^*(p; N, \{e\})$ is the connected subgraph of $\mathcal{G}^*$ corresponding to the genus with these invariants. Let $\mathcal{M}^*(N, \{e\})$ be the corresponding Hecke submodule of $\mathcal{M}^*(N)$; then our pious hope is that

$$? \ \mathcal{M}^*(N, \{e\}) \otimes \mathbb{Q} \cong \mathbb{Q} \oplus S_2(N, \{e\}) \ ?$$

where the extra $\mathbb{Q}$ corresponds to the 'trivial' eigenvector **Eis** with all its components 1.

Certainly, this is false, even when we restrict N to be squarefree, though in that case it appears that the 'new' part is correct. We recollect that

$$S_2(N, \{e\}) \cong \bigoplus_{M|N} \{ \ \alpha \text{ copies of } S_2^{new}(M, \{f\}) \ \} \ , \tag{3.2}$$

where the sign f_p is conventionally taken as $+$ if p does not divide M. Here, the multiplicity $\alpha = \prod_{p|N} \alpha_p$, where α_p depends on e_p, f_p and the exponent r_p of p dividing N/M. Precisely, α_p is the integer part of $\frac{1}{2}(r_p + \chi_p + 1)$ where χ_p is 1 or 0 according as $e_p = f_p$ or $e_p \neq f_p$. The most important part of $S_2(N, \{e\})$ is of course the 'new' part $S_2^{new}(N, \{e\})$; the rest has already appeared at lower level.

In contrast, it appears that if $(N, \{e\})$ is *good* in the sense that

$$\prod_{p^r||N} e_p^r = \prod_{p|N} e_p = -1 \tag{3.3}$$

where in the first product the e_p are raised to an exponent the same as the exponent of p in N, while the second product is simply the product of the e_p's, then

$$\mathcal{M}^*(N, \{e\}) \otimes \mathbb{Q} \cong \mathbb{Q} \oplus \bigoplus_{M|N} \{ \ \beta \text{ copies of } S_2^{new}(M, \{e\}) \ \} \ , \tag{3.4}$$

where again the multiplicity $\beta = \prod_{p|N} \beta_p$. This time, the multiplicity β is zero unless the sign-sequence at the level M is the same as that at level N, and in this case β_p depends on e_p and r_p. Experiment and a little bit of reason suggests that if $e_p = +1$ then $\beta_p = 1 + [\frac{1}{2}r_p]$ while if $e_p = -1$ then β_p is 1 or 0 according as r_p is even or odd. Note that only *good* $(M, \{e\})$ occur on the right hand side with non-zero multiplicity. (The union of the $\mathcal{M}^*(N, \{e\})$ is very similar to the space $\mathcal{M}_2^-(N)$ of Skoruppa and Zagier [18], but I do not think it is quite the same when N is not square free.) When $(N, \{e\})$ is not good, I do not know the interpretation of $\mathcal{M}^*(N, \{e\})$.

In the rest of this section, I will give some justification for these assertions in case N is squarefree; the more general case remains a guess. We need a lemma; note that in what follows the discriminant D is usually the same as the level N.

Lemma 3.1. *Let $F(x, y, z) = ax^2 + \ldots + gzx$ be as usual an integral ternary quadratic form of discriminant D, and suppose that D is squarefree. Then there is a 'flipped' form $F^\vee(x, y, z)$ with discriminant D^2 and $c_p(F^\vee) = c_p$ for all p, and a quaternary form $G_F(t, x, y, z)$ with determinant D^2, $G_F(1, 0, 0, 0) = 1$ and $G_F(0, x, y, z) = F^\vee(x, y, z)$ such that G_F is the norm form of an order (with Eichler invariant dividing D) of a quaternion algebra ramified at ∞ and the primes with $c_p = -1$, and of appropriate level, and $F^\vee$ is the restriction of G_F to the elements whose trace is zero. The forms $G_F, G_{F'}$ are in the same class if and only if F, F' are in the same class.*

This 'flipping' transformation is the inverse of the transformation at the end of section 1.1; I think it is the same as the δ-transformation of Ponomarev [14]. The class of G_F determines the *type* but not the *class* of the quaternion order. We have thus a somewhat roundabout 1-1 correspondence between classes of ternary forms of discriminant D, and types of quaternion orders ramified at the correct primes and with the correct Eichler invariant; so our Hecke operators can be viewed as operators on types of quaternion orders. In a famous paper, Eichler [7] has shown how the Hecke algebra $S_2(\Gamma_0(N))$ can be described by means of Brandt matrices, coming from classes of ideals of quaternion orders; any trusting person will believe, and Ponomarev has proved that our action is essentially the same as that described by the Brandt matrices, except that we have lost information since our ternary form determines only the type of the quaternion order, not the class. So (3.4) really is provable!

There is another approach that we might have taken, which seems at first sight to be simpler but which turns out to be inadequate. We have already remarked that if $F \in \mathcal{T}^*(N)$ then

$$\theta_F(\tau) = \sum_{(x,y,z)\in\mathbb{Z}^3} e^{2\pi F(x,y,z)\tau} \tag{3.5}$$

is a modular form of weight 3/2. So $\theta_F \in S_{3/2}(N, \chi)$, where the level really is N because $F \in \mathcal{T}^*(N)$ rather than just $F \in \mathcal{T}(N)$, and where χ is a suitable character.

It is reasonable to suppose that our Hecke action T_p on F is essentially the same as Shimura's T_{p^2} acting on $S_{3/2}(\Gamma_0(N), \chi)$; and so, after Shimura's lift from

$S_{3/2}$ to S_2, essentially the same as the action of T_p on S_2. This seems a nice simple argument: the image of $S_{3/2}$ is only a subspace of S_2, but it is very nearly the correct W_N-eigenspace.

It seems to follow from Ponomarev's work that this argument is essentially correct, though not in the least simple (Ponomarev works with Brandt matrices, and then verifies that the Shimura lifts are correct; and the details of his calculations are horrid). Moreover, there is a major snag which means that any argument on these lines is necessarily inadequate. One wastes information (particularly interesting information, as we will see in section 4) by going from quadratic forms to theta series, since it is possible for the theta series corresponding to the set of distinct classes of a genus to be linearly dependent, so that the space spanned by the θ_F has dimension less than $dim(\mathcal{M}(N))$. Accordingly, it seems essential to use the theory of Brandt matrices, theta series aren't good enough! On the other hand, the two approaches are not far separated, since the theta series formed from G_F is the Shimura lift of θ_F.

4 Numerical results

4.1 Generalities

Our module $\mathcal{M}^*(D)$ has a very rich structure: it is a module, and it has a Hecke action. In addition, it is a quadratic space with the inner product

$$<\sum \lambda_F(F), \sum \mu_F(F)>= 2\sum w_F\lambda_F\mu_F \tag{4.1.1}$$

where of course the weight w_F is $|Aut(F)|^{-1}$; it has a distinguished basis; and, since ternary forms are rather recognisable, the graph $\mathcal{G}^*(D)$ even has distinguished vertices. As we have explained, it contains (when D is squarefree) the information contained in $S_2^-(\Gamma_0(D))$, so, according to the standard yoga explained in the introduction to Antwerp IV [2], it tells us all about $J_0^-(D)$, where $J_0(D)$ is the Jacobian of the modular curve $X_0(D)$. So of course do other Hecke modules — the homology $H_1(X_0(D))$ tells us about the whole of $J_0(D)$ — but our module $\mathcal{M}^*$ is quite a bit easier to calculate, and maybe its distinguished basis will prove a valuable advantage.

Accordingly, there is a great deal of interesting information to be calculated; since the program is very fast, it is possible for anyone who owns it to generate interesting numbers much faster than it is possible to read them. Accordingly, I will summarise the sort of results that have been found, but I will give very few tables.

According to the standard yoga, the Jacobian $J_0(D)$ is isogenous to a product of simple abelian varieties, and the splitting of $J_0^-(D)$ corresponds (more or less: not quite, because there is a difference between the right hand side of (3.2) and the right hand side of (3.4)) to the splitting of $\mathcal{M}^*(D) \otimes \mathbb{Q}$ as a sum of rational eigenspaces of the Hecke algebra. More precisely, if $G(D, \{e\})$ is a genus,

$$\mathcal{M}^*(D, \{e\}) \otimes \mathbb{Q} \cong \mathbb{Q} \oplus \bigoplus_{\rho=1}^{R} M_\rho \otimes \mathbb{Q} \ , \tag{4.1.2}$$

where each $M_\rho \otimes \mathbb{Q}$ is irreducible, and the modules M_ρ are chosen so that $\mathbf{Eis} \oplus \bigoplus M_\rho$ has minimal index in $\mathcal{M}^*$. In particular, rational eigenvectors correspond to elliptic curves over $\mathbb{Q}$ of level D, that is to say, equipped with a Weil parametrisation $\phi : X_0(D) \to E$. In general, the modules M_ρ turn out to be orders of rings of integers of totally real number fields K_ρ, and the corresponding eigenvectors come as complete sets of $K_\rho/\mathbb{Q}$ conjugates.

One's first instinct, when presented with a module with a commutative algebra of operators acting on it, is to calculate the characteristic values of T_p for suitable small p and hence calculate the corresponding primitive integral eigenvectors, particularly those defined over $\mathbb{Z}$; and considerations like the above confirm that this instinct is correct, and suggest interpretations of the numbers we get.

So far, most of the information that people have been interested in obtaining from the module $\mathcal{M}^*$, and from its relatives such as $H_1(X_0(D))$, has been in one of three areas. First, some of the most interesting information to be calculated is the splitting (4.1.2), and the determination of the fields K_ρ; this was done long ago for $H_1(X_0(D))$ by Tingley [20], for levels $D \leq 300$, and the splittings are available in Antwerp IV [2]. Information about the fields K_ρ for these levels may be extracted from the tables in Tingley's thesis, and Mestre has tables which go a lot further -- but very little about these fields has been published. (For the most part, the fields K_ρ have very small discriminant for their degree, but there are occasional surprises; the largest discriminant found for a quadratic K_ρ is 37, at level 571. There are reasons to suppose that excessively large discriminants occur only finitely often: it is an interesting problem!) Second, it is very desirable to calculate the elliptic curves corresponding to the rational eigenvectors - this is almost a separate computation, and in fact Cremona [5] has recently computed a table of all elliptic curves over $\mathbb{Q}$ with a Weil parametrisation of level at most 1000. Thirdly, there is information carried by the integer arithmetic of the primitive eigenvectors, and that is what I will describe in the next section.

Before leaving a section described as 'Generalities', I should remind the reader again about newforms and oldforms. The orthogonal splitting (4.1.1) is a refinement of (3.4), so several factors M_ρ are 'old' in the standard sense that they have already appeared at lower level, and a fair amount of the information in $\mathcal{M}^*$ is about oldforms, or about how to split oldforms from new. [An important part of Cremona's program is devoted to splitting off the old part of $H_1(X_0(D))$, so that he does not compute the same curve many times over.]

4.2 The primitive eigenvectors

In this section, I denote by $\mathbf{Eis}, u_1, ..., u_\gamma$ the primitive integral eigenvectors of the algebra $\mathcal{M}^*(D, \{e\})$, and $\lambda_{i,p}$ will denote the eigenvalue of T_p acting on u_i:

I may occasionally write u_0 instead of **Eis**, and I may enumerate the classes of the genus $G(D,\{e\})$ as $C_0,...,C_\gamma$. The components of u_i with respect to the standard basis $\{C\}$ will be $u_i(C)$, so that in particular $\mathbf{Eis}(C)$ is always 1. The u_i may be either new of level D, or old. There are a number of easy relations: the weight formula (1.3.1) can be written as $<\mathbf{Eis},\mathbf{Eis}>=2w(G)$, and there are orthogonality relations

$$<\mathbf{Eis},u_i>=<u_i,u_j>=0 \text{ for } 1\le i\ne j\le\gamma \qquad (4.2.1)$$

which may be written as

$$\sum_{C\in G} w_C u_i(C)=\sum_C w_C u_i(C)u_j(C)=0 \text{ for } 1\le i\ne j\le\gamma. \qquad (4.2.2)$$

There are also some more exciting relations that the calculations confirm.

First, *congruence primes*. Let v_1,v_2 be a pair of primitive eigenvectors defined respectively over fields K_1,K_2; suppose that there are integers $\alpha_1\in K_1$, $\alpha_2\in K_2$ (usually α_1,α_2 are units) and a prime π of the compositum K_1K_2, with π not dividing $\alpha_1\alpha_2$, such that

$$\pi \text{ divides } (\alpha_1v_1(C)-\alpha_2v_2(C)) \text{ for all } C\ ; \qquad (4.2.3)$$

then we call π a *congruence prime joining* v_1,v_2. As Mazur remarked, it follows from Torelli's theorem that the set of eigenforms spanning $S_2(\Gamma_0(D))$ is connected via congruence primes; this may also be true of the set of classes in $G^*(D,\{e\})$, that is to say, of the subset of eigenvectors that span $\mathcal{M}^*(D,\{e\})$, but it is less obvious. At any rate, when one calculates the eigenvectors u_i the congruence primes become conspicuous.

[The congruence prime phenomenon is certainly arithmetically very pretty. Let us recollect two reasons why it is important. *First*, in Mazur's work (originally in the case D prime), the eigenvectors v for which there is a congruence prime joining v to **Eis** are particularly interesting; such eigenvectors span the so-called 'Eisenstein part'. Suppose that v is joined to **Eis** by π, that p is the rational prime below π, and that A is the abelian factor of $J_0(D)$ corresponding to the submodule of $\mathcal{M}^*$ spanned by the conjugates of v; then it appears that $A(\mathbb{Q})$ has p-torsion; in the case where D is prime, this 'handle' enabled Mazur to prove that $A(\mathbb{Q})$ was finite. *Second*, Ribet [15] has given an almost complete explanation of the congruence primes joining old forms onto the 'main part' $S_2^{new}(\Gamma_0(D))$ of $S_2(\Gamma_0(D))$, and this line of investigation led him to his beautiful theory, deducing Fermat's Last Theorem, and more, from the Taniyama-Weil conjecture.]

Next, we consider the norms $<u_i,u_i>$ of the primitive eigenvectors. These have lots of properties; we know what to expect, as Zagier [21] gives a very good exposition of what happens when we consider the Hecke module $H_1(X_0(D))$ rather than $\mathcal{M}^*$. First, suppose that u_i is defined over $\mathbb{Q}$ — then it corresponds to an elliptic curve E with a Weil mapping $\phi: X_0(D)\to E$. It appears that

$$<u_i,u_i>\doteq deg(\phi)\ , \qquad (4.2.4)$$

where as usual $\doteq$ means 'equals apart from junk factors'. [In this case, Gross tells me that the correct junk factors are the usual $\prod_{p|D}c_p$ on the right hand side, together with a power of 2 to make up for factoring by some W-involutions in going from $S_2(\Gamma_0(D))$ to $\mathcal{M}^*(D,\{e\})$. If $E(\mathbb{Q})$ has torsion, it appears in one of the c_p's.] It also appears that, in general,

$$(< u_i, u_i >) \doteq (\prod \pi) \tag{4.2.5}$$

where the product is over the congruence primes joining u_i to other primitive eigenvectors. [This time, we are talking about an approximate equality of ideals. Note that if u_i is rational and corresponds to E as above then any torsion of $E(\mathbb{Q})$ occurs on the right hand side as a congruence prime connecting u_i to **Eis**.] Because both $\{C_k\}$ and $\{u_k\}$ are orthogonal bases, both (4.2.4) and (4.2.5) seem consistent with

$$|det_{0\leq j,k\leq\gamma} < u_j, u_k > | \doteq deg(\psi).w(G) \tag{4.2.6}$$

where ψ is the isogeny $J_0^-(D) \to \prod A_\rho$ referred to above (4.1.2), and of course $w(G)$ is the weight of the genus.

Finally, there are some extraordinary things that seem to happen when u is a primitive eigenvector corresponding to an elliptic curve E over $\mathbb{Q}$ with Mordell-Weil rank($E(\mathbb{Q})$) equal to 2. These seem particularly interesting, since (post Kolyvagin) the outstanding problem in the arithmetic of elliptic curves is to find a replacement for the Heegner point construction in the rank ≥ 2 case. If one doesn't know what to do, having an extraordinary phenomenon to look at may help!

Most of these phenomena seem to have been first noticed by Dick Gross. Suppose for the rest of this section that u is a primitive eigenvector with integral coordinates $u(C)$ corresponding to an elliptic curve E of rank ≥ 2. If F is a form in the class C, let $C^\vee$ be the class of $F^\vee$, and let $\theta_{C^\vee}(\tau)$ be the corresponding theta function of weight 3/2, as in (3.5); then it appears to be a 'moral consequence' of the Gross-Zagier theorem that

$$\sum_{C\in G} w_C u(C)\theta_{C^\vee} \text{ is identically zero.} \tag{4.2.7}$$

More precisely, Gross ([9] (see also Kramer [12]) justifies this in the Oesterlé-Mestre set-up, and experiment shows that the similar relation, with θ_C replacing $\theta_{C^\vee}$, stays true in ours. (The relation was actually used to predict curves with rank 2 now in Cremona's tables, before Cremona found them). As Schulze-Pillot [16] has shown, rather more is true. Let G_F be a quaternary form corresponding to $F \in C$ as in Lemma 3.1; then there is a degree 2 theta series $\theta_C^{(2)}$ that counts the representations of binary integral forms by G_F, and we have the extraordinary linear relation, that

$$\sum_{C\in G} w_C u(C)\theta_C^{(2)} \text{ is identically zero.} \tag{4.2.8}$$

The tissue of proof leading to (4.2.8) is more than a little fragile (I haven't proved that my Hecke action is related to the normal one, though Ponomarev has, Gross justifies (4.2.7) only in the prime case, and Schulze-Pillot only claims (4.2.8) for conductor 389), but it seems pretty certain that (4.2.8) is true for eigenforms corresponding to curves of rank 2 in fair generality.

Subsequently, even more extraordinary relations have been found (this time, they were spotted by Zagier [22], and proved by Gross [10], again in the Oesterlé-Mestre set-up.) Suppose that u is as above, and that v is *any* primitive eigenvector of $\mathcal{M}^*(D, \{e\})$, v may or may not be different from u. Then

$$\sum_{C \in G} w(C)u(C)v^2(C) = 0 \; ! \qquad (4.2.9)$$

(The relation actually noted by Zagier, using the Oesterlé-Mestre tables for prime levels less than 1000, was $\sum_C w(C)u^{2n+1}(C) = 0$ for all integer $n \geq 0$, but there is not enough evidence to convince me that this is true for $n \geq 2$.)

5 Possible extensions

A notion of neighbourliness, leading to a Hecke graph, may be defined in a wide variety of contexts. The classical Hecke operators and the associated graphs arise from the neighbourliness of classes of 2-dimensional lattices, or lattices with level structure; and there are Hecke graphs of higher valency that correspond to neighbourliness of lattices of higher dimension. In a similar way, there is no reason to restrict ourselves to Hecke graphs of classes of ternary quadratic forms, that is, to quadratic spaces of dimension 3; after all, Eichler's Anzahlmatrizen are defined in a very general context. We may form Hecke graphs of classes of n-ary quadratic forms for any n, using the p-neighbourliness relation

$$f(\frac{1}{p}x_1, px_2, x_3, ..., x_n) \text{ is a } p\text{-neighbour of } f(x_1, x_2, x_3, ..., x_n),$$

at least when both forms have $\mathbb{Z}$ coefficients.

The case $n = 2$ is amusing but wrong-headed: the theory is essentially what is left of Gauss' theory of composition (c.f. Shanks [ ]) when one forgets to distinguish between the classes of *im*properly equivalent forms. For $n \geq 4$ the theory of section 1.2 remains valid, but I don't know what the associated graph and its eigenvectors are good for. A possible application is as a good computational method of making a self-checking list of all 'semi-stable' definite quaternary quadratics of given discriminant (in particular of all definite quaternary quadratics of given squarefree discriminant), but I have not yet programmed it (a complete normalisation procedure for quaternary forms is not easy to write).

For $n = 4$ one may define a different neighbourliness relation by

$$f(\frac{1}{p}x_1, \frac{1}{p}x_2, px_3, px_4) \text{ is a } p\text{-neighbour of } f(x_1, ..., x_4)$$

if both forms are integral.

With this relation, a form will usually have $2p + 2$ neighbours, corresponding to the $2p + 2$ lines on the quadric $f \equiv 0 \; mod(p)$; and these neighbours will sort themselves into two sets of $p + 1$, as the lines on the quadric belong to 2 reguli. With a little care, one obtains two $(p + 1)$-valent graphs, one for each regulus. If

N is squarefree, we may act in this way on the set $\mathcal{Q}^*(N^2)$ of classes of definite quaternary forms of determinant N^2 whose mod p reduction is always either of rank 2 or rank 4; when I tried this for $p = 2$ and $N = 105$ and 165, I obtained *all* the information given by the Brandt matrices. However, though this method seems to work in principle, it seems of little value in practice. The set $\mathcal{Q}^*(N^2)$ is large and the corresponding graph splits into a large number of connected subgraphs, most of them carrying essentially the same information. Accordingly, one has in effect computed the same thing a large number of times -- but it is not clear how to avoid this redundancy.

6 A table

For a selection of elliptic curves E corresponding to eigenvectors $\{u(C)\}_{C\in\mathcal{C}}$, I list $M_2 = 2\sum w(C)u^2(C)$ and $M_3 = 2\sum w(C)u^3(C)$; I have restricted myself to eigenvectors of ternary form graphs whose discriminant D is square free. I have listed all cases with $D < 200$, and a few more. In most cases, the discriminant D is simply the level of the curve E, and I have used the Antwerp [2] or Cremona [5] name of the curve to label the entry. In the other cases, I have calculated M_2, M_3 using oldforms, that is to say by using a ternary form graph with discriminant D that is properly divisible by the level of the curve, and then I have used as label the discriminant D followed by a colon and the name of the curve. All the curves with level less than 200 have Mordell-Weil rank zero, and so do most of the rest; a few curves with rank 2 are marked with an asterisk.

As the table was not computer generated, there are almost certainly a few transscription errors, but it gives some idea of the highly erratic way in which M_2, M_3 vary for newforms. For oldforms, the variation appears to be systematic. Suppose that E is an elliptic curve of level N, and that MN is square free; for $p \not| N$ suppose that the modulo p reduction of E has $p + 1 - a_p$ points. Then the theorem of Ribet [15] on congruence primes tells us that

$$M_2(MN : E) \doteq \prod_{p|M} (p + 1 + a_p) M_2(E) ,$$

where this time $\doteq$ denotes equality apart from a predictable power of 2; and for M_3 our tables appear consistent with the rather odd formula

$$M_3(MN : E) \doteq \prod_{p|M} (p + 1 + 3a_p) M_3(E) .$$

Table 1

Curve	M_2	M_3	Curve	M_2	M_3	Curve	M_2	M_3
11B	5	5	14C	$\frac{3}{2}$	$\frac{3}{2}$	15C	2	4
17C	4	8	19B	3	3	21B	2	4
22:11B	5	15	26B	3	3	26D	7	7
30A	$\frac{3}{2}$	$\frac{3}{2}$	30:15C	1	0	33B	3	4
33:11B	15	5	34A	$\frac{3}{2}$	$\frac{3}{2}$	34:17C	2	0
35B	3	3	37C	3	3	38A	5	5
38D	9	9	38:19B	9	9	39B	2	0
42A	2	4	42:14C	3	3	42:21B	1	0
46A	$\frac{5}{2}$	$\frac{3}{2}$	51A	3	3	51:17C	4	4
55B	2	0	55:11B	35	45	57B	3	4
57F	15	5	57:19B	6	6	58B	5	15
62A	4	8	66A	3	3	66E	1	0
66I	$\frac{5}{2}$	$\frac{1}{2}$	66:11B	15	15	66:33B	3	3
67A	5	3	69A	9	0	70A	1	0
70:14C	9	9	70:35B	9	9	73B	6	0
74:37C	18	18	77A	6	8	77D	15	21
77:11B	30	10	78A	10	6	78:26B	15	21
78:26D	7	35	78:39B	2	0	85A	2	0
85:17C	4	0	89A	10	8	94A	2	0
95:19B	27	45	102A	9	9	102G	2	4
102:17C	2	0	102:34A	3	3	102:51A	9	9
105A	1	0	105:15C	4	4	105:21B	2	0
105:35B	15	21	106B	9	9	106D	5	3
106E	9	9	109A	4	0	110A	3	3
110C	5	5	110E	21	33	110:11B	35	135
110:55B	2	0	111:37C	30	42	113B	12	8
114A	$\frac{3}{2}$	$\frac{3}{2}$	114E	5	3	114G	3	4
114:19B	18	18	114:38A	15	5	114:38D	45	63
114:57B	3	3	114:57F	15	15	115A	5	3
118B	15	15	118D	3	3	118E	19	9
119:17C	12	20	129B	5	4	130A	1	0
130J	$\frac{5}{2}$	$\frac{3}{2}$	130:26B	9	9	130:26D	35	21
133:19B	21	15	134:67A	25	27	138A	2	0
138G	6	6	138:46A	10	6	138:69A	2	0
139A	6	0	141A	12	24	141G	6	4
141H	6	6	142A	$\frac{9}{2}$	$\frac{3}{2}$	142G	6	4
142C	6	6	143:11B	90	130	146:73B	6	0
154A	2	0	154E	2	0	154:11B	30	30
154:14C	18	18	154:77A	6	6	154:77D	45	63
155A	4	0	158B	60	60	158F	6	8
158H	30	10	161B	10	12	165:11B	105	45
165:15	4	0	165:33B	3	0	165:55B	2	0

Table 1 (continued)

Curve	M_2	M_3	Curve	M_2	M_3	Curve	M_2	M_3
170C	5	3	170D	9	9	170F	9	9
170H	120	312	170:17C	2	0	170:34A	9	9
170:85A	8	0	174A	$\frac{5}{2}$	$\frac{3}{2}$	174E	13	9
174F	3	3	174G	7	35	174I	165	627
174:58B	15	15	178C	14	12	178A	12	6
178:89A	10	6	179A	9	3	182A	3	3
182D	9	5	182E	9	4	182I	35	27
182J	77	69	182:14C	15	3	182:26B	21	15
182:26D	63	77	185:37C	36	36	186B	5	15
186A	7	3	186D	11	3	186:62A	4	4
187A	24	12	187C	15	3	187:11B	80	60
187:17C	12	12	190A	3	3	190:19B	81	135
190:38A	10	30	190:38D	54	54	194A	$\frac{7}{2}$	$\frac{3}{2}$
195A	6	12	195I	3	3	195J	21	59
195K	7	3	195:15C	6	4	195:39B	4	0
222E	621	573	222C	9	6	222D	13	15
222A	3	3	222B	11	29	222:37C	90	126
246E	3	4	246B	15	5	246A	7	3
246G	11	3	246F	15	3	246C	35	27
258D	5	4	256E	95	331	258F	21	21
258G	$\frac{3}{2}$	$\frac{3}{2}$	258B	49	27	258:129B	5	3
291A	506	378	291B	120	72	291D	6	0
389A*	40	0	433A*	28	0	446D*	56	0
446C	12	8						

References

[1] A.O.L. Atkin and J. Lehner, *Hecke Operators on* $\Gamma_0(m)$, Math. Annalen **185** (1970), 134-160.

[2] B.J. Birch and W. Kuyk, *Modular Forms of One Variable IV*, Springer Lecture Notes **476**, 1975.

[3] Burton W. Jones, *The Arithmetic Theory of Quadratic Forms*, Carus Mathematical Monographs **10**, John Wiley 1950.

[4] J.W.S. Cassels, *Rational Quadratic Forms*, London Math. Soc. Monographs **13**, Academic Press 1978.

[5] J.E. Cremona, *Computation of modular elliptic curves and the Birch-Swinnerton-Dyer conjecture*, To be published.

[6] M. Eichler, *Quadratische Formen und Orthogonale Gruppen*, Grundlehren d. Math. Wiss. LXIII, Springer 1952.

[7] M. Eichler, *The basis problem for modular forms and the traces of the Hecke operator*, In *Modular Forms of One Variable I*, Springer Lecture Notes **320**, pp 75-151, 1973.

[8] B. Gross and D.B. Zagier, *Heegner points and derivatives of L-series*, Invent. Math. **84**, 225-320 (1986).

[9] B. Gross, *Heights and the special values of L-series*, Sem. Math. Sup., Presses Univ. Montreal.

[10] B. Gross. Informal lecture delivered at the conference on L-functions in Durham, July 1989.

[11] V.A. Kolyvagin, *Finiteness of $E(\mathbb{Q})$ and Ш for a class of Weil curves*, Izv. Akad. Nauk. SSSR Ser. Mat. 1988 (Russian), and subsequent papers. For good expositions, q.v. Karl Rubin *'The work of Kolyvagin on the arithmetic of elliptic curves'* to appear in Springer Lecture Notes, and Benedict Gross *'Kolyvagin's work on modular elliptic curves'* to appear in the proceedings of the 1989 Durham conference on L-functions.

[12] J. Kramer, *Jacobiformen und Thetareihen*, Inauguraldissertation Basel 1985, and *'On the linear independence of certain theta series'*, Preprint Max-Planck-Institut Bonn MPI-SFB 85/40.

[13] J-F. Mestre, *La Méthode des graphes. Examples et applications*, Proc. Int. Conf. on Class Numbers and Units of Algebraic Number Fields, Katata (1986) 217-242.

[14] P. Ponomarev, *Ternary quadratic forms and Shimura's correspondence*, Nagoya Math. J. **81** (1981) 123-151.

[15] K.A. Ribet, *Congruence Relations between Modular Forms*, Proc. Int. Congress of Mathematicians, Warszawa (1983) 503-514.

[16] R. Schulte-Pillot, *A linear relation between theta series of degree and weight 2*. Proceedings of the Journées Arithmétiques in Ulm 1987, 197-201, Springer Lecture Notes **1380** (1989).

[17] D. Shanks, *On Gauss and Composition I, II*, Number Theory and Applications (Banff conference 1988) 163-204, Kluwer 1989.

[18] N-P. Skoruppa and D.B. Zagier, *Jacobi forms and a certain space of modular forms*, Invent. Math. **94** (1988) 113-146.

[19] H.J.S. Smith, *On the Orders and Genera of Ternary Quadratic Forms*, Phil. Trans. CLVII (1867), in Collected Mathematical Papers 455-509, Oxford 1894.

[20] D.J. Tingley, *Elliptic curves uniformised by modular functions*, Oxford D.Phil. thesis 1975.

[21] D.B. Zagier, *Modular parametrisations of elliptic curves*, Canad. Math. Bull. **28** (1985) 372-384.

[22] D.B. Zagier, Letter to Gross, April 1989.

Computation of singular moduli by multi-valued modular equations

Harvey Cohn *

Abstract. A highlight of Fricke's *Algebra* is the development of two-valued modular equations for the computation of singular moduli. For example, for 37 cases of N ranging from $N = 2$ to 71 the two values $j(z/N)$ and $j(z)$ are the branches of a Riemann Surface. They satisfy a quadratic equation whose discriminant displays the singular moduli. The computations require an algebraic system, and the author completed the program using MACSYMA; the results appear in Math. of Comp. **51** (1988) and in Proc. Symp. Pure Math. **49** (1989). The fundamental domain for symmetric functions of the branches is defined by the locus of the real singular moduli.

By methods essentially described by Fricke in Acta.Arith.**52** (1929), it is possible to consider cases $N = nm = 30, 42, 70, 105$ where $j(z/nm)$, $j(z/n)$, $j(z/m)$, $j(z)$ are four branches of a Riemann Surface. His results were limited by computational considerations. Here we shall make a more complete study of the simpler cases $N = 22, 33, 34, 46$ (not in previous computations), leading to singular moduli of order nm and also nm^2. Again the location of these points determines the fundamental domain of symmetric functions of the branches. A new type of singular modulus is seen here, a common value of all four branches at once. Attention is also called to the anomalous case $N = 65$.

1 Introduction

Even for small values of N the modular equation of order N relating $j(z)$ and $j(z/N)$ has coefficients of such magnitude that it is more of a challenge to obtain useful information than to perform the computation (see [9]).

Fricke [5] showed for 37 values of N, namely

$$N = 2, ..., 21, 23, ..., 27, 29, 31, 32, 35, 36, 39, 41, 47, 49, 50, 59, 71, \tag{1.1}$$

that $j(z)$ and $j(z/N)$ are defined on a *two-valued* Riemann Surface of "at worst" hyperelliptic type (genus 6 for $N = 71$),

$$s^2 = P_N(t). \tag{1.2}$$

Here $P_N(t)$ is a monic polynomial in $\mathbb{Z}[t]$, (a fortuitous result which seems possible to prove only by verifying all 37 cases individually!). The main results of Fricke's

*Research supported by NSF Grant DMS-8700878.

Computational Number Theory

method (as carried to completion in [3], [4]), are that by symmetry conditions there exists functions $M_N(t), R_N(t)$ in $\mathbb{Q}[t]$ such that

$$j(z)j(z/N) = M_N(t), \quad j(z) - j(z/N) = R_N(t)P_N^{(1/2)}(t). \tag{1.3}$$

Accordingly, the *singular moduli* which belong to N (roots of $j(z) - j(z/N)$) are displayed by factorization of the polynomials $R_N(t), P_N(t)$ into *class equations* which determine ring class fields (see [2]). Furthermore, the real roots can be made to correlate in a simple fashion to the boundary of the fundamental domain for symmetric functions of $j(z)$ and $j(z/N)$.

Most important for Fricke's method is that the coefficients in the functions appearing in (1.3) are reasonably small under factorization, seldom exceeding seven digits.

Subsequently Fricke [5], [6], [7] devised a four-valued generalization to deal with the cases $N = 30, 42, 70, 105$, but again, the analysis was not computationally completed.

In this paper, we shall apply Fricke's method to cases omitted in the original list (1.1), but where N has only two prime factors,

$$N = 22, 33, 34, 46. \tag{1.4}$$

This is by no means a complete list, nor are our methods theoretically most general, since we shall use the discriminant function but not theta-functions. Indeed, completeness is not the only theoretical challenge, (see the anomalous case $N = 65$).

2 The Fricke involutions

We are dealing with the *Weber modular invariant*

$$j(z) = E_4^3(z)/\Delta(z). \tag{2.1a}$$

Here the *Eisenstein series* and the *discriminant* are given , respectively by

$$E_4(z) = 1 + 240\sum_1^\infty h^3q^h/(1-q^h), \tag{2.1b}$$

$$\Delta(z) = q\prod_1^\infty(1-q^h)^{24}, \tag{2.1c}$$

$$q = \exp 2\pi i z \tag{2.1d}$$

defined in the upper half-plane (see [13])

$$H : \Im z > 0, \quad |q| < 1. \tag{2.1e}$$

The function $j(z)$ is invariant under the *modular group* acting on H,

$$\Gamma = PSL_2(\mathbb{Z}) = \{M = \pm\begin{pmatrix} a & b \\ c & d \end{pmatrix} \; a,b,c,d \in \mathbb{Z}, ad - bc = 1\}, \tag{2.2a}$$

(where $Mz = (az+b)/(cz+d)$). The function $\Delta(z)$ is an invariant form of weight 12, i.e.,

$$\Delta(Mz) = \Delta(z)(cz+d)^{12}. \tag{2.2b}$$

The subgroup $\Gamma^0 \subset \Gamma$ under which $j(z/N)$ is invariant is

$$\Gamma^0 = \{PSL_2(\mathbb{Z}),\ b \equiv 0 \bmod N\}, \tag{2.2c}$$

a subgroup of Γ of index

$$m = \prod\nolimits_{p|N} (1 + \frac{1}{p}). \tag{2.2d}$$

For any group containing Γ we define the fundamental domain H/Γ (suitably compactified at the cusps).

Fricke's original method [5] was to introduce the involution

$$W = -N/z \tag{2.3a}$$

which he used to enlarge the group Γ^0 to

$$\Gamma^* = \Gamma^0 + W\Gamma^0 \tag{2.3b}$$

a group of order $2m$. For precisely the values in (1.1), the fundamental domain for Γ^* is of genus zero. In [6], however, Fricke introduces additional involutions (attributed to unpublished work of Bessel-Hagen, also see [1]). We think of N as fixed in each problem and we let T equal 1 or any *primary* divisor (so $(T, N/T) = 1$). Then we define cosets of integral matrices (acting always on H)

$$S_T = \{\pm \begin{pmatrix} A & B \\ C & D \end{pmatrix},\ N|B,\ T|(A,D),\ AD - BC = T\}. \tag{2.4a}$$

Thus the S_1 make up Γ^0 while W is in S_N. It is seen that

$$S_T S_U = S_V \tag{2.5a}$$

where TU and V differ by a square multiplier, so for some integer E

$$TUV = E^2. \tag{2.5b}$$

Thus each matrix in S_T acts as an involution on Γ^0.

The important property of the involution is that for

$$M = \begin{pmatrix} D/T & -B/T \\ -C & A \end{pmatrix}, \tag{2.6a}$$

$M \in \Gamma$ and $MS_T z = z/T$. Therefore

$$j(S_T z) = j(z/T). \tag{2.6b}$$

Furthermore, using (2.4a), we find

$$\Delta(S_T z) = \Delta(z/T)(Cz+D)^{12}/T^{12}. \tag{2.6c}$$

We define the largest possible augmented group (see (2.3b))

$$\Gamma^c = \Gamma^0 \cup \{S_T : T|N, (T, N/T) = 1\},\ |\Gamma^c/\Gamma^0| = 2^\nu \tag{2.7}$$

where there are ν distinct prime divisors of N. Now in many of Fricke's cases (1.1), the fundamental domain for Γ^* had cusps. We shall, for simplicity assume that N is squarefree so that there is only the cusp at ∞. For this cusp (of width N) there is the local parameter

$$r = q^{1/N} = \exp 2\pi i z/N. \tag{2.8a}$$

Hence, precisely when H/Γ^c is of genus zero, there will be a one-valued function with pole at $z = \infty$

$$t = 1/r + a_0 + a_1 r + a_2 r^2 \cdots. \tag{2.8b}$$

3 Determination of single-valued functions

We narrowed our cases to new ones where N factors into two primes ($N = nm$). We let $\{T, U, V\}$ be one of six permutations of $\{1, n, m\}$.Then making use of the discriminant function (see [11]), we define three functions, one for each choice of T,

$$\Lambda_T(z) = \frac{\Delta(zU/N)\Delta(zV/N)}{\Delta(zT/N)\Delta(z)}. \tag{3.1a}$$

We verify from (2.6c) and (2.5) that

$$\Lambda_T(S_1 z) = \Lambda_T(S_{N/T} z) = \Lambda_T(z), \tag{3.1b}$$

$$\Lambda_T(S_{N/U} z) = \Lambda_T(S_{N/V} z) = T^{24}/\Lambda_T(z). \tag{3.1c}$$

The pole of $\Lambda_T(z)$ is of order r^{-p} where

$$p = \begin{cases} (m-1)(n-1) & \text{if } T = 1 \\ (m+1)(n-1) & \text{if } T = n. \end{cases}$$

Now if for integers g, h,

$$gp = 24h \tag{3.2a}$$

we can use "principal root definitions" to define a function single-valued in H,

$$D_{T,g}(z) = \Lambda_T(z)^{g/24} + T^g/\Lambda_T(z)^{g/24}. \tag{3.2b}$$

It has a pole r^{-h}, but we are no longer sure that it is single-valued over the fundamental domain H/Γ^c.

The worst that can happen is that the function in (3.2b) experiences a change in sign (as the S_T are involutory over Γ^0), but this possibility can be precluded in most cases by the knowledge of enough fixed-points or "mirror points" of S_T. For example, a very general form of $z' = S_T z$ is given by

$$(z - qT/s)(z' - q'T/s) = T/s^2 \tag{3.3a}$$

where the reduced fractions $r = q/s$, $r' = q'/s$ satisfy

$$(1 - qq'T)/s \equiv 0 \bmod N/T. \tag{3.3b}$$

Thus $z' = S_T z$ where

$$z = \frac{qT + \sqrt{-T}}{s}, \quad z' = \frac{q'T + \sqrt{-T}}{s}. \tag{3.3c}$$

It is not hard to find special values for (3.3b) where $r = r'$ so a fixed-point occurs ($z = z'$). This precludes a sign change in (3.2b). Likewise if $r = -r'$ (mirror points) then $-z$ and z' are conjugates and this also works the same way as a fixed-point. In this way we verify the table below for our main cases (the desired transformations of (3.3a) actually occur in the description of the fundamental domains H/Γ^c below). For our new cases we have the following table:

N	22	33	34	34	46
T	2	3	1	17	2
g	2	1	3	1	1
h	1	1	2	2	1

For $N = 22, 33, 46$, $h = 1$ and $t = D_{T,g}(z)$ as shown, while for $N = 34$ the power series in r will establish

$$t = (D_{1,3}(z) - D_{17,1}(z))/4.$$

To illustrate, we expand the last expression into power series:

$$\Lambda_1(z) = \frac{\Delta(z/17)\Delta(z/2)}{\Delta(z/34)\Delta(z)} = \frac{\prod(1-r^{2h})^{24}(1-r^{17h})^{24}}{r^{16}\prod(1-r^h)^{24}(1-r^{34h})^{24}},$$

$$D_{1.3}(z) = \Lambda_1^{1/8}(z) + \Lambda_1^{-1/8}(z) = 1/r^2 + 3/r + 6 + 13r + \cdots;$$

$$\Lambda_{17}(z) = \frac{\Delta(z/17)\Delta(z/34)}{\Delta(z/2)\Delta(z)} = \frac{\prod(1-r^{2h})^{24}(1-r^h)^{24}}{r^{48}\prod(1-r^{17h})^{24}(1-r^{34h})^{24}},$$

$$D_{17,1}(z) = \Lambda_{17}^{1/24}(z) + 17\Lambda_{17}^{-1/24}(z) = 1/r^2 - 1/r - 2 + r + \cdots;$$

$$t = 1/r + 2 + 3r + \cdots.$$

4 Four-valued representation

We are considering cases where N $(= nm)$ the product of primes and a simple pole exists in H/Γ^c, given by

$$t = 1/r + a_0 + a_1 r + a_2 r^2 + \cdots. \tag{4.1}$$

Then the symmetric functions of the four functions

$$j_T = j(z/T) = j(S_T z), \quad T = 1, n, m, N \tag{4.2}$$

are polynomials in t. We define the invariants

$$k_1 = j_1 + j_n + j_m + j_N \tag{4.3a}$$
$$k_n = j_1 + j_n - j_m - j_N \tag{4.3b}$$
$$k_m = j_1 - j_n + j_m - j_N \tag{4.3c}$$
$$k_N = j_1 - j_n - j_m + j_N \tag{4.3d}$$

For practical purposes we start with

$$j_N = j(z/N) = 1/r + 744 + 196884r^2 + \cdots, \tag{4.4}$$

the well-known expansion, and successively substitute for r the powers r^n, r^m, r^N to obtain j_m, j_n, j_1 respectively. For later purposes we truncate each Taylor series to error $O(r^{N+1})$.

Let us consider the action of the involutions S_T on the above invariants. We see k_n for instance is invariant under S_n but transforms into its negative under the other generators of Γ^c, namely S_m and S_N. All of the k_T^2, (for $T = n,\ m,\ N$) are invariant (hence polynomials of degree $2N$ in t), while k_1 more easily is invariant hence a polynomial of degree N. We therefore have enough accuracy in the Taylor expansions to calculate each of the polynomials by the nonvanishing terms of the series for $k_1,\ k_n^2,\ k_m^2,\ k_N^2$ by the "reversion of series".

We factor out the perfect squares to obtain

$$k_1 = p_1(t),\ k_n = p_n(t)v_n^{1/2}(t),\ k_m = p_m(t)v_m^{1/2}(t),\ k_N = p_N(t)v_N^{1/2}(t) \tag{4.4}$$

where the p_T and the v_T are polynomials. As far as their degrees are concerned, for now it can only be said that the total degree (at infinity) is N for each k_T. The radicands are square-dependent, i.e.,

$$v_n(t)v_m(t)v_N(t) = v^2(t) \tag{4.5}$$

for some polynomial $v(t)$. We now have the equations

$$j_1 = (p_1(t) + p_n(t)v_n^{1/2}(t) + p_m(t)v_m^{1/2}(t) + p_N(t)v_N^{1/2}(t))/4 \tag{4.6a}$$
$$j_n = (p_1(t) + p_n(t)v_n^{1/2}(t) - p_m(t)v_m^{1/2}(t) - p_N(t)v_N^{1/2}(t))/4 \tag{4.6b}$$
$$j_m = (p_1(t) - p_n(t)v_n^{1/2}(t) + p_m(t)v_m^{1/2}(t) - p_N(t)v_N^{1/2}(t))/4 \tag{4.6c}$$
$$j_N = (p_1(t) - p_n(t)v_n^{1/2}(t) - p_m(t)v_m^{1/2}(t) + p_N(t)v_N^{1/2}(t))/4 \tag{4.6d}$$

A complete listing of equations for $N = 22$ appears in the Appendix.

To summarize the invariance theory, for $T = n,\ m,\ N$ we can say that k_T is invariant under S_T and belongs to the group $\Gamma^0 \cup \Gamma^0 S_T$ and the corresponding fundamental domain $H/\Gamma^0 \cup \Gamma^0 S_T$ is the "Riemann Surface of $v_T^{1/2}(t)$". Likewise k_1 is invariant under Γ^c and the corresponding fundamental domain H/Γ^0 is the "Riemann Surface of $v_n^{1/2}(t),\ v_m^{1/2}(t),\ v_N^{1/2}(t)$". Illustrations in our special cases follow.

Case 22 genus $H/\Gamma^0 = 2$
$$\begin{aligned} v_{22} &= (t+4)(t^3+12t^2+44t+52)\\ v_2 &= (t-4)(t^3+12t^2+44t+52)\\ v_{11} &= (t+4)(t-4). \end{aligned}$$

Case 33 genus $H/\Gamma^0 = 3$
$$\begin{aligned} v_{33} &= (t+1)(t^3+7t^2+19t+17)\\ v_3 &= (t+1)(t^3+7t^2+19t+17)(t^2-12)\\ v_{11} &= (t^2-12). \end{aligned}$$

Case 34 genus $H/\Gamma^0 = 3$
$$\begin{aligned} v_{34} &= (t-1)(t^2-t-4)\\ v_2 &= (t^2-t-4)(t^2-9t+16)\\ v_{17} &= (t-1)(t^2-9t+16). \end{aligned}$$

Case 46 genus $H/\Gamma^0 = 5$
$$\begin{aligned} v_{46} &= (t^3+5t^2+4t+1)(t^3+5t^2+8t+5)\\ v_2 &= (t^3+5t^2+4t+1)(t^3+5t^2+8t+5)(t^2-8)\\ v_{23} &= (t^2-8). \end{aligned}$$

It is probably no coincidence that the cubics in Cases 22 and 33 both have discriminant $-16 \cdot 11$, while those of Case 46 have discriminant -23. Likewise both quadratics in Case 34 have discriminant 17.

5 Singular moduli

An unexpectedly complicated matter is the search for singular moduli belonging to the order N

$$j(z) = j(z/N) \tag{5.1}$$

They generally produce quadratic fields of with orders of discriminant

$$d_N = -(4N - g^2)\ (< 0),\ g = 1, 2, 3, \cdots. \tag{5.2}$$

From the equations (4.a-d), we see they arise from $j_1 = j_N$, or $k_n + k_m = 0$. Yet to solve this equation in polynomial terms (see (4.4)), we have to solve $k_n^2 = k_m^2$, producing the extraneous roots $j_n = j_m$. Actually, we are dealing with the classical case of the resolvent cubic of a biquadratic with three discriminant factors:

$$s_N = (k_n^2 - k_m^2)/4 \quad \to j(z) = j(z/N),\ j(z/n) = j(z/m) \tag{5.3a}$$
$$s_n = (k_m^2 - k_N^2)/4 \quad \to j(z) = j(z/n),\ j(z/m) = j(z/N) \tag{5.3b}$$
$$s_m = (k_n^2 - k_N^2)/4 \quad \to j(z) = j(z/m),\ j(z/n) = j(z/N) \tag{5.3c}$$

We therefore solve the three equations $s_T = 0$ for t and evaluate the appropriate values of j to obtain a panoply of singular moduli. They will correspond to the discriminants in (5.2) and additionally

$$\begin{aligned} d_n &= -(4n - g^2), & d'_n &= -m^2(4n - g^2); \\ d_m &= -(4m - g^2), & d'_m &= -n^2(4m - g^2). \end{aligned} \tag{5.4}$$

(Here g takes integer values subject to the discriminant remaining negative.) The "extra discriminants" come from the fact that if $j(z) = j(z/n)$ for $z = \zeta$ then $j(z/m) = j(z/N)$ for $z = m\zeta$, introducting a factor of m^2 in the discriminant.

At this point we naturally substitute the roots t of each $s_T = 0$ into (4.6a-d) to find the moduli, but a further surprise awaits us. The field of irrationality of j is not always determined by that of t, (unlike the two-valued modular equations [3]). The degree could be doubled owing to the radicals $v_T^{1/2}$ or it could be lower than that of t owing to fortuitous cancellations of radicals in (4.6a-d).

A complete listing of s_{22}, s_2, s_{11} for Case 22 is in the Appendix. We analyze just the factors of s_2 for brevity (showing discriminants d and class numbers h associated with the singular moduli [2], [8]).

$(t+3)^2 \cdots d = -8,\ h = 1;$
$(t+4) \cdots d = -7,\ h = 1;$
$(t^3 + 32t^2 + 180t + 288)^2 \cdots d = -4 \cdot 11^2,\ h = 6;$
$(t^5 - 68t^4 - 1160t^3 - 7456t^2 - 20720t - 20800) \cdots d = -8 \cdot 11^2,\ h = 10;$
$(t^{10} + 41t^9 + 4827t^8 + 158648t^7 + 2548693t^6 + 23700126t^5 + 136492029t^4$
$+496032660t^3 + 1110597075t^2 + 1402292250t + 765905625) \cdots d = -7 \cdot 11^2,$
$h = 10.$

The correlation of these polynomials with discriminants is far from systematic (unlike the analogous problem for two-valued modular functions). A typical class field theoretical result emerging from the above data is that if the tenth-degree polynomial splits mod p and $(-7/p) = 1$, $(p > 11)$, then we have the integral representation

$$4p = x^2 + 7(11y)^2.$$

We finally note that of the three quantities s_N, s_n, s_m in (5.3a-c) any root t common to two must be common to all three, and indeed three or four of the branches j_1, j_n, j_m, j_N are equal. These common values will be called *strong singular moduli.* We list some of the strong singular moduli.

Case 22	$t = -4$,	$z = 8 + \sqrt{-2}$,	$j = 20^3$
	$t = -3$,	$z = (9 + \sqrt{-7})/2$,	$j = -15^3$
Case 33	$t = -1$,	$z = (11 + \sqrt{-11})/2$,	$j = -2^{15}$
	$t = -2$,	$z = 8 + \sqrt{-2}$,	$j = 20^3$
Case 34	$t = 0$,	$z = 13 + i$,	$j = 12^3$
	$t = 1$,	$z = 10 + \sqrt{-2}$,	$j = 20^3$
Case 46	$t = -1$,	$z = (27 + \sqrt{-7})/2$,	$j = -15^3$.

In each case $j_1 = j_n = j_m = j_N$ as we can verify from the usual rules that $j(z) = j(z+1)$ and that, when $cc' = a^2 + b$,

$$j\left(\frac{a + \sqrt{-b}}{c}\right) = j\left(\frac{-a + \sqrt{-b}}{c'}\right).$$

6 Fundamental domains

The fundamental domains for H/Γ^c are found next by the method of Fricke [5], (also see [4]). First of all there is only one cusp (at ∞ of width N), so we need find only the floor, indeed (by symmetry) only the positive real half from 0 to $N/2$. In all four of our special cases we are (fortunately) able to find enough isometric circles C_T $(T = N, n, m)$ to piece segments together to make the floor. These circles are of the type

$$|z - qT/s| = \sqrt{T}/s. \tag{6.1a}$$

They correspond to the conditions (3.3ab) for S_T,

$$(z - qT/s)(z' \pm qT/s) = -T/s^2. \tag{6.1b}$$

If the $\pm$ is negative, then C_T collapses about its fixed-point $z = (qT + \sqrt{-T})/s$, while if we have a positive sign, then C_T maps into its mirror image (under $\Re z \to -\Re z$) so that $z \to z' = (-qT + \sqrt{-T})/s$. In any case the floor folds up simply, so the genus is zero. The strong singular moduli are important here as points where the segments of the floor join. Case 22 is illustrated in Figure 1.

The proof that the region matched onto itself by Γ^c is really the fundamental domain comes from Poincare's Theorem once we verify by a computation (in trigonometry) that the noneuclidean area $\int_R dxdy/y^2$ over R the region above the (half) floor is $|\Gamma^c/\Gamma|/8$ times $\pi/3$, (the area of Γ, see (2.2a-d)).

We list the (half) floors of the fundamental domains. They all start at $z = \sqrt{-N}$, (corresponding to $zz' = -N$), and end at $(N + \sqrt{-N})/2$ for N odd or at $N/2 + \sqrt{-N/2}$ for N even. (Note the transformations (6.1b) are listed instead of the circles (6.1a) because they indicate whether a fixed-point or a mirroring of arcs is involved).

Case 22	S_{22}	$zz' = -22$ to $(9 + \sqrt{-7})/2$
	S_{11}	$(z - 11/2)(z' - 11/2) = -11/4$ to $(13 + \sqrt{-7})/2$
	S_{22}	$(z - 22/3)(z' + 22/3) = -22/9$ to $8 + \sqrt{-2}$
	S_{11}	$(z - 11)(z' - 11) = -11$ to $11 + \sqrt{-11}$.
Case 33	S_{33}	$zz' = -33$ to $(11 + \sqrt{-11})/2$
	S_3	$(z - 6)(z' + 6) = -3$ to $(22 + \sqrt{-8})/3$
	S_{33}	$(z - 33/4)(z' + 33/4) = -33/16$ to $8 + \sqrt{-2}$
	S_{11}	$(z - 11)(z' - 11) = -11$ to $14 + \sqrt{-2}$
	S_{33}	$(z - 33/2)(z' - 33/2) = -33/4$ to $(33 + \sqrt{-33})/2$.
Case 34	S_{34}	$zz' = -34$ to $(17 + \sqrt{-17})/3$
	S_2	$(z - 6)(z' + 6) = -2$ to $7 + \sqrt{-2}$
	S_{17}	$(z - 17/2)(z' - 17/2) = -17/4$ to $11 + \sqrt{-2}$
	S_{34}	$(z - 34/3)(z' + 34/3) = -34/9$ to $13 + i$
	S_{17}	$(z - 17)(z' - 17) = -17$ to $17 + \sqrt{-17}$.
Case 46	S_{46}	$zz' = -46$ to $(20 + \sqrt{-14})/3$
	S_{23}	$(z - 23/3)(z' - 23/3) = -23/9$ to $(26 + \sqrt{-14})/3$
	S_{46}	$(z - 46/5)(z + 46/5) = -46/25$ to $(19 + \sqrt{-7})/2$
	S_{23}	$(z - 23/2)(z' - 23/2) = -23/4$ to $(27 + \sqrt{-7})/2$
	S_2	$(z - 18)(z' + 18) = -2$ to $(92 + \sqrt{-46})/5$
	S_{23}	$(z - 23)(z' - 23) = -23$ to $23 + \sqrt{-23}$.

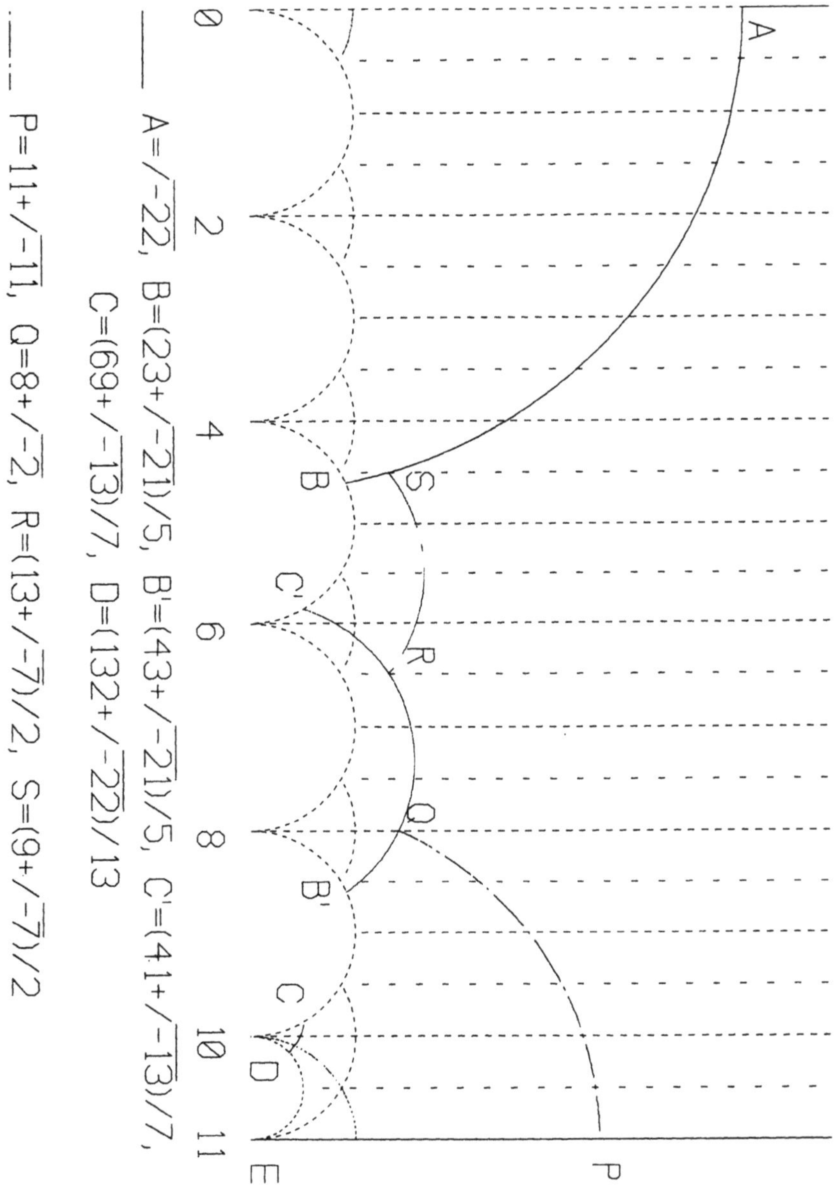

Figure 1. Fundamental domain for Case 22.

The (half) floor of H/Γ^0 is the semicircles connecting $0, 2, 4, 6, 8, 10, 11$.
The (half) floor of H/Γ^* is $ASBC'RQB'CDE$.
The (half) floor of H/Γ^c is $ASRQP$. (See §2 and §6).

7 Concluding remarks

There are many formal similarities of four-valued modular equations to two-valued modular equations, but the plethora of singular moduli (not necessarily located on the boundary of the fundamental domain) makes the computations harder to use. The simple cases in this report are not exhaustive. A single-valued function on H/Γ^c also exists for $N = 55, 69, 94$ but only one two-valued function is available in each case by the method of §3.

The case of $N = 65$ is far more fascinating. We can construct two functions single-valued on H but not on H/Γ^c by the methods of §3, namely

$$\Lambda_1(z) = \frac{\Delta(z/13)\Delta(z/5)}{\Delta(z/65)\Delta(z)} = \frac{\prod(1-r^{5h})^{24}(1-r^{13h})^{24}}{r^{48}\prod(1-r^h)^{24}(1-r^{65h})^{24}},$$

$$u = D_{1,1}(z) = \Lambda_1^{1/24}(z) + \Lambda_1^{-1/24}(z) = 1/r^2 + 1/r + \cdots;$$

$$\Lambda_{13}(z) = \frac{\Delta(z/13)\Delta(z/65)}{\Delta(z/5)\Delta(z)} = \frac{\prod(1-r^{5h})^{24}(1-r^h)^{24}}{r^{72}\prod(1-r^{13h})^{24}(1-r^{65h})^{24}},$$

$$v = D_{13,1}(z) = \Lambda_{13}^{1/24}(z) + 13\Lambda_{13}^{-1/24}(z) = 1/r^3 - 1/r^2 - 1/r + \cdots.$$

Here we find that u and v are not single-valued on H/Γ^c. Actually S_{65} causes no problem because of the fixed-point $z = \sqrt{-65}$, but not so with S_{13}. We can not find a fixed-point or a mirror point as desired in §3. For instance, (3.3b) is not satisfied with $q = \pm q'$, $T = 13$, $s = 1$, because the quadratic residue character $(\pm 13/5) = -1$. If we take S_{13} as $(z+13)(z'-26) = -13$ then two corresponding points are

$$z = (-13 + \sqrt{-13}), \quad z' = (26 + \sqrt{-13}).$$

We can use a trigonometric (arctangent) routine to compare the arguments of $\Lambda_1(z)$, $\Lambda_{13}(z)$ for

$$r = \exp 2\pi i(-13 + \sqrt{-13})/65, \quad r' = \exp 2\pi i(26 + \sqrt{-13})/65.$$

We find that S_{13} transforms u and v into their negatives.

Now we restart the process with new variables defined in H/Γ^c, namely

$$z = u^2, \quad x = uv.$$

Since z is of order four and x is of order five, then $z^5 - x^4$ can be successively expanded into powers of the type $z^a x^b$ with order $4a + 5b < 20$, (actually there is only one monomial for each order). The Laurent series in r then lead to the polynomial

$$F(z,x) = z^5 - 106z^4 + 329z^3 + x(160z^2 - 100z^3) - 260z^2 \\ + x^2(32z - 45z^2) - 10x^3 z - x^4 = 0.$$

This is a curve of genus one equivalent to

$$s^2 = (4t+1)(t^2+4)$$

under the transformation

$$z = 4 + t^2, \quad x = (ts - 2s - 20 - 5t^2)/2.$$

We have the remarkable fact that H/Γ^0 is of genus five but it reduces to genus one under the involution S_{65}. Yet it is not reduced further under the second involution S_{13}, (see [10], [12]).

We might finally remark that there is a viable theory for an eight-valued modular equation to accomodate values of N with three different prime factors (although Fricke [5] has made a four-valued theory work by permitting more than one cusp). At present an eight-valued theory seems too difficult computationally.

Appendix. Data for Case 22.

$$\begin{aligned} p_1(t) = {} & t^{22} + 44t^{21} + 814t^{20} + 7788t^{19} + 33297t^{18} - 63932t^{17} \\ & -1508496t^{16} - 6621164t^{15} + 56144t^{14} + 110192016t^{13} \\ & +391474754t^{12} + 28164273t^{11} - 3379762474t^{10} - 8608599659t^{9} \\ & +48775496t^{8} + 38340977004t^{7} + 66534406290t^{6} + 3418406640t^{5} \\ & -116625182256t^{4} - 129981360960t^{3} - 23762250720t^{2} \\ & +34441459200t + 13793760000 \end{aligned}$$

$$\begin{aligned} p_{22}(t) = {} & (t+3)(t+5)(t+6)(t^{17} + 22t^{16} + 141t^{15} - 264t^{14} - 6588t^{13} \\ & -17710t^{12} + 78368t^{11} + 464288t^{10} + 71808t^{9} - 3964224t^{8} \\ & -6568704t^{7} + 11173935t^{6} + 36586496t^{5} + 5873048t^{4} - 58171392t^{3} \\ & -41354640t^{2} + 13271040t + 10627200) \end{aligned}$$

$$\begin{aligned} p_2(t) = {} & (t-2)t(t+2)(t+3)(t+4)(t+5)(t+6)(t^2-12) \\ & (t^{11} + 22t^{10} + 165t^{9} + 264t^{8} - 2772t^{7} - 14542t^{6} - 11792t^{5} \\ & +80080t^{4} + 209088t^{3} + 85536t^{2} - 207360t - 166095) \end{aligned}$$

$$\begin{aligned} p_{11}(t) = {} & (t-2)t(t+2)(t+3)(t^2-12)(t^{15} + 41t^{14} + 715t^{13} + 6699t^{12} \\ & +33168t^{11} + 42484t^{10} - 478132t^{9} - 3046960t^{8} - 6782192t^{7} \\ & +3671808t^{6} + 55621442t^{5} + 124595241t^{4} + 93780063t^{3} \\ & -66997098t^{2} - 155455290t - 71864550) \end{aligned}$$

$$\begin{aligned} s_{22}(t) = {} & (t-4)(t-2)^2t^2(t+2)^2(t+3)^2(t+4)(t^2-12)^2 \\ & (t^4 + 12t^3 + 42t^2 + 15t - 75)(t^4 + 13t^3 + 61t^2 + 117t + 69) \\ & (t^5 + 16t^4 + 76t^3 + 15t^2 - 666t - 1161) \\ & (t^6 + 19t^5 + 112t^4 + 66t^3 - 1548t^2 - 5220t - 5175) \end{aligned}$$

$$\begin{aligned} s_2(t) = {} & (t+3)^2(t+4)(t^3 + 32t^2 + 180t + 288)^2 \\ & (t^5 - 68t^4 - 1160t^3 - 7456t^2 - 20720t - 20800) \\ & (t^{10} + 41t^9 + 4827t^8 + 158648t^7 + 2548693t^6 + 23700126t^5 \\ & +136492029t^4 + 496032660t^3 \\ & +1110597075t^2 + 1402292250t + 765905625) \end{aligned}$$

$$\begin{aligned} s_{11}(t) = {} & (t+3)^2(t+4)(t+5)^2(t+6)^2(t^2+9)(t^2+6t+10) \\ & (t^3-4t^2-36t+180)(t^3+2t^2-28t-82) \\ & (t^3+8t^2+24t+36)(t^3+12t^2+44t+52) \\ & (t^4-2t^3-14t^2+60t+180) \\ & (t^6+12t^5+44t^4+88t^3+576t^2+2480t+3280) \end{aligned}$$

References

[1] A.O.L. Atkin and J. Lehner, *Hecke operators on* $\Gamma_0(m)$, Math. Ann. **185** (1970), 134–160.

[2] H. Cohn, "Introduction to the Construction of Class Fields", Cambridge University Press 1985.

[3] H. Cohn, *Fricke's Two-Valued Modular Equation*, Math. of Comput. **51** (1988), 787–807.

[4] H. Cohn, *Singular moduli and modular equations for Fricke's cases*, Proc. Symp. Pure Math. **49(2)** (1989), Proc. Bowdoin AMS Summer Institute 1987, 233–249.

[5] R. Fricke, *Lehrbuch der Algebra III (Algebraische Zahlen)*, Vieweg, Braunschweig 1928.

[6] R. Fricke, *Über die Berechnung der Klasseninvarianten*, Acta Arithmetica **52** (1929), 257–279.

[7] R. Fricke, *Transformation der elliptischen Funktionen bei zusammengesetzem Transformationsgrade*, Math. Ann. **101** (1929), 316–341.

[8] B. Gross and D. Zagier, *On singular moduli*, J. Reine und angew. Math. **355** (1985), 191–220.

[9] E. Kaltofen and N. Yui, *Explicit construction of the Hilbert class fields of imaginary quadratic fields with class numbers* 7 *and* 11, Lecture Notes in Computer Science **174** (1984), Eurosam 84, 310–320

[10] G. Ligozat, *Courbes Modulaires de genre* 1, Bull. de la Soc. Math. de France **43** Supplement (1975), 1–80.

[11] M. Newman, *Construction and application of a class of modular functions*, Proc. London Math. Soc. **7** (1957), 334–350.

[12] A. P. Ogg, *Hyperelliptic modular curves*, Bull. de la Soc. Math. de France **102** (1974), 107–148.

[13] H. Weber "Elliptische Funktionen und algebraische Zahlen" Vieweg, Braunschweig 1891.

Congruent numbers and elliptic curves

P. Serf

Abstract. The problem of determining whether or not a given positive integer is a *congruent number*, i.e. the area of a right triangle with rational sides, is very old. Only recently, in 1983, it has been solved almost completely by J.B. Tunnell ([12]), who gave a simple equivalence criterion, one direction of which however depends on the truth of a weak version of the famous Birch and Swinnerton-Dyer conjecture.

The aim of the present paper is threefold: We construct infinite series of congruent and non-congruent numbers, provide a method of calculating an upper bound for the rank of the corresponding elliptic curves and produce a table of all square-free congruent numbers between 2000 and 3000.

1 The connection between congruent numbers and elliptic curves

From the definition given above, we know that a positive integer n is a congruent number iff there exist positive rational numbers X, Y and Z such that

$$\left\{ \begin{array}{rcl} \frac{1}{2} \cdot X \cdot Y & = & n \\ X^2 + Y^2 & = & Z^2 \end{array} \right\}.$$

First of all, it is clear that it suffices to consider square-free numbers, because for $n' := \alpha^2 n$ ($\alpha \in \mathbb{N}$)

$$\begin{array}{ll} & X,\ Y,\ Z \text{ are sides of a triangle belonging to } n \\ \text{iff} & \alpha X,\ \alpha Y,\ \alpha Z \text{ are sides of a triangle belonging to } n'. \end{array}$$

The following theorem, whose proof is elementary (cf. e.g. [10], Ch.2), shows the relation between congruent numbers and elliptic curves.

Theorem 1.1. *Let n be a square-free positive integer and E^n be the elliptic curve over $\mathbb{Q}$ defined by the equation*

$$E^n :\ y^2 = x^3 - n^2 x.$$

Then

$$n \textit{ is a congruent number} \iff \operatorname{rank}_{\mathbb{Q}}(E^n) > 0.$$

Computational Number Theory

2 An infinite series of congruent numbers

In 1984, J.S. Chahal ([2]) showed (by using the identity of A. Desboves given below) that all positive integers of the form

$$n = m(1+4m^2) \qquad (m \in \mathbb{N})$$

are congruent numbers.

His result can be generalized as follows:

Theorem 2.1. *All positive integers of the series*

$$n = \tfrac{1}{2} m_1 m_2 (m_1^2 + m_2^2) \qquad \textit{for } m_1, m_2 \in \mathbb{N},\ \gcd(m_1, m_2) = 1,\ m_1 \cdot m_2 > 1$$

are congruent numbers.

The following two lemmas are needed in the proof.

Lemma 2.2. *Let $a, b \in \mathbb{Z}$ with $\gcd(a,b) = 1$, $\frac{a}{b} \neq -\frac{1}{2}$ and*

$$A := 8ab(2a-b)^2 \neq 0 \qquad (i.e.\ a \neq 0,\ b \neq 0,\ 2a \neq b).$$

Then, the elliptic curve

$$E:\ y^2 = x^3 + Ax$$

over $\mathbb{Q}$ has positive rank, and the rational point $P = (\tilde{x}, \tilde{y})$ with coordinates

$$\tilde{x} = \frac{(b^2 - 12ab + 4a^2)^2}{4(b+2a)^2},$$

$$\tilde{y} = \frac{(b^2 - 12ab + 4a^2)(b^4 + 40ab^3 - 104a^2b^2 + 160a^3b + 16a^4)}{8(b+2a)^3}$$

belongs to E and has infinite order.

The fact that P lies on E can be verified by choosing $u = b - 2a$ and $v = 4a$ in Desboves' identity

$$\begin{aligned}&(v^2 + 2uv - u^2)^4 + (2u^3v + u^2v^2)(2(u+v))^4 \\ =\ &(u^4 + 12u^3v + 10u^2v^2 + 4uv^3 + v^4)^2\end{aligned}$$

(cf. [4]).
To prove that P is a non-torsion point, one simply shows that the numerator of $\tilde{y}$ is non-zero by considering the first factor modulo 3 and the second modulo 5. Here, we use the fact that every non-trivial torsion point on E must have y-coordinate 0 (cf. e.g. [6], I, Th.3.2).

Lemma 2.3. *Let A be as in Lemma 2.2. Then*

$$A = -n^2 \iff \begin{array}{c} n \textit{ is of the form } n = 8m_1m_2(m_1^2 + m_2^2) \textit{ for} \\ m_1, m_2 \in \mathbb{N},\ \gcd(m_1, m_2) = 1,\ 2 \nmid m_1,\ m_1 \cdot m_2 > 1. \end{array}$$

The proof of Lemma 2.3 is just an examination of different cases for a and b such that $A = 8ab(2a - b)^2$ (cf. [10], Ch.3).

Now, it is easy to deduce Theorem 2.1:
A combination of Lemma 2.2, Lemma 2.3 and Theorem 1.1 shows, that integers of the form

$$n \;=\; 8m_1m_2(m_1^2+m_2^2) \quad \text{for } m_1, m_2 \in \mathbb{N},\ \gcd(m_1,m_2)=1,\ 2\nmid m_1,\ m_1\cdot m_2>1$$

are congruent numbers. The product $m_1m_2(m_1^2 + m_2^2)$ is divisible by 2, because either m_1m_2 or $m_1^2 + m_2^2$ is even. Therefore, the expression $8m_1m_2(m_1^2 + m_2^2)$ can be divided by the square 16, since congruent numbers may be assumed to be square-free. This means that n is of the form

$$n \;=\; \tfrac{1}{2}m_1m_2(m_1^2+m_2^2) \quad \text{for } m_1, m_2 \in \mathbb{N},\ \gcd(m_1,m_2)=1,\ 2\nmid m_1,\ m_1\cdot m_2>1.$$

Finally, the condition "$2\nmid m_1$" may be dropped, because $\frac{1}{2}m_1m_2(m_1^2 + m_2^2)$ is symmetric in m_1 and m_2, and because m_1 and m_2 must be relatively prime.

Remark 2.4. Taking

$$m_1 = 1 \quad \text{and} \quad m_2 = 2m \ (m \in \mathbb{N})$$

in Theorem 2.1 yields

$$n = m(1+4m^2),$$

which is Chahal's series.

3 An upper bound for $\text{rank}_{\mathbb{Q}}(E^n)$

In this section, we shall describe a method of determining an upper bound for $\text{rank}_{\mathbb{Q}}(E^n)$. It is based on the next theorem, which is a special case of the complete 2-descent for elliptic curves over algebraic number fields (cf. [11], X, §1).

Theorem 3.1 (Complete 2-descent for E^n over $\mathbb{Q}$) *Let*

$$n = 2^{\epsilon}p_1\cdot\ldots\cdot p_l$$

be a square-free positive integer with $\epsilon \in \{0,1\}$, $l \in \mathbb{N}_0$, $p_1, \ldots, p_l \in \mathbb{P}\setminus\{2\}$ (where $\mathbb{P}$ is the set of all prime numbers), E^n be the elliptic curve over $\mathbb{Q}$ defined by the equation

$$E^n:\ y^2 = x^3 - n^2x = x(x-n)(x+n),$$

$$S = \{\infty, 2, p_1, \ldots, p_l\}$$

be a finite subset of $M_{\mathbb{Q}}$, the set of all places of $\mathbb{Q}$, and

$$\mathbb{Q}(S,2) \;:=\; \{\, a \in \mathbb{Q}^{\times}/\mathbb{Q}^{\times 2} \mid v_p(a) \equiv 0(2) \quad \forall\, p \in M_{\mathbb{Q}}\setminus S\,\}.$$

(1) *There exists an injective homomorphism*

$$b: E^n(\mathbb{Q})/2E^n(\mathbb{Q}) \hookrightarrow \mathbb{Q}(S,2)\times\mathbb{Q}(S,2)$$

$$P \mapsto \left\{ \begin{array}{ll} (1,1) & P=\mathcal{O}=(\infty,\infty) \\ (-1,-n) & P=(0,0) \\ (n,2) & P=(n,0) \\ (x,x-n) & P=(x,y)\neq\mathcal{O},(0,0),(n,0) \end{array} \right\}.$$

(2) *If* $(b_1,b_2)\in\mathbb{Q}(S,2)\times\mathbb{Q}(S,2)\setminus\{(1,1),(-1,-n),(n,2)\}$, *then*

$$(b_1,b_2)\in \text{image}(b) \iff \exists\ (z_1,z_2,z_3)\in\mathbb{Q}^\times\times\mathbb{Q}^\times\times\mathbb{Q} \therefore$$

$$\left\{ \begin{array}{lcl} b_1z_1^2-b_2z_2^2 & = & n \\ b_1z_1^2-b_1b_2z_3^2 & = & -n \end{array} \right\} \quad (*).$$

In this case

$$(b_1,b_2)=b(P)$$

for

$$P=(b_1z_1^2,b_1b_2z_1z_2z_3)=(b_2z_2^2+n,b_1b_2z_1z_2z_3).$$

Remark 3.2. On choosing square-free integers as representatives for $\mathbb{Q}(S,2)$, one obtains the set

$$\{\pm 2^\epsilon p_1^{\epsilon_1}\cdot\ldots\cdot p_l^{\epsilon_l}\mid \epsilon,\epsilon_1,\ldots,\epsilon_l\in\{0,1\}\} =: R,$$

showing in particular, that the cardinality of $\mathbb{Q}(S,2)$ is

$$\#\mathbb{Q}(S,2)=2^{l+2}.$$

Now, we shall demonstrate how to calculate the rank r of E^n over $\mathbb{Q}$, if the image of the homomorphism b is known:
We have

$$E^n_{\text{tors}}(\mathbb{Q})\cong\mathbb{Z}/2\mathbb{Z}\times\mathbb{Z}/2\mathbb{Z}$$

(cf. e.g. [6], I, Th.3.2).
The theorem of Mordell-Weil states that $E^n(\mathbb{Q})$ is of isomorphism type

$$E^n(\mathbb{Q})\cong\mathbb{Z}/2\mathbb{Z}\times\mathbb{Z}/2\mathbb{Z}\times\mathbb{Z}^r.$$

Therefore,

$$E^n(\mathbb{Q})/2E^n(\mathbb{Q})\cong(\mathbb{Z}/2\mathbb{Z})^{r+2}$$

and

$$\text{image}(b)\cong(\mathbb{Z}/2\mathbb{Z})^{r+2},$$

since b is injective. Hence,

$$r=\log_2\left(\frac{\#\text{image}(b)}{4}\right).$$

For determining the rank r, one must decide whether a given pair $(b_1,b_2)\in\mathbb{Q}(S,2)\times\mathbb{Q}(S,2)$ (respectively $R\times R$) is or is not contained in image(b), i.e. whether the system of equations $(*)$ is or is not solvable.
In general, it is difficult to establish solvability; but it is relatively easy to derive sufficient criteria for unsolvability by considering $(*)$ modulo p ($p\in\mathbb{P}$) for $p=2$ and $p\mid n$ (cf. [10], Ch.5).

Theorem 3.3 (Unsolvability-conditions) *Let*

$$n = 2^{\epsilon} p_1 \cdot \ldots \cdot p_l$$

be a square-free positive integer with $\epsilon \in \{0,1\}$, $l \in \mathbb{N}_0$, $p_1, \ldots, p_l \in \mathbb{P} \setminus \{2\}$ *and let* $(b_1, b_2) \in R \times R$. *The system of equations*

$$\left\{ \begin{array}{lcl} b_1 z_1^2 - b_2 z_2^2 & = & n \\ b_1 z_1^2 - b_1 b_2 z_3^2 & = & -n \end{array} \right\}$$

has no solution $(z_1, z_2, z_3) \in \mathbb{Q}^\times \times \mathbb{Q}^\times \times \mathbb{Q}$ *in the following cases:*

General unsolvability-condition:

U(I) $\quad b_1 \cdot b_2 < 0$

Unsolvability-conditions modulo 2:

U(II.1) $\quad 2 \nmid n \ \wedge \ 2 \mid b_1$

Let $n = 2\tilde{n}$, $b_1 = 2\widetilde{b_1}$ *(if* $2 \mid b_1$*) and* $b_2 = 2\widetilde{b_2}$ *(if* $2 \mid b_2$*).*

U(II.2.a) $\quad (v_2(b_1), v_2(b_2)) = (0,0) \ \wedge \ b_1 \equiv 3(4)$
U(II.2.b) $\quad (v_2(b_1), v_2(b_2)) = (0,1) \ \wedge \ b_1 \equiv 1(4)$
U(II.2.c) $\quad (v_2(b_1), v_2(b_2)) = (1,0) \ \wedge \ \tilde{n}\widetilde{b_1} \equiv 1(4)$
U(II.2.d) $\quad (v_2(b_1), v_2(b_2)) = (1,1) \ \wedge \ \tilde{n}\widetilde{b_1} \equiv 3(4)$

Unsolvability-conditions modulo p **for** $p \in \{p_1, \ldots, p_l\}$**:**
Let $p \in \{p_1, \ldots, p_l\}$, $n = p \cdot n'$, $b_1 = p \cdot b_1'$ *(if* $p \mid b_1$*),* $b_2 = p \cdot b_2'$ *(if* $p \mid b_2$*) and denote by* $\left(\frac{a}{p}\right)$ *for* $a \in \mathbb{Z}$ *with* $p \nmid a$ *the Legendre-symbol.*

U(III.a) $\quad (v_p(b_1), v_p(b_2)) = (0,0)$ *and* $\left(\frac{b_1}{p}\right) = -1 \vee \left(\frac{b_2}{p}\right) = -1$
U(III.b) $\quad (v_p(b_1), v_p(b_2)) = (0,1)$ *and* $\left(\frac{-b_1}{p}\right) = -1 \vee \left(\frac{-n'b_2'}{p}\right) = -1$
U(III.c) $\quad (v_p(b_1), v_p(b_2)) = (1,0)$ *and* $\left(\frac{n'b_1'}{p}\right) = -1 \vee \left(\frac{2b_2}{p}\right) = -1$
U(III.d) $\quad (v_p(b_1), v_p(b_2)) = (1,1)$ *and* $\left(\frac{-n'b_1'}{p}\right) = -1 \vee \left(\frac{-2n'b_2'}{p}\right) = -1$

Corollary 3.4. *An upper bound for the rank of* E^n *over* $\mathbb{Q}$ *is given by*

$$\begin{aligned} ubr \ &:= \ \log_2\left(\tfrac{1}{4} \cdot \text{number of pairs } (b_1, b_2) \in R \times R, \text{ for which no unsolvability-conditions hold}\right) \\ &= \ \log_2\left(\tfrac{1}{4} \cdot \text{number of “possibly solvable systems”}\right). \end{aligned}$$

Remark 3.5. We know that

$$\#(R \times R) \ = \ \#\big(\mathbb{Q}(S,2) \times \mathbb{Q}(S,2)\big) \ = \ 2^{2l+4},$$

so that, for each n, a large set of pairs $(b_1, b_2) \in R \times R$ must be examined in order to find out whether the corresponding systems of equations are unsolvable. For $l \in \{2,3\}$ the use of a computer is advisable, and for $l \geq 4$ it is absolutely necessary.

In 1975, Lagrange ([8]) deduced the following conjecture from the well-known conjecture according to which the rank and the number of the first descent have the same parity.

Conjecture 3.6.
The rank of E^n over $\mathbb{Q}$ is $\left\{ \begin{array}{ll} \textit{even} & \textit{for } n \equiv 1,2,3(8) \\ \textit{odd} & \textit{for } n \equiv 5,6,7(8) \end{array} \right\}$.

Remark 3.7. (a) If $ubr = 0$, then $r = 0$ and n is a non-congruent number (cf. Theorem 1.1).
(b) If $ubr = 1$ and $n \equiv 5,6,7(8)$, then $r = 1$ and n is a congruent number (provided Conjecture 3.6 is true).

The following table contains examples illustrating Remark 3.7.(a) and (b) in the case of $l = 1,2$ or 3 odd prime factors. They have been found on the basis of Theorem 3.3 and Corollary 3.4.
It should be pointed out that all of them were already known previously - except for the congruent numbers of the form $n = pqr$ and $n = 2pqr$ (cf. e.g. [8] and [16]).

Table 3.8. Non-congruent numbers

$n = p$	with	$p \equiv 3(8)$
$n = 2p$	with	$p \equiv 5(8)$
$n = pq$	with	$p \equiv 1(8),\ q \equiv 3(8),\ \left(\frac{p}{q}\right) = -1$
	or	$p \equiv 3(8),\ q \equiv 3(8)$
	or	$p \equiv 5(8),\ q \equiv 7(8),\ \left(\frac{p}{q}\right) = -1$
$n = 2pq$	with	$p \equiv 1(8),\ q \equiv 5(8),\ \left(\frac{p}{q}\right) = -1$
	or	$p \equiv 3(8),\ q \equiv 3(8)$
	or	$p \equiv 3(8),\ q \equiv 7(8),\ \left(\frac{p}{q}\right) = -1$
	or	$p \equiv 5(8),\ q \equiv 5(8)$
e.g. $n = pqr$	with	$p \equiv 1(8),\ q \equiv 3(8),\ r \equiv 3(8)$ and $\left(\frac{p}{q}\right) = -\left(\frac{p}{r}\right)$
	or	$p \equiv 3(8),\ q \equiv 5(8),\ r \equiv 7(8)$ and $\left(\frac{q}{r}\right) = -1$
	or	$p \equiv 3(8),\ q \equiv 7(8),\ r \equiv 7(8)$ and $\left(\frac{p}{q}\right) = -\left(\frac{p}{r}\right) = \left(\frac{q}{r}\right)$
e.g. $n = 2pqr$	with	$p \equiv 1(8),\ q \equiv 3(8),\ r \equiv 3(8)$ and $\left(\frac{p}{q}\right) = -\left(\frac{p}{r}\right)$
	or	$p \equiv 1(8),\ q \equiv 5(8),\ r \equiv 5(8)$ and $\left(\frac{p}{q}\right) = -\left(\frac{p}{r}\right)$
	or	$p \equiv 3(8),\ q \equiv 5(8),\ r \equiv 7(8)$ and $\left(\frac{p}{r}\right) = -\left(\frac{q}{r}\right)$
	or	$p \equiv 5(8),\ q \equiv 7(8),\ r \equiv 7(8)$ and $\left(\frac{p}{q}\right) = -\left(\frac{p}{r}\right) = \left(\frac{q}{r}\right)$

Table 3.9. Congruent numbers n with $\text{rank}_{\mathbb{Q}}(E^n) = 1$
(modulo the truth of Lagrange's conjecture)

$n = p$	with	$p \equiv 5, 7(8)$
$n = 2p$	with	$p \equiv 3, 7(8)$
$n = pq$	with	$p \equiv 1(8),\ q \equiv 5(8),\ \left(\frac{p}{q}\right) = -1$
	or	$p \equiv 1(8),\ q \equiv 7(8),\ \left(\frac{p}{q}\right) = -1$
	or	$p \equiv 3(8),\ q \equiv 5(8)$
	or	$p \equiv 3(8),\ q \equiv 7(8)$
$n = 2pq$	with	$p \equiv 1(8),\ q \equiv 3(8),\ \left(\frac{p}{q}\right) = -1$
	or	$p \equiv 1(8),\ q \equiv 7(8),\ \left(\frac{p}{q}\right) = -1$
	or	$p \equiv 3(8),\ q \equiv 5(8)$
	or	$p \equiv 5(8),\ q \equiv 7(8)$
e.g. $n = pqr$	with	$p \equiv 3(8),\ q \equiv 3(8),\ r \equiv 5(8)$
	or	$p \equiv 3(8),\ q \equiv 3(8),\ r \equiv 7(8)$
	or	$p \equiv 7(8),\ q \equiv 7(8),\ r \equiv 7(8)$ and $\left(\frac{p}{q}\right) = -\left(\frac{p}{r}\right) = \left(\frac{q}{r}\right)$
e.g. $n = 2pqr$	with	$p \equiv 3(8),\ q \equiv 3(8),\ r \equiv 7(8)$
	or	$p \equiv 3(8),\ q \equiv 5(8),\ r \equiv 5(8)$
	or	$p \equiv 5(8),\ q \equiv 5(8),\ r \equiv 7(8)$
	or	$p \equiv 7(8),\ q \equiv 7(8),\ r \equiv 7(8)$ and $\left(\frac{p}{q}\right) = -\left(\frac{p}{r}\right) = \left(\frac{q}{r}\right)$

We also obtained a lot of results for $l \in \{4, 5, 6\}$, but it is almost impossible to exhibit them in a reasonable way, because the number of Legendre-symbols involved is equal to $(l-1)l/2$, so that there are $2^{(l-1)l/2}$ possibilities for these symbols ($2^6 = 64$ for $l = 4$, $2^{10} = 1024$ for $l = 5$, $2^{15} = 32768$ for $l = 6$). Therefore, we confine ourselves here to give just 4 examples for $l = 4$.

Table 3.10. Non-congruent numbers

$n = pqrs$	with	$p \equiv 5(8),\ q \equiv 5(8),\ r \equiv 7(8),\ s \equiv 7(8)$
	and	$1 = \left(\frac{p}{r}\right) = -\left(\frac{p}{s}\right) = -\left(\frac{q}{r}\right)$
	or	$1 = -\left(\frac{p}{r}\right) = \left(\frac{p}{s}\right) = -\left(\frac{q}{s}\right)$
	or	$1 = -\left(\frac{p}{r}\right) = -\left(\frac{p}{s}\right),\ \left(\frac{q}{r}\right) = -\left(\frac{q}{s}\right)$
$n = 2pqrs$	with	$p \equiv 1(8),\ q \equiv 1(8),\ r \equiv 3(8),\ s \equiv 3(8)$
	and	$\left(\frac{p}{q}\right) = 1,\ \left(\frac{p}{r}\right) = -\left(\frac{p}{s}\right),\ \left(\frac{q}{r}\right) = -\left(\frac{q}{s}\right)$
	or	$\left(\frac{p}{q}\right) = -1,\ \left(\frac{p}{r}\right) = \left(\frac{p}{s}\right),\ \left(\frac{q}{r}\right) = -\left(\frac{q}{s}\right)$
	or	$\left(\frac{p}{q}\right) = -1,\ \left(\frac{p}{r}\right) = -\left(\frac{p}{s}\right)$

Table 3.11. Congruent numbers n with $\operatorname{rank}_{\mathbb{Q}}(E^n) = 1$ (modulo the truth of Lagrange's conjecture)

$n = pqrs$	with $p \equiv 1(8),\ q \equiv 3(8),\ r \equiv 3(8),\ s \equiv 5(8)$ and $1 = \left(\frac{p}{q}\right) = \left(\frac{p}{r}\right) = \left(\frac{p}{s}\right)$ or $1 = -\left(\frac{p}{q}\right) = -\left(\frac{p}{r}\right) = \left(\frac{p}{s}\right),\ \left(\frac{q}{s}\right) = \left(\frac{r}{s}\right)$
$n = 2pqrs$	with $p \equiv 1(8),\ q \equiv 3(8),\ r \equiv 5(8),\ s \equiv 5(8)$ and $1 = \left(\frac{p}{q}\right) = \left(\frac{p}{r}\right) = \left(\frac{p}{s}\right)$ or $1 = \left(\frac{p}{q}\right) = -\left(\frac{p}{r}\right) = -\left(\frac{p}{s}\right),\ \left(\frac{q}{r}\right) = \left(\frac{q}{s}\right)$

Since this method yields only an upper bound *ubr* for the rank r, it would be interesting to know to what extent *ubr* differs from r.
Therefore, we have compared our upper bounds for r with the lower bounds obtained by A. Wiman in 1944-1948 ([13], [14], [15]) for some numerical examples.

Table 3.12

number n	factorization	$r \geq \dots$	$r \leq \dots$
2	2	0	0
3	3	0	0
5,13,29,37,61	5,13,29,37,61	1	1
7,23,31,47,71	7,23,31,47,71	1	1
41,137,761	41,137,761	2	2
6,22,38,86	$2 \cdot\ 3, 11, 19, 43$	1	1
14,46,62,94,158,254	$2 \cdot\ 7, 23, 31, 47, 79, 127$	1	1
34,194,226,386,514,674	$2 \cdot\ 17, 97, 113, 193, 257, 337$	2	2
15	$3 \cdot 5$	1	1
65	$5 \cdot 13$	2	2
145	$5 \cdot 29$	2	2
1241	$17 \cdot 73$	2	2
2329	$17 \cdot 137$	$\rightarrow$ 2 $\leftarrow$	$\rightarrow$ 4 $\leftarrow$
3281	$17 \cdot 193$	2	2
210	$2 \cdot 3 \cdot 5 \cdot 7$	2	2
330	$2 \cdot 3 \cdot 5 \cdot 11$	2	2
1254	$2 \cdot 3 \cdot 11 \cdot 19$	3	3
42486	$2 \cdot 3 \cdot 73 \cdot 97$	$\rightarrow$ 3 $\leftarrow$	$\rightarrow$ 5 $\leftarrow$
112326	$2 \cdot 3 \cdot 97 \cdot 193$	$\rightarrow$ 3 $\leftarrow$	$\rightarrow$ 5 $\leftarrow$
1155	$3 \cdot 5 \cdot 7 \cdot 11$	2	2
12155	$5 \cdot 11 \cdot 13 \cdot 17$	2	2

Table 3.12 (continued)

number n	factorization	$r \geq \ldots$	$r \leq \ldots$
2730	$2 \cdot 3 \cdot 5 \cdot 7 \cdot 13$	2	2
3570	$2 \cdot 3 \cdot 5 \cdot 7 \cdot 17$	2	2
22134	$2 \cdot 3 \cdot 7 \cdot 17 \cdot 31$	3	3
29274	$2 \cdot 3 \cdot 7 \cdot 17 \cdot 41$	4	4
153510	$2 \cdot 3 \cdot 5 \cdot 7 \cdot 17 \cdot 43$	3	3
163590	$2 \cdot 3 \cdot 5 \cdot 7 \cdot 19 \cdot 41$	3	3
588126	$2 \cdot 3 \cdot 7 \cdot 11 \cdot 19 \cdot 67$	3	3
1978086	$2 \cdot 3 \cdot 11 \cdot 17 \cdot 41 \cdot 43$	3	3
21677754	$2 \cdot 3 \cdot 7 \cdot 17 \cdot 97 \cdot 313$	$\rightarrow$ 3 $\leftarrow$	$\rightarrow$ 4 $\leftarrow$
87636486	$2 \cdot 3 \cdot 7 \cdot 23 \cdot 257 \cdot 353$	$\rightarrow$ 3 $\leftarrow$	$\rightarrow$ 5 $\leftarrow$
510510	$2 \cdot 3 \cdot 5 \cdot 7 \cdot 11 \cdot 13 \cdot 17$	3	3
881790	$2 \cdot 3 \cdot 5 \cdot 7 \cdot 13 \cdot 17 \cdot 19$	3	3
1009470	$2 \cdot 3 \cdot 5 \cdot 7 \cdot 11 \cdot 19 \cdot 23$	3	3
3233230	$2 \cdot 5 \cdot 7 \cdot 11 \cdot 13 \cdot 17 \cdot 19$	3	3
13123110	$2 \cdot 3 \cdot 5 \cdot 7 \cdot 11 \cdot 13 \cdot 19 \cdot 23$	3	3

In the most cases, the upper and lower bounds coincide, i.e. they are exact. There are only 5 cases (marked by "$\rightarrow$ $\leftarrow$"), in which they differ; and for $n = 21677754 \equiv 2(8)$, r equals $ubr = 4$ if Conjecture 3.6 is true.

4 The L-function $L(E^n, s)$

In this final section, we shall use the L-series $L(E^n, s)$ of E^n for computing the rank r of E^n over $\mathbb{Q}$.

The famous conjecture of Birch and Swinnerton-Dyer states that

$$\text{ord}_{s=1} L(E, s) = \text{rank}_{\mathbb{Q}}(E)$$

for an elliptic curve E over $\mathbb{Q}$.

In 1987, K. Rubin ([9]) applied theorems of Coates and Wiles ([3]) as well as Gross and Zagier ([5]) to prove one part of this conjecture:

Theorem 4.1. *Let E be an elliptic curve over $\mathbb{Q}$ with complex multiplication and $r := \text{rank}_{\mathbb{Q}}(E)$. Then, the following implication is true:*

$$\text{ord}_{s=1} L(E, s) \leq 1 \implies \text{ord}_{s=1} L(E, s) = r.$$

In particular,

$$L(E, 1) \neq 0 \implies r = 0$$

$$\textit{and} \quad L(E, 1) = 0 \wedge L'(E, 1) \neq 0 \implies r = 1.$$

For square-free numbers $n < 3000$, we computed $L(E^n, 1)$ (if $n \equiv 1, 2, 3(8)$) and $L'(E^n, 1)$ (if $n \equiv 5, 6, 7(8)$) using the following formulas, which can be derived (cf. [10], Ch.8) from the general formula for $L^{(i)}(E, 1)$ given in [1]. (In [1], E is a modular curve defined over $\mathbb{Q}$ and $i \in \mathbb{N}_0$.)

Proposition 4.2. *Let n be a square-free positive integer. Then*

$$L(E^n, 1) = 2 \sum_{\substack{m=1 \\ (m,2n)=1}}^{\infty} \frac{a_m^{(1)}}{m} \cdot \left(\frac{n}{m}\right) \cdot e^{-2\pi m/\sqrt{N}} \qquad \text{for } n \equiv 1, 2, 3(8)$$

and

$$L'(E^n, 1) = 2 \sum_{\substack{m=1 \\ (m,2n)=1}}^{\infty} \frac{a_m^{(1)}}{m} \cdot \left(\frac{n}{m}\right) \cdot \int_{2\pi m/\sqrt{N}}^{\infty} \frac{e^{-t}}{t} dt \qquad \text{for } n \equiv 5, 6, 7(8),$$

where

$$N = \left\{ \begin{array}{ll} 2^5 n^2 & 2 \nmid n \\ 2^4 n^2 & 2 \mid n \end{array} \right\}$$

is the conductor of E^n and

$$L(E^1, s) = \sum_{m=1}^{\infty} a_m^{(1)} m^{-s}.$$

In this way, we were able to extend to 3000 an earlier table of G. Kramarz ([7]) giving all congruent numbers less than 2000. For most of these numbers, it was even possible to determine the rank of E^n over $\mathbb{Q}$ (cf. [10], Ch.8).

It should be pointed out that for the 4 numbers

2306 2818 2833 2953,

we only knew that $L(E^n, 1) = 0$, so that their congruent number property depends on the truth of a weak version of the Birch and Swinnerton-Dyer conjecture stating that

$$L(E, 1) = 0 \iff \text{rank}_{\mathbb{Q}}(E) > 0.$$

An application of Kramarz' method (cf. [7]) showed that 2306 and 2818 are congruent numbers, but this method failed for 2833 and 2953.
For $n = 2953$, we found positive integers m_1 and m_2 such that n is the square-free part of $\frac{1}{2}m_1 m_2(m_1^2 + m_2^2)$, namely $m_1 = 22201$ and $m_2 = 625$. Hence, 2953 is a congruent number by Theorem 2.1.
However, for $n = 2833$, this latter method was not successful either. We believe that the smallest possible solution (m_1, m_2), if at all it can be obtained by Kramarz' method or the method given in Theorem 2.1, is rather large.

Table 4.3. All square-free congruent numbers between 2000 and 3000

2005	2006	2013	2014	2015	2021	2022	2029	2030	2031
2035	2037	2038	2039	2045	2046	2047	2053	2054	2055
2059	2062	2063	2069	2071	2077	2078	2085	2086	2087
2093	2094	2095	2101	2102	2103	2109	2110	2111	2113
2117	2118	2119	2126	2127	2130	2134	2135	2139	2141
2143	2145	2149	2154	2157	2158	2159	2165	2167	2170
2173	2174	2181	2182	2183	2189	2190	2191	2195	2198
2199	2201	2206	2207	2213	2215	2221	2222	2229	2230
2231	2237	2238	2239	2245	2246	2247	2249	2253	2255
2257	2261	2262	2263	2269	2270	2271	2273	2278	2279
2282	2285	2287	2293	2294	2298	2301	2302	2306	2309
2310	2311	2317	2318	2319	2326	2327	2329	2333	2334
2335	2337	2341	2342	2343	2351	2357	2359	2365	2373
2374	2379	2381	2382	2383	2389	2390	2391	2397	2398
2399	2405	2406	2407	2413	2414	2415	2418	2422	2423
2429	2431	2434	2437	2438	2445	2446	2447	2453	2454
2455	2461	2462	2463	2465	2469	2470	2471	2477	2478
2479	2485	2486	2487	2494	2495	2501	2503	2509	2510
2517	2518	2519	2526	2530	2533	2534	2542	2543	2549
2551	2557	2558	2559	2566	2567	2569	2573	2581	2582
2589	2590	2591	2594	2598	2599	2605	2606	2607	2611
2613	2614	2615	2621	2622	2623	2629	2630	2631	2638
2639	2647	2653	2654	2661	2663	2669	2670	2671	2677
2678	2679	2685	2686	2687	2693	2694	2698	2701	2702
2703	2706	2710	2711	2717	2719	2726	2730	2733	2734
2735	2741	2742	2743	2749	2751	2757	2758	2759	2765
2766	2767	2773	2774	2777	2782	2789	2791	2797	2798
2805	2806	2807	2810	2813	2814	2815	2818	2821	2822
2823	2829	2830	2831	2833	2837	2838	2839	2845	2846
2847	2849	2854	2855	2861	2863	2865	2869	2870	2877
2878	2879	2885	2886	2887	2893	2894	2895	2901	2902
2903	2909	2910	2911	2914	2917	2918	2919	2926	2927
2929	2933	2935	2941	2942	2945	2949	2951	2953	2957
2958	2959	2965	2966	2967	2973	2974	2981	2982	2983
2990	2991	2995	2998	2999					

Acknowledgment. I would like to thank Prof. H.G. Zimmer for drawing my attention to this topic and Prof. G. Frey for his suggestion to bring in L-series.

References

[1] S.A. Abramov, M.Y. Rozenblyum, *Approximate calculation of the first term of the Taylor expansion at the identity of the canonical L-function of Weyl curves*, Math. Notes **26** (1979), 960-964.

[2] J.S. Chahal, *On an identity of Desboves*, Proc. Japan Acad. **60**, Ser.A (1984), 105-108.

[3] J. Coates, A. Wiles, *On the conjecture of Birch and Swinnerton-Dyer*, Invent. math. **39** (1977), 223-251.

[4] A. Desboves, Comptes Rendus Paris **87** (1878), 159-161.

[5] B. Gross, D. Zagier, *Heegner points and derivatives of L-series*, Invent. math. **84** (1986), 225-320.

[6] D. Husemöller, "Elliptic curves", Springer, Berlin, 1987.

[7] G. Kramarz, *All congruent numbers less than 2000*, Math. Ann. **273** (1986), 337-340.

[8] J. Lagrange, *Nombres congruents et courbes elliptiques*, Sém. Delange-Pisot-Poitou (Théorie des nombres), 16e anneé, 1974/75, no.16.

[9] K. Rubin, *Tate-Shafarevich groups and L-functions of elliptic curves with complex multiplication*, Invent. math. **89** (1987), 527-559.

[10] P. Serf, *Kongruenzzahlen und elliptische Kurven*, Diploma Thesis, Saarbrücken, 1989.

[11] J.H. Silverman, "The arithmetic of elliptic curves", Springer, New York, 1986.

[12] J.B. Tunnell, *A classical diophantine problem and modular forms of weight* 3/2, Invent. math. **72** (1983), 323-334.

[13] A. Wiman, *Über rationale Punkte auf Kurven dritter Ordnung vom Geschlechte eins*, Acta math. **80** (1948), 223-257.

[14] A. Wiman, *Über den Rang von Kurven* $y^2 = x(x+a)(x+b)$, Acta math. **76** (1944), 225-251.

[15] A. Wiman, *Über rationale Punkte auf Kurven* $y^2 = x(x-c)(x+c)$, Acta math. **77** (1945), 281-320.

[16] H.G. Zimmer, *Congruent numbers - From elementary to algebraic number theory*, ACM-SIGSAM, Bulletin 18-19, 1985.

The rank of elliptic curves upon quadratic extension *

U. Schneiders and H.G. Zimmer

Abstract. The rank of elliptic curves over the rationals $\mathbb{Q}$ is conjectured to be unbounded. To date, however, only curves of ranks up to 14 are known (see [19]). It is therefore natural to pose the question as to how the rank grows upon transition from $\mathbb{Q}$ to an algebraic number field K. We study this question for quadratic fields $K = \mathbb{Q}(\sqrt{d})$. For an infinite series of rank zero elliptic curves over $\mathbb{Q}$ and prime values $d = q$, the rank is shown to go up by at most 0, 1 or 2 depending on the residue behaviour of q. Furthermore, some computer–assisted experiments yield jumps of the rank from 6 to 10, 7 to 10, 8 to 11, 9 to 11, 10 to 12 and 11 to 12 as one passes from $\mathbb{Q}$ to a suitable quadratic field K. These examples provide some numerical evidence for a boundedness conjecture enunciated by Honda (see [8], [24]). At the same time two elliptic curves over $\mathbb{Q}$ of rank 11 and with nontrivial torsion group are found. One of these curves completes the numerical evidence for the truth of a refined unboundedness conjecture enunciated by Kretschmer (see [12]).

Introduction

The purpose of this paper is to study the behavior of the rank of an elliptic curve upon transition from $\mathbb{Q}$ to a quadratic number field $K_d := \mathbb{Q}(\sqrt{d})$. The first part contains a theoretical description of methods for computing the rank of an elliptic curve upon quadratic extension. In the second part an algorithm for computing the rank, based on a method of Tate, is presented. The third part provides some information concerning the choice of the coefficients of an elliptic curve E and the generator d of the quadratic extension K_d of $\mathbb{Q}$ in order to achieve that the rank of E over K_d becomes large. Our investigation is based on two types of curves E over $\mathbb{Q}$:

(1) A list of curves of high rank
(2) A series of curves of rank zero.

In both cases we try to obtain big jumps of the rank upon transition from $\mathbb{Q}$ to $K_d = \mathbb{Q}(\sqrt{d})$. In the last part we illustrate our investigation by choosing some

*A report on this work was delivered to the conference by the second author.

Computational Number Theory

examples of elliptic curves and comparing their rank over $\mathbb{Q}$ to their rank over $K_d = \mathbb{Q}(\sqrt{d})$ for several d's.

In our search for curves of high rank we succeeded in finding two elliptic curves of rank 11 with non trivial torsion group over $\mathbb{Q}$ and one elliptic curve over $\mathbb{Q}$ whose rank jumps at least by four upon transition from $\mathbb{Q}$ to K_d.

Acknowledgment. We thank the referees for some very interesting supplementary remarks.

1 Theoretical background

The famous *Mordell–Weil Theorem* states the following: Let K be an arbitrary algebraic number field and let E denote any elliptic curve defined over K. Then the group of rational points $E(K)$ is a finitely generated group of isomorphism type

$$E(K) \cong E_{tors}(K) \oplus \mathbb{Z}^{r_K}$$

where

$$\begin{array}{rll} & E_{tors}(K) & \text{is the (finite) } \textit{torsion group} \\ \text{and} & r_K = rk_K(E) & \text{is the } \textit{rank} \text{ of } E \text{ over K.} \end{array}$$

Now we restrict ourselves to the case in which K is a quadratic number field. Let $K_d := \mathbb{Q}(\sqrt{d})$, $d \in \mathbb{Z}$ squarefree, denote a quadratic number field and

$$E: \; y^2 = x^3 + Ax + B \qquad (A, B \in \mathbb{Z})$$

be an elliptic curve over $\mathbb{Q}$ with *discriminant*

$$\Delta := -16(4A^3 + 27B^2) \neq 0.$$

Denote by

$$E'_d: \; dy^2 = x^3 + Ax + B$$

the *d-twist* of E over $\mathbb{Q}$. Then E'_d is birationally equivalent to the elliptic curve

$$E_d: \; y^2 = x^3 + Ad^2x + Bd^3$$

over $\mathbb{Q}$ with discriminant

$$\Delta_d = d^6 \Delta.$$

We consider the groups of rational points $E(K_d)$, $E(\mathbb{Q})$ and $E_d(\mathbb{Q})$. For an elliptic curve E defined over $\mathbb{Q}$, the rank of E over a quadratic number field $K_d = \mathbb{Q}(\sqrt{d})$ can be computed by virtue of the following theorem.

Theorem 1.1.

$$rk_{K_d}(E) = rk_{\mathbb{Q}}(E) + rk_{\mathbb{Q}}(E_d).$$

Hence the rank of E over K_d is obtained by computing the ranks of E over $\mathbb{Q}$ and of E_d over $\mathbb{Q}$.

Proof. The result is classical (see [1], [10], [14]). It follows from the fact that, if $Gal(K_d/\mathbb{Q}) = \{e, \sigma\}$ is the Galois group of $K_d/\mathbb{Q}$, the map

$$\begin{array}{rcl} E(K_d) & \longrightarrow & E(\mathbb{Q}) \oplus E_d(\mathbb{Q}) \\ P & \longmapsto & (P + P^\sigma, P - P^\sigma) \end{array}$$

has finite kernel and cokernel. □

If one assumes the truth of the conjecture of Birch and Swinnerton–Dyer, the result can also be obtained by looking at L-series. Since L-series will also be used in Section 4, we outline here the corresponding proof. The *L-series* of E over K can be defined as an *Euler product* of the form:

$$L_K(E,s) \quad = \prod_{\substack{\wp \in M_K^0 \\ E \text{ has good reduction at } \wp}} \frac{1}{1 - a_\wp q_\wp^{-s} + q_\wp^{1-2s}} \cdot \prod_{\substack{\wp \in M_K^0 \\ E \text{ has bad reduction at } \wp}} \frac{1}{1 - e_\wp q_\wp^{-s}},$$

converging for $\mathrm{Re}(s) > \frac{3}{2}$. Here M_K^0 is the set of non–archimedian absolute values on K, $q_\wp = \#k_\wp$, $a_\wp = \#k_\wp - (1 + \#\widehat{E}(k_\wp))$, $k_\wp$ the residue class field mod $\wp$, $\widehat{E}(k_\wp)$ the reduction of $E(K)$ mod $\wp$ and

$$e_\wp = \begin{cases} 1, & \text{if } E \text{ has split multiplicative reduction at } \wp \\ -1, & \text{if } E \text{ has non–split multiplicative reduction at } \wp \\ 0, & \text{if } E \text{ has additive reduction at } \wp. \end{cases}$$

At this point we assume the truth of the conjecture of Birch and Swinnerton–Dyer:

Conjecture. *The L-series $L_K(E,s)$ of E over a number field K has an analytic continuation beyond the line $Re(s) = \frac{3}{2}$ and vanishes at $s = 1$ of order $r_K(E) =$ rank of E over K.*

Then one establishes the product formula

$$L_{K_d}(E,s) = L_{\mathbb{Q}}(E,s) \cdot L_{\mathbb{Q}}(E_d, s)$$

which is true up to a finite number of factors, by showing that the following holds:

a) If p splits completely in K_d, i.e. i $p \cong \wp\wp'$, then:
$a_\wp = a'_\wp = a_p = a_p^{(d)}$.

b) If p is inert in K_d, i.e. $p \cong \wp$, then:
i) $a_\wp = a_p^2 - 2p$
ii) $a_p = -a_p^{(d)}$.

Here $a_\wp$ and $a'_\wp$ denote the coefficients corresponding to the places $\wp$ and $\wp'$ in the Euler product of $L_{K_d}(E,s)$, and a_p, $a_p^{(d)}$ are analogously defined with respect to $L_{\mathbb{Q}}(E,s)$, $L_{\mathbb{Q}}(E_d,s)$, respectively. The product formula for the L-series now implies the desired relation $rk_{K_d}(E) = rk_{\mathbb{Q}}(E) + rk_{\mathbb{Q}}(E_d)$.

2 An algorithm for computing the rank of an elliptic curve

We now describe the method of Tate [26], as modified by Penney & Pomerance [22], [23] and by Kretschmer [11], [12]. Let us consider the elliptic curve

$$E: \; y^2 = x(x^2 + ax + b) \qquad (a, b \in \mathbb{Z})$$

over $\mathbb{Q}$, having a torsion group $E_{tors}(\mathbb{Q}) \supseteq \mathbb{Z}/2\mathbb{Z}$. The discriminant of E is

$$\Delta = 2^4 b^2 (a^2 - 4b) \neq 0$$

and its d-twist for $d \in \mathbb{Z}, d \neq 1$ squarefree, is

$$E_d: \; y^2 = x(x^2 + adx + bd^2)$$

over $\mathbb{Q}$ with discriminant $\Delta_d = d^6 \Delta$. The torsion point of order 2

$$P_0 = (0,0) \in E_{tors}(\mathbb{Q})$$

gives rise to the 2-isogenous elliptic curve

$$\widetilde{E}: \; \widetilde{y}^2 = \widetilde{x}(\widetilde{x}^2 + \widetilde{a}\widetilde{x} + \widetilde{b}) \qquad (\widetilde{a}, \widetilde{b} \in \mathbb{Z})$$

over $\mathbb{Q}$ with coefficients $\widetilde{a} = -2a$, $\widetilde{b} = a^2 - 4b \in \mathbb{Z}$ via $\widetilde{E} = E/ < P_0 >$. The isogeny is explicitly given as the homomorphism

$$\begin{array}{rcl} \iota: \; E(\mathbb{Q}) & \longrightarrow & \widetilde{E}(\mathbb{Q}) \\ P = (x,y) & \longmapsto & \widetilde{P} = (\widetilde{x}, \widetilde{y}) \quad \text{with} \quad \widetilde{x} = \dfrac{y^2}{x^2}, \quad \widetilde{y} = y\dfrac{x^2 - b}{x^2} \end{array}$$

and has kernel Ker $\iota = \{\mathcal{O}, P_0\}$, # Ker $\iota = 2$. The 2-descent homomorphism

$$\alpha: \; E(\mathbb{Q}) \longrightarrow \mathbb{Q}^\times / \mathbb{Q}^{\times^2}$$

is defined by setting, for $P = (x,y) \in E(\mathbb{Q})$,

$$\alpha(P) := \begin{cases} 1\mathbb{Q}^{\times^2} & \text{if } P = \mathcal{O} \\ b\mathbb{Q}^{\times^2} & \text{if } P = P_0 \\ x\mathbb{Q}^{\times^2} & \text{if } P \neq \mathcal{O}, P_0. \end{cases}$$

Since $2P = (x_2, y_2) \in E(\mathbb{Q})$ has x-coordinate

$$x_2 = \left(\frac{x^2 - b}{2y}\right)^2 \in \mathbb{Q}^{\times^2},$$

we have Ker $\alpha \supseteq 2E(\mathbb{Q})$. We define analogously

$$\widetilde{\alpha}: \; \widetilde{E}(\mathbb{Q}) \longrightarrow \mathbb{Q}^\times / \mathbb{Q}^{\times^2}.$$

The rank $r = rk_{\mathbb{Q}}(E)$ is given by the following formula (see [26], [4]):
Theorem 2.1.

$$2^r = \frac{\#\alpha E(\mathbb{Q}) \cdot \#\widetilde{\alpha}\widetilde{E}(\mathbb{Q})}{2^2}. \qquad (*)$$

Hence it remains to determine the images $\alpha E(\mathbb{Q})$, $\widetilde{\alpha}\widetilde{E}(\mathbb{Q})$. The image $\alpha E(\mathbb{Q})$ is computed as follows (see [26], [4]):

Theorem 2.2. *Let*

$$T := \{1, b\} \quad \cup \quad \{\ b_1 \in \mathbb{Z} \ \mid\ b = b_1 b_2 \text{ and } W^2 = b_1 U^4 + aU^2V^2 + b_2 V^4 \text{ has a solution } (U, V, W) \in \mathbb{Z}^3, \text{ such that } UV \neq 0\ \}.$$

Then $\alpha E(\mathbb{Q}) = < T\mathbb{Q}^{\times 2} >$.

Corollary 2.3. *Let*

$$S := \{1, b\} \cup \{\ b_1 \in \mathbb{Z} \ \mid\ b \ = \ b_1 b_2 \text{ and } b_1 + a + b_2 \in \mathbb{Q}^{\times^2}\ \}.$$

Then $\alpha E(\mathbb{Q}) \supseteq B := < S\mathbb{Q}^{\times^2} >$.

Hence, we obtain the estimate $\#\alpha E(\mathbb{Q}) \geq \#B =: 2^s$ $(s \in \mathbb{N})$ and, analogously, $\#\widetilde{\alpha}\widetilde{E}(\mathbb{Q}) \geq \#\widetilde{B} =: 2^{\widetilde{s}}$ $(\widetilde{s} \in \mathbb{N})$. Furthermore, we consider the prime decompositions of b and $\widetilde{b}$:

$$b = \pm \prod_{i=1}^{t} p_i^{\beta_i}, \qquad \widetilde{b} = \pm \prod_{j=1}^{\widetilde{t}} q_j^{\widetilde{\beta}_j}.$$

Then, in the rank formula $(*)$ of Theorem 2.1, the images can be estimated according to

$$\begin{array}{ccccccc} 2^s & = & \#B & \leq & \#\alpha E(\mathbb{Q}) & \leq & 2^{t+1} \\ 2^{\widetilde{s}} & = & \#\widetilde{B} & \leq & \#\widetilde{\alpha}\widetilde{E}(\mathbb{Q}) & \leq & 2^{\widetilde{t}+1}. \end{array}$$

Therefore, we end up with the following inequalities:

$$s + \widetilde{s} - 2 \ \leq \ r \ \leq \ t + \widetilde{t}.$$

Special case: $\widetilde{b} = a^2 - 4b \notin \mathbb{Q}^{\times^2}$.
In this case the torsion group is (cf. [11]) $E_{tors}(\mathbb{Q}) \cong \mathbb{Z}/2\mathbb{Z}$, $\mathbb{Z}/4\mathbb{Z}$ or $\mathbb{Z}/6\mathbb{Z}$. Then, since Ker $\alpha \supseteq 2E(\mathbb{Q})$,

$$2^{r+1} \ = \ \#E(\mathbb{Q})/2E(\mathbb{Q}) \ \geq \ \#E(\mathbb{Q})/\mathrm{Ker}\alpha \ = \ \#\alpha E(\mathbb{Q})$$

so that we have $r \geq \log \#\alpha E(\mathbb{Q}) - 1$; hence the inequality $\#\alpha E(\mathbb{Q}) \geq \#B = 2^s$ entails the upper estimate

$$s - 1 \ \leq \ r.$$

Special subcase: $\widetilde{b} = a^2 - 4b = p$.
is a prime, which means that $\widetilde{t} = 1$. This leads to the estimates

$$s - 1 \leq r \leq t + 1.$$

By determing the set S we obtain a lower bound for the rank.

3 Conditions on the coefficients

The algorithm for computing the rank $r = rk_{\mathbb{Q}}(E)$ consists of the following steps:

(1) Choose $b = \prod_{i=1}^{t} p_i^{\beta_i} \in \mathbb{Z}$ to have a large number t of prime factors.

(2) Choose $a \in \mathbb{Z}$ in such a way that the set

$$S = \{1, b\} \cup \{b_1 \in \mathbb{Z} \mid b = b_1 b_2,\ b_1 + a + b_2 \in \mathbb{Q}^{\times^2}\}$$

and hence the group $B = < S\mathbb{Q}^{\times^2} >$ is large. In particular, ensure that the conditions i) – iv) given below are satisfied.

(3) Estimate the rank from below according to $r \geq s - 1$, where $2^s = \#B$.

In the sequel,

$$\widehat{E}:\ \widehat{y}^2 = \widehat{x}(\widehat{x}^2 + \widehat{a}\widehat{x} + \widehat{b}) \qquad (\widehat{a}, \widehat{b} \in \mathbb{F}_p)$$

denotes the modulo p reduced elliptic curve, where p is to be chosen in such a way that $\widehat{\Delta} \neq 0$, i.e. $p \not| \Delta$.

Empirical conditions on the coefficients (see [11], [17])

i) $a \equiv 1 \bmod 24$

ii) $\widetilde{b} = a^2 - 4b = p'$ is a prime (if possible)

iii) $q \mid b \implies q \not| a$ for primes $q > 3$

iv) $N_p := \#\widehat{E}(\mathbb{F}_p)$ is maximal for a set of small primes $p > 2$, so that $p \not| b$, $p \not| \widetilde{b} = a^2 - 4b$.

N_p should be maximal for a set of small prime numbers not dividing Δ. This is true because, as was observed by Mestre [18], the rank of an elliptic curve is likely to be large if the numbers N_p are maximal for a large set of small prime numbers $p \not| \Delta$. This follows from the conjecture of Birch and Swinnerton–Dyer. We now explain what it means for N_p to be maximal and we first state the (analogue of the) *Riemann hypothesis*:

$$|N_p - (p+1)| \leq 2\sqrt{p}.$$

Now, we put

$$N_{p,max} = p + 1 + \lfloor 2\sqrt{p} \rfloor.$$

Since $\mathbb{Z}/2\mathbb{Z} \subseteq E_{tors}(\mathbb{Q})$ it follows that $2 \mid N_p$. Hence the maximal value attained by N_p is

$$N_p = \begin{cases} N_{p,max} & \text{if } N_{p,max} \text{ is even} \\ N_{p,max} - 1 & \text{if } N_{p,max} \text{ is odd} \end{cases}.$$

Therefore, being given the coefficient b, we have to choose $a \bmod p$ in such a way that N_p is maximal in the above sense. Empirically, it turned out to be advisable to choose the absolute value of the discriminant

$$|\Delta| = 2^4 b^2 |a^2 - 4b|$$

to be small; more precisely, one takes

$$\left\{ \begin{array}{ll} a \approx 2\sqrt{b} & \text{if } b > 0 \\ a \approx 0 & \text{if } b < 0 \end{array} \right\}.$$

Note that if b contains many or large prime factors, $|\Delta|$ cannot be chosen too small. This observation casts some doubt on Kretschmer's conjecture (cf. also [16]). In any case, if the above conditions *i)* and *iv)* are satisfied, we must choose a to satisfy the congruence relation

$$a \equiv a_{min} \pmod{c},$$

where a_{min} and c are computed in accordance with the Chinese Remainder Theorem. Then we make sure that $|\Delta|$ is small.

Now we recall Theorem 1.1 telling us that

$$rk_{K_d}(E) = rk_{\mathbb{Q}}(E) + rk_{\mathbb{Q}}(E_d),$$

where $K_d = \mathbb{Q}(\sqrt{d})$. We want to know under which condition the rank of E goes up by at least 1 as one passes from $\mathbb{Q}$ to K_d as the basic field of E. Put $N_p := \#\widehat{E}(\mathbb{F}_p)$ as before, and let $N_{p,d} := \#\widehat{E_d}(\mathbb{F}_p)$. First it should be made sure that $N_{p,d}$ is maximal for a set of small primes $p > 2$ satisfying $p \nmid b$, $p \nmid \widetilde{b} = a^2 - 4b$, i.e. $p \nmid \Delta$.

Proposition 3.1. *Suppose $a, b \in \mathbb{Z}$ are chosen in such a way that N_p is maximal for a prime $p \geq 3$. Then*

$$N_{p,d} \text{ is maximal} \iff \left(\frac{d}{p}\right) = 1.$$

Sketch of Proof. We have

$$a_p^{(d)} = \left(\frac{d}{p}\right) a_p, \quad N_p = p - 1 - a_p.$$

□

Theorem 3.2. *Suppose $d \in \mathbb{Z}$, $d \neq 1$ squarefree, is chosen in such a way that*

$$b = b_1 b_2 \quad \text{and} \quad b_1 + a + b_2 \in d\mathbb{Q}^{\times 2},$$

where $b_1 \not\equiv b_2 \ mod^{\times} \mathbb{Q}^{\times 2}$, $b_1, b_2 \not\equiv d \ mod^{\times} \mathbb{Q}^{\times 2}$, $b, a^2 - 4b \notin \mathbb{Q}^{\times 2}$. Then

$$rk_{\mathbb{Q}}(E_d) \geq 1.$$

Sketch of Proof. We have $\{1, bd^2, b_1 d, b_2 d\}\mathbb{Q}^{\times 2} \subseteq \alpha E_d(\mathbb{Q})$. □

Application of Theorem 1.1 then leads to the inequality

$$rk_{K_d}(E) \geq rk_{\mathbb{Q}}(E) + 1, \quad \text{where } K_d = \mathbb{Q}(\sqrt{d}).$$

4 A series of examples

According to Kretschmer [11], for a prime $p \equiv 5 \pmod 8$, the elliptic curve

$$E:\ y^2 = x^3 + p^2 x$$

has rank $r = rk_{\mathbb{Q}}(E) = 0$. Its d-twist is

$$E_d:\ y^2 = x^3 + p^2 d^2 x$$

and the corresponding 2-isogenous elliptic curve

$$\widetilde{E_d}:\ \widetilde{y}^2 = \widetilde{x}^3 - 2^2 p^2 d^2 \widetilde{x}.$$

In Theorem 2.1 on the rank, we have the inclusions

$$\begin{array}{rcll} \alpha E_d(\mathbb{Q}) & \supseteq & B_d =< S_d \mathbb{Q}^{\times 2} >, & \text{where } \# B_d = 2^s, \\ \text{and } \widetilde{\alpha}\widetilde{E_d}(\mathbb{Q}) & \supseteq & \widetilde{B_d} =< \widetilde{S_d} \mathbb{Q}^{\times 2} >, & \text{where } \# \widetilde{B_d} = 2^{\widetilde{s}}. \end{array}$$

Hence, we obtain the inequality $2^{r_d} \geq 2^{s+\widetilde{s}-2}$ leading to the lower estimate

$$r_d \geq s + \widetilde{s} - 2.$$

On the other hand, the inequality $2^{r_d} \leq 2^{t+\widetilde{t}}$ leads to the upper estimate

$$r_d \leq t + \widetilde{t}.$$

If d is a prime, the decompositions $b = p^2 d^2$ and $\widetilde{b} = 2^2 p^2 d^2$ show that $t = 2$ and $\widetilde{t} = 3$, respectively. Local considerations show that in all cases, rank 2 is the highest possible rank we can reach provided that d is a prime number. For determing the image $\alpha E_d(\mathbb{Q})$, we have to solve the equation

$$(i)\quad d(U^4 + p^2 V^4) = W^2 \text{ in } \mathbb{Z} \text{ with } UV \neq 0$$

and for determing $\widetilde{\alpha}\widetilde{E_d}(\mathbb{Q})$, we have to solve the equations

$$\left.\begin{array}{lrcl} (ii) & \pm d(U^4 - 2^2 p^2 V^4) & = & W^2 \\ (iii) & \pm 2p(U^4 - d^2 V^4) & = & W^2 \end{array}\right\} \text{ in } \mathbb{Z} \text{ with } UV \neq 0.$$

Here, p is a prime satisfying $p \equiv 5 \bmod 8$. Since $E_d(\mathbb{Q}) = E_{-d}(\mathbb{Q})$ and $rk_{\mathbb{Q}}(E) = 0$, we know that $rk_{K_d}(E) = rk_{K_{-d}}(E)$. In what follows we choose $d = q$ to be a prime.

Theorem 4.1. *Let $K_d := \mathbb{Q}(\sqrt{q})$ and $p \equiv 5$ mod 8.*

(a)	*If $q = p$,*	*then $rk_{K_{\pm p}}(E) = 0$.*
(b)	*If $q \neq p, q \equiv 5$ mod 8, according to Genocchi.*	*then $rk_{K_{\pm q}}(E) = 0$*
(c)	*If $q \neq p, q \equiv 3, 7$ mod 8,*	*then $rk_{K_{\pm q}}(E) = 1$.*
(d)	*If $q \neq p, q \equiv 1$ mod 8, $\left(\frac{q}{p}\right) = -1$ according to Lagrange.*	*then $rk_{K_{\pm q}}(E) = 0$*
(e)	*If $q \neq p, q \equiv 1$ mod 8, $\left(\frac{q}{p}\right) = 1$*	*then $rk_{K_{\pm q}}(E) = 0$ or 2.*

Proof. Besides Theorem 1.1 we use the following result (see [9]). □

Theorem 4.2. *The Hasse–Weil L-function* $L_{\mathbb{Q}}(E^{(n)}, s)$ *for the elliptic curve*

$$E^{(n)} : \ y^2 = x^3 - n^2 x,$$

which for $Re(s) > \frac{3}{2}$ *is definied, extends analytically to an entire function on* $\mathbb{C}$ *. In addition, let*

$$N \ = \ \begin{cases} 32n^2 & \textit{if } n \textit{ is odd} \\ 16n^2 & \textit{if } n \textit{ is even.} \end{cases}$$

Let

$$\Lambda(s) := \left(\frac{\sqrt{N}}{2\pi}\right)^s \Gamma(s) L_{\mathbb{Q}}(E^{(n)}, s).$$

Then $L_{\mathbb{Q}}(E^{(n)}, s)$ *satisfies the functional equation*

$$\Lambda(s) = \omega\Lambda(2 - s)$$

$$\textit{where} \quad \omega = \pm 1 \quad \textit{and} \quad \omega = \begin{cases} 1 & \textit{if } n \equiv 1, 2, 3 \textit{ mod } 8 \\ -1 & \textit{if } n \equiv 5, 6, 7 \textit{ mod } 8. \end{cases}$$

In general it is conjectured that

$$\omega = (-1)^{rk_{\mathbb{Q}}(E)}. \tag{+}$$

According to Theorem 2.1, the rank of

$$\widetilde{E_d} : \ y^2 = x^3 - 4d^2p^2x$$

equals $rk_{\mathbb{Q}}(E_d) = rk_{\mathbb{Q}}(\widetilde{E_d})$. Since $p \equiv 5$ mod 8, the sign of the functional equation of $\widetilde{E_d}$ is

$$\omega = \begin{cases} 1 & \text{if } d \equiv 1, 5 \text{ mod } 8 \\ -1 & \text{if } d \equiv 3, 7 \text{ mod } 8. \end{cases}$$

From the conjecture (+), we have that

$$rk_{\mathbb{Q}}(E_d) \text{ is } \begin{cases} \text{even} & \text{for } d \equiv 1, 5 \text{ mod } 8 \\ \text{odd} & \text{for } d \equiv 3, 7 \text{ mod } 8. \end{cases}$$

Remark 4.3. Without referring to Theorem 4.1 and conjecture (+) we have only the following estimates:

(c') If $d = q \neq p, q \equiv 3, 7$ mod 8, then $rk_{K_{\pm q}}(E) \leq 1$.
(e') If $d = q \neq p, q \equiv 1$ mod 8, $\left(\frac{d}{p}\right) = 1$ then $rk_{K_{\pm q}}(E) \leq 2$.

Modulo conjecture (+) and conjecture of Birch and Swinnerton-Dyer, the rank of E over $K_{\pm d}$ is exactly 2 for the primes $p < 500$ and $q < 1000$ in table 1 below, whereas for the primes within this range not listed in table 1 the rank is 0 by the Theorem of Coates and Wiles. The results are obtained by checking whether or not the L-series of $\widetilde{E_d}$ vanishes at $s = 1$. Then one can decide if $rk_{\mathbb{Q}}(E_d)$ is zero or strictly positive. The corresponding L-series calculations are performed by virtue of a computer program developed by P. Serf [25].

Table 1

p	q
5	41, 89, 281, 401, 449, 601, 809
13	17, 233, 313, 337, 521, 569, 601, 641, 673, 809, 857, 953
29	313, 353, 401, 857
37	593, 673, 929, 937
53	89, 113, 281, 857, 937
61	41, 73, 137, 569, 601, 809, 857
101	17, 137, 233, 977
109	89, 113, 137, 193, 353, 409, 433, 457, 953
149	17, 73, 601, 769
157	193
173	41, 73, 769
181	73, 137, 233, 241, 449, 953
197	41, 193, 233, 457, 769
229	97, 433, 449, 457
269	233, 353, 433
277	193, 761, 769
293	449, 641
317	577, 977
349	41, 257, 457, 577, 937
373	17, 89, 457
389	337, 641
397	353, 409, 761
421	97, 337, 401
461	73, 97

Remark 4.4. It turned out that for a given prime p with $p \equiv 5 \bmod 8$ and $p < 500$, there always exists a prime q with $q \equiv 1 \bmod 8$, $\left(\frac{q}{p}\right) = 1$ and $q < 1000$ such that $rk_{K_{\pm q}}(E) = 2$ in accordance with conjecture (+).

In this connection we note the following: Murty & Murty [20] showed that for a given modular rank zero elliptic curve E over $\mathbb{Q}$ there exist infinitely many $d < 0$ such that, on the one hand, all primes dividing the conductor of E over $\mathbb{Q}$ split completely in K_d and, on the other hand, $rk_{K_d}(E) = 1$. The same theorem was proved independently by Bump, Friedberg and Hoffstein.

Examples. Consider the elliptic curve

$$E: \; y^2 = x^3 + p^2 x$$

for prime numbers p satisfying $p \equiv 5 \bmod 8$ and $p < 500$. Let $d = q$ be a prime number, satisfying $d \equiv 1 \bmod 8$, $\left(\frac{d}{p}\right) = 1$ and $d < 1000$. Then for the pairs (p, d) in table 2 below, the rank $rk_{K_{\pm d}}(E)$ of E over $K_{\pm d}$ is exactly equal to 2. Table 2 is not based on any conjecture.

Table 2

p	d	equations (*i*)		(*ii*)		(*iii*)	
5	41	U =	2	U =	9		
		V =	1	V =	2		
5	89	U =	1	U =	7		
		V =	10	V =	2		
5	449	U =	5	U =	131		
		V =	28	V =	22		
13	17	U =	4	U =	11		
		V =	1	V =	2		
13	569	U =	43			U =	187
		V =	8			V =	49
13	641	U =	4			U =	601
		V =	35			V =	31
37	929	U =	20			U =	39
		V =	29			V =	1
53	89	U =	128			U =	623
		V =	11			V =	731
61	137	U =	646			U =	53
		V =	413			V =	7
101	17	U =	8	U =	99		
		V =	1	V =	2		
109	137	U =	1	U =	101		
		V =	4	V =	4		
149	17	U =	19	U =	299		
		V =	2	V =	2		
197	41	U =	433			U =	229
		V =	4			V =	7
229	97	U =	316			U =	247
		V =	3			V =	17
349	41	U =	62			U =	149
		V =	5			V =	119
461	97	U =	632			U =	479
		V =	99			V =	41

Here, U and V denote the solutions found for the equations (*i*), (*ii*) or (*iii*) within the range ≤ 1000.

We list some authors who determined elliptic curves of high rank over $\mathbb{Q}$:

Wiman	(1948) :	r	≥ 4
Penney & Pomerance	(1974) :	r	≥ 6
Penney & Pomerance	(1975) :	r	≥ 7
Grunewald & Zimmert	(1977) :	r	≥ 8
Brumer & Kramer	(1977) :	r	≥ 9
Nakata	(1979) :	r	≥ 9
Mestre	(1982) :	r	≥ 12
Kretschmer*	(1983) :	r	$= 10$
Mestre	(1986) :	r	≥ 14
Quer**	(1987) :	r	$= 12$
Schneiders*	(1989) :	r	$= 11$

* These curves have torsion groups $E_{tors}(\mathbb{Q}) \neq \mathcal{O}$.

** These curves are CM–curves of the form $y^2 = x^3 + D$.

As a consequence of his numerical search for elliptic curves of high rank over $\mathbb{Q}$, Kretschmer enunciated the following conjecture:

Conjecture. *For every $n \in \mathbb{N}$, there exists an elliptic curve E over $\mathbb{Q}$ and a set of distinct primes $S = \{p_1, \cdots, p_n, 2, 3, p\}$ such that*

(*a*) $E_{tors}(\mathbb{Q}) \supseteq \mathbb{Z}/2\mathbb{Z}$

(*b*) *E has good reduction away from S*

(*c*) $rk_{\mathbb{Q}}(E) = n$.

This conjecture was numerically verified for $1 \leq n \leq 9$ by Kretschmer. By our tables 6 and 8 below this numerical verification is extended to the range $10 \leq n \leq 11$. In our search for elliptic curves of high rank, we investigated 22660 curves of ranks up to 11 over $\mathbb{Q}$ of which 22 are listed in tables 6 – 24 below. The distribution of ranks for these 22660 curves is shown in the subsequent tables 3 – 5. Table 3 gives the number of curves of ranks up to 11 over $\mathbb{Q}$.

Table 3

Rank over $\mathbb{Q}$ at least	Number of curves	Per cent
0	0	0.00
1	0	0.00
2	0	0.00
3	3	0.01
4	572	2.52
5	3476	15.34
6	6418	28.32
7	7475	32.99
8	4113	18.15
9	547	2.41
10	54	0.24
11	2	0.01

Tables 4 and 5 show for a random selection of these curves how often rank jumps by 0, 1, 2, 3 and 4 were obtained upon transition from $\mathbb{Q}$ to suitable quadratic extensions $K_d = \mathbb{Q}(\sqrt{d})$. The fields K_d were chosen at random too but in compliance with Theorems 3.1 and 3.2.

Table 4

Rank over $\mathbb{Q}$ at least	Number of curves with rank jump of at least				
	0	1	2	3	4
4	186	3	0	0	0
5	1225	49	4	1	0
6	12693	509	46	5	1
7	33784	1336	87	10	0
8	13261	424	26	4	0
9	9858	200	6	0	0
10	4212	729	3	0	0
11	154	24	0	0	0

Table 5

Rank over $\mathbb{Q}$ at least	Percentage of curves with rank jump of at least				
	0	1	2	3	4
4	98.41	1.59	0.00	0.00	0.00
5	95.78	3.83	0.31	0.08	0.00
6	95.77	3.84	0.35	0.04	0.01
7	95.93	3.79	0.25	0.03	0.00
8	96.69	3.09	0.19	0.03	0.00
9	97.95	1.99	0.06	0.00	0.00
10	85.19	14.75	0.06	0.00	0.00
11	86.52	13.48	0.00	0.00	0.00

Our numerical results appear to support the following conjecture of Honda (see [8], [24]).

Conjecture. *For an elliptic curve E over $\mathbb{Q}$, there is a constant*

$$C = C(E)$$

such that $rk_K(E) \leq C[K : \mathbb{Q}]$ for every algebraic number field K.

This conjecture implies that, for a fixed curve E over $\mathbb{Q}$, the d-twists E_d of E have bounded rank over $\mathbb{Q}$. Honda's conjecture thus contradicts the following "folklore" conjecture.

Conjecture. *For a fixed elliptic curve E over $\mathbb{Q}$,*

$$\sup_{d\in\mathbb{Z}} rk_{\mathbb{Q}}(E_d) = \infty.$$

The "folklore" conjecture is not in compliance with the heuristics of Section 3 either. According to that heuristics, for the rank to be *large*, the discriminant of an elliptic curve should be *small* in absolute value. The d-twist E_d of E, however, has discriminant $\Delta_d = d^6\Delta$ which tends to infinity as $|d| \to \infty$.

Moreover, F. Gouvea and B. Mazur [6] conjecture that the number of quadratic twists E_d of a fixed elliptic curve E over $\mathbb{Q}$ which have rank $rk_{\mathbb{Q}}(E_d) \geq 2$ is of zero density. More precisely, they conjecture that the number of positive integers $d \leq x$ such that $rk_{\mathbb{Q}}(E_d) \geq 2$ is $O(x^{\frac{3}{4}+\varepsilon})$, and they show that this number is $>> x^{\frac{1}{2}-\varepsilon}$. In contrast to this conjecture of Gouvea and Mazur, it is conjectured by Kramarz and Zagier [27] that, for the elliptic curves of CM-type

$$E_m: \; y^2 = x^3 - 2^4 3^3 m^2$$

parametrized by cubefree positive integers m, the number of $m \leq x$ such that $rk_{\mathbb{Q}}(E_m) \geq 2$ is a positive proportion.

Now we exhibit some examples of elliptic curves

$$E: \; y^2 = x^3 + ax^2 + bx \qquad (a, b \in \mathbb{Z})$$

of various ranks over $\mathbb{Q}$ and $K_d = \mathbb{Q}(\sqrt{d})$. The notation

moduli	a	b
$(m_1 \cdots m_l)$	$(a_1 \cdots a_l)$	$(b_1 \cdots b_l)$

means that

$$\left.\begin{array}{rcl} a & \equiv & a_i \bmod m_i \\ b & \equiv & b_i \bmod m_i \end{array}\right\} \quad \forall\, 1 \leq i \leq l.$$

The tables show rank jumps by 0, 1, 2, 3 or 4.

Table 6

moduli	a	b
(8 3 5 7 11)	(1 1 0 6 9)	(7 2 3 4 6)
b	$= 13 \cdot 17 \cdot 23 \cdot 29 \cdot 31 \cdot 37 \cdot 41 \cdot 43 \cdot 47 \cdot 53 \cdot 59$ $= 43808620418186663$	

a	exact rank over $\mathbb{Q}$	rank jump at least 1 when $d =$
1797856465	11	370441
		144994489 $= 101 \cdot 1435589$
		-211151

Table 7

moduli	a	b
(8 3 5 7 11 13)	(5 1 0 1 5 3)	(3 2 3 5 10 10)

b	$=$	$-\ 19 \cdot 23 \cdot 29 \cdot 31 \cdot 37 \cdot 47 \cdot 53 \cdot 59 \cdot 67 \cdot 79 \cdot 83 \cdot 97$
	$=$	$-\ 9103749782424487077$

a	exact rank over $\mathbb{Q}$	rank jump at least 1 when $d =$
8532707485	11	$707041 = 113 \cdot 6257$
		$1632409 = 43 \cdot 37963$
		-335999

Table 8

moduli	a	b
(32 3 5 11)	(2 1 0 1)	(29 2 3 7)

b	$=$	$-\ 7 \cdot 13 \cdot 17 \cdot 19 \cdot 23 \cdot 31 \cdot 37 \cdot 41 \cdot 43 \cdot 53$
	$=$	$-\ 72454164114787$

a	exact rank $\mathbb{Q}$ over $\mathbb{Q}$	rank jump at least 1 when $d =$
6342370	10	$346 = 2 \cdot 173$
		370186111
		-453896231

Table 9

moduli	a	b	
(8 3 5 7 11 13)	(1 1 0 5 9 1)	(0 0 3 2 6 4)	

b	$=$	$2^{10} \cdot 3^{6} \cdot 17 \cdot 19 \cdot 23 \cdot 29 \cdot 31 \cdot 37 \cdot 41 \cdot 43 \cdot 53$
	$=$	17236434803911308288

a	rank over $\mathbb{Q}$ at least	rank jump at least 1 when $d =$	rank jump at least 2 when $d =$
12273038545	10	$394 = 2 \cdot 197$	-11399
		-101	

Table 10

moduli	a	b	
(8 3 5 7 13)	(1 1 0 1 2)	(0 0 3 5 11)	
b	$= - 2^{10} \cdot 3^8 \cdot 11 \cdot 17 \cdot 23 \cdot 29 \cdot 31 \cdot 37 \cdot 41 \cdot 47 \cdot 53$ $= - 98165399195348001792$		
a	rank over $\mathbb{Q}$ at least	rank jump at least 1 when d =	rank jump at least 2 when d =
4481541625	10	79	$406 = 2 \cdot 7 \cdot 29$

Table 11

moduli	a	b
(64 3 5 7 11)	(18 1 0 1 1)	(1 2 3 5 7)
b	$= - 19 \cdot 23 \cdot 29 \cdot 31 \cdot 37 \cdot 41 \cdot 43 \cdot 59 \cdot 61 \cdot 71 \cdot 79 \cdot 89$ $= - 46041817427223966847$	
a	rank over $\mathbb{Q}$ at least	rank jump at least 1 when d =
18695818450	10	$106 = 2 \cdot 53$ 211 $646 = 2 \cdot 17 \cdot 19$ $-374 = - 2 \cdot 11 \cdot 17$

Table 12

moduli	a	b
(8 3 5 7 11 13)	(1 1 0 4 9 4)	(0 0 3 3 6 12)
b	$= 2^9 \cdot 3^4 \cdot 17 \cdot 19 \cdot 23 \cdot 29 \cdot 31 \cdot 37 \cdot 43 \cdot 47 \cdot 53 \cdot 59$ $= 64765086329601704448$	
a	rank over $\mathbb{Q}$ at least	rank jump at least 1 when d =
18926510425	10	$466 = 2 \cdot 233$

Table 13

moduli	a	b
(8 3 5)	(1 1 0)	(0 0 3)
b	$= 2^8 \cdot 3^6 \cdot 7 \cdot 11 \cdot 13 \cdot 17 \cdot 19 \cdot 23 \cdot 29 \cdot 31 \cdot 37 \cdot 41$ $= 18926800393500067968$	
a	rank over $\mathbb{Q}$ at least	rank jump at least 2 when d =
2792554705	9	$-26 = - 2 \cdot 13$ -59

Table 14

moduli	a	b
(64 3 5)	(42 1 0)	(41 2 3)
b	$= -7 \cdot 11 \cdot 13 \cdot 17 \cdot 19 \cdot 23 \cdot 29 \cdot 31 \cdot 37 \cdot 43 \cdot 53 \cdot 59$	
	$= -33259995678156247$	

a	rank over $\mathbb{Q}$ at least	rank jump at least 2 when d =
65290090	9	61

Table 15

moduli	a	b
(8 3 5 11)	(1 1 0 9)	(0 0 3 6)
b	$= 2^8 \cdot 3^5 \cdot 7 \cdot 13 \cdot 17 \cdot 19 \cdot 23 \cdot 29 \cdot 31 \cdot 37 \cdot 53$	
	$= 741404597823751168$	

a	rank over $\mathbb{Q}$ at least	rank jump at least 2 when d =
1021882585	9	1061
		$2509 = 13 \cdot 193$

Table 16

moduli	a	b
(8 3 5)	(1 1 0)	(0 0 3)
b	$= 2^8 \cdot 3^5 \cdot 7 \cdot 11 \cdot 13 \cdot 17 \cdot 19 \cdot 23 \cdot 67 \cdot 73 \cdot 79$	
	$= 178745628330517248$	

a	rank over $\mathbb{Q}$ at least	rank jump at least 1 when d =
3240270025	9	-29
		$-119 = -7 \cdot 17$
		-269

Table 17

moduli	a	b
(8 3 5)	(1 1 0)	(7 2 3)
b	$= 7 \cdot 11 \cdot 13 \cdot 17 \cdot 19 \cdot 23 \cdot 29 \cdot 31 \cdot 41 \cdot 43 \cdot 47 \cdot 53$	
	$= 29359602231702743$	

a	rank over $\mathbb{Q}$ at least	rank jump at least 3 when d =
353838145	8	$-119 = -7 \cdot 17$

Table 18

moduli	a	b	
(8 3 5)	(1 1 0)	(0 0 3)	

$b = -2^{10} \cdot 3^8 \cdot 7 \cdot 11 \cdot 13 \cdot 17 \cdot 19 \cdot 37 \cdot 43 \cdot 47 \cdot 53$
$= -8608956262203128832$

a	rank over $\mathbb{Q}$ at least	rank jump at least 1 when $d =$	3 when $d =$
42068305	8	-59 -89	$-119 = -7 \cdot 17$

Table 19

moduli	a	b	
(8 3 5)	(1 1 0)	(0 0 3)	

$b = -2^{10} \cdot 3^8 \cdot 7 \cdot 11 \cdot 13 \cdot 17 \cdot 19 \cdot 23 \cdot 31 \cdot 37$
$= -57305703462239232$

a	rank over $\mathbb{Q}$ at least	rank jump at least 1 when $d =$	3 when $d =$
57059065	8	349 $-254 = -2 \cdot 127$	241

Table 20

moduli	a	b	
(8 3 5)	(5 1 0)	(3 2 3)	

$b = 7 \cdot 11 \cdot 13 \cdot 17 \cdot 23 \cdot 31 \cdot 37 \cdot 43 \cdot 47 \cdot 59$
$= 53529424952003$

a	rank over $\mathbb{Q}$ at least	rank jump at least 1 when $d =$	3 when $d =$
19639045	8	19 $-314 = -2 \cdot 157$	-71

Table 21

moduli	a	b	
(32 3 5)	(2 1 0)	(29 2 3)	
b	$= 7 \cdot 11 \cdot 13 \cdot 17 \cdot 23 \cdot 29 \cdot 37 \cdot 41 \cdot 47 \cdot 53$		
	$= 42891194479133$		

a	rank over $\mathbb{Q}$ at least	rank jump at least 1 when $d =$	2 when $d =$	3 when $d =$
13674370	7	19	$214 = 2 \cdot 107$	$-119 = - 7 \cdot 17$
		$34 = 2 \cdot 17$	$301 = 7 \cdot 43$	
		$46 = 2 \cdot 23$	$-86 = - 2 \cdot 43$	
		$-14 = - 2 \cdot 7$		
		$-206 = - 2 \cdot 103$		
		$-299 = - 13 \cdot 23$		
32300770	7	31		139
		$-14 = - 2 \cdot 7$		
50902690	7	31		19
		$34 = 2 \cdot 17$		
		61		
		-41		
		$-86 = - 2 \cdot 43$		
		$-119 = - 7 \cdot 17$		
		$-146 = - 2 \cdot 73$		
		$-209 = - 11 \cdot 19$		

Table 22

moduli	a	b	
(64 3 5)	(26 1 0)	(25 2 3)	
b	$= - 7 \cdot 11 \cdot 13 \cdot 17 \cdot 19 \cdot 23 \cdot 31 \cdot 37 \cdot 47 \cdot 53 \cdot 59$		
	$= - 1253584440155047$		

a	rank over $\mathbb{Q}$ at least	rank jump at least 1 when $d =$	3 when $d =$
76940890	7	109	19
		$274 = 2 \cdot 137$	
		349	
		-41	
		$-86 = - 2 \cdot 43$	
		-149	
		$-161 = - 7 \cdot 23$	
		-191	
		-251	

Table 23

moduli	a	b	
(8 3 5)	(1 1 0)	(0 0 3)	
b	$= 2^{10} \cdot 3^8 \cdot 7 \cdot 11 \cdot 13 \cdot 17 \cdot 19 \cdot 37 \cdot 41 \cdot 43 \cdot 59$ $= 8360122520381119488$		

a	rank over $\mathbb{Q}$ at least	rank jump at least 1 when d =	 3 when d =
5789481505	7	$274 = 2 \cdot 137$ -191	-71

Table 24

moduli	a	b	
(64 3 5)	(2 1 0)	(49 2 3)	
b	$= -7 \cdot 11 \cdot 13 \cdot 17 \cdot 23 \cdot 29 \cdot 31 \cdot 37 \cdot 41 \cdot 43 \cdot 53$ $= -1216467281716687$		

a	rank over $\mathbb{Q}$ at least	rank jump at least 1 when d =	 3 when d =	 4 when d =
134603650	7	79 $106 = 2 \cdot 53$ -179	$-14 = -2 \cdot 7$	
131502850	6	151 181 $-14 = -2 \cdot 7$ -29 $-209 = -11 \cdot 19$ $-314 = -2 \cdot 157$		$301 =$ $7 \cdot 43$

References

[1] B.J. Birch, *Elliptic curves and modular functions*, London and New York: Sympos. Math., vol. IV, INDAM Rome, Academic Press 1970, 27-32.

[2] D. Bump, S. Friedberg, J. Hoffstein, *Nonvanishing theorems for L-functions of modular forms and their derivatives*, Preprint (1988).

[3] A. Brumer, K. Kramer, *The rank of elliptic curves*, Duke Math. Journal **44** (1977), 715-743.

[4] J. Coates, *Elliptic curves and Iwasawa theory*, In: Chichester: Modular Forms, Ed.: R.A. Rankin, Ellis Horwood Ltd. 1984, 51-73.

[5] J. Coates, A. Wiles, *On the conjecture of Birch and Swinnerton-Dyer*, Invent. Math. **39** (1977), 223-251.

[6] F. Gouvea, B. Mazur, *The square-free sieve and the rank of Mordell-Weil*, Preprint (1989).

[7] F.J. Grunewald, R. Zimmert, *Über einige rationale elliptische Kurven mit freiem Rang* $\geq$ 8, Journal Reine u. Angew. Math. **296** (1977), 100-107.

[8] T. Honda, *Isogenies, rational points and section points of group varieties*, Jap. J. Math. **30** (1960), 84-101.

[9] N. Koblitz, *Introduction to elliptic curves and modular forms*, New York: Springer 1984.

[10] K. Kramer, *Arithmetic of elliptic curves upon quadratic extensions*, Trans. of the Am. Math. Soc. **264** (1981), 121-135.

[11] T.J. Kretschmer, *Konstruktion elliptischer Kurven von hohem Rang*, Diplomarbeit, Saarbrücken 1983.

[12] T.J. Kretschmer, *Construction of elliptic curves with large rank*, Math. of Comp. **174** (1986), 627-635.

[13] J. Lagrange, *Nombres congruents et courbes elliptiques*, Séminaire Delange-Pisot-Poitou (Théorie des Nombres) 16 (1974/75).

[14] M. Laska, *Punkte auf elliptischen Kurven über* $\mathbb{Q}$ *in quadratischen Zahlkörpern*, Bonn: Universität Bonn, Sonderforschungsbereich 40 und Max-Planck-Institut für Mathematik 1983.

[15] M. Laska, *Punkte auf elliptischen Kurven über* $\mathbb{Q}$ *in quadratischen Zahlkörpern*, Arch. Math. **44** (1985), 159-167.

[16] J.F. Mestre, *Majoration du rang d'une courbe elliptique définie sur* $\mathbb{Q}$, Grenoble: Séminaire de Théorie des Nombres 1977.

[17] J.F. Mestre, *Construction d' une courbe elliptique de rang* $\geq$ 12, C. R. Acad. Sci. Paris **295** (1982), 643-644.

[18] J.F. Mestre, *Courbes elliptiques et formules explicites*, Grenoble: Séminaire de Théorie des Nombres 1982.

[19] J.F. Mestre, *Formules explicites et minorations de conducteurs de variétés algébriques*, Comp. Math. **58** (1986), 209-232.

[20] M.R. Murty, V.K. Murty, *Mean values of derivatives of modular L-series*, Preprint (1989).

[21] K. Nakata, *On some elliptic curves defined over* $\mathbb{Q}$ *of free rank* $\geq$ 9, Manu. math. **29** (1979), 183-194.

[22] D.E. Penney, C. Pomerance, *A search for elliptic curves with large rank*, Math. of Comp. **28** (1974), 851-853.

[23] D.E. Penney, C. Pomerance, *Three elliptic curves with rank at least seven*, Math. of Comp. **29** (1975), 965-967.

[24] J.P. Serre, *Lectures on the Mordell-Weil theorem*, Braunschweig: Vieweg 1989.

[25] P. Serf, *Congruent numbers and elliptic curves*, These Proceedings.

[26] J.T. Tate, *Rational points on elliptic curves*, Phillips Lectures, Haverford College 1961.

[27] D. Zagier, G. Kramarz, *Numerical investigations related to the L-series of certain elliptic curves*, Journal Indian Math. Soc. **52** (1987), 51-69.

On the resolution of some diophantine equations

I. Gaál *

Abstract. In this survey our main purpose is to give an impression to the reader about the application of the reduction algorithm, introduced by Baker and Davenport [5], to Thue equations and index form equations. Until today this is the only general method which makes possible to solve completely such equations. We also include several computational results, connected with these equations, and add numerical examples.

Introduction

In the present paper we give a survey about the methods for the resolution of Thue equations and index form equations. The occasion to consider these types of diophantine equations together are given by two facts. First, our main purpose is to illustrate the so-called reduction method, which was applied to Thue equations and some types of index form equations in a very similar manner. Secondly, the index form equations in the most simple case of cubic number fields are in fact cubic Thue equations.

Our further purpose is to give several connected computational results on Thue equations and index form equations, and to include some numerical examples.

I. Thue equations

Let $F \in \mathbb{Z}[X,Y]$ be an irreducible binary form of degree ≥ 3 and let m be a non-zero rational integer. Consider the well known equation

$$F(x,y) = m \qquad \text{in } x, y \in \mathbb{Z} \tag{1}$$

which is called *Thue equation* in honour of the Norwegian mathematician Thue who proved in 1909 [48] that the above equation has only finitely many solutions.

*Research supported in part by Hungarian National Foundation for Scientific Research Grant No. 273/86

His theorem gave a finiteness assertion for a whole class of equations and it had a very deep influence for the further development of the theory of diophantine equations. This theorem of Thue was extended and improved by several authors, (see e.g. [40] chapters 5,7), among others by Siegel [41] to the relative case and by Mahler [27] to the p-adic case.

All these results are ineffective in the sense that they assert only the finiteness of the number of solutions, but do not give any further information about them. It was Baker [3] in 1966 who introduced a deep effective method in number theory, giving effective lower estimates for linear forms in the logarithms of certian algebraic numbers. Using his method it became possible to give effective upper bounds for the solutions of several types of diophantine equations. It was Baker himself who gave the first effective bounds for the solutions of Thue equations:

Theorem 1.1 (Baker [4], 1968) *Under the above conditions for all solutions, of equation* (1) *we have*

$$\max(|x|, |y|) < exp\{n^{\nu^2} H^{\nu n^3} + (log\,|m|)^{2n+2}\},$$

where $n = deg(F)$, H *denotes the height of* F *(the maximum absolute value of its coefficients) and* $\nu = 128n(n+1)$.

This celebrated result of Baker was the first one, which made possible, at least in principle, to determine all solutions of Thue equations. The theorems of this type are of basic importance, since formerly there was no hope for determining the solutions of such equations. Since there are only finitely many candidates for x and y satisfying the above inequality, hence substituting these values into (1) we can select all solutions of it. Unfortunately the bounds obtained by Baker's method are usually so high, that the above algorithm can not be carried out even with the fastest computer. It means that in practice these bounds themselves do not make possible really to find all solutions.

The reduction algorithm for finding all solutions

An algorithm, which enables one to reduce these bounds for specific equations in a numerical way was invented by Baker and Davenport [5] in 1969, originally for cubic Thue equations, and was further extended and generalized by several authors.

In the following our purpose is to illustrate this method on a simple example. First we remark, that if $F \in \mathbb{Z}[X,Y]$ is an irreducible form of degree n with leading coefficient 1, then it can be written as a norm form in two variables, that is

$$F(X,Y) = N_{K/\mathbb{Q}}(X+\alpha Y) = \prod_{i=1}^{n}(X+\alpha^{(i)}Y),$$

where $K = \mathbb{Q}(\alpha)$ is the field generated by a root α of $F(X,1) = 0$, and $\alpha^{(i)}$ $(i = 1,\dots,n)$ denote the conjugates of α. For simplicity we consider the cubic

Thue equation

$$N_{K/\mathbb{Q}}(x+\alpha y)=1 \qquad \text{in } x,y\in\mathbb{Z}, \tag{2}$$

where $K=\mathbb{Q}(\alpha)$ is a totally real cubic field. Denote by η_1,η_2 the fundamental units of K of norm $+1$. Let x,y be an arbitrary but fixed solution of (2), and set $X=\max(|x|,|y|)$. Let $\beta=x+\alpha y$ and denote by $\gamma^{(i)}$ $(i=1,2,3)$ the conjugates of any $\gamma\in K$.

We can see from (2) that β is a unit of K, whence

$$\beta=\eta_1^{b_1}\eta_2^{b_2}, \tag{3}$$

with some $b_1,b_2\in\mathbb{Z}$. In the sequel let $H=\max(|b_1|,|b_2|)$. We can choose the numeration of the conjugates so that

$$\left|\beta^{(1)}\right|\le\left|\beta^{(2)}\right|\le\left|\beta^{(3)}\right|.$$

Using standard arguments we get

$$\left|\beta^{(i)}\right|\ge c_1X \qquad (i=2,3), \tag{4}$$

where c_1 (and in the following $c_2,c_3,\ldots$) denote effectively computable positive constants depending only on α and the parameters of K. By our original equation (2) we get from (4) that

$$\left|\beta^{(1)}\right|=\frac{1}{\left|\beta^{(2)}\right|\left|\beta^{(3)}\right|}\le c_2X^{-2}. \tag{5}$$

Taking logarithms in the conjugates of β, (3) implies

$$log\left|\beta^{(i)}\right|=b_1log\left|\eta_1^{(i)}\right|+b_2log\left|\eta_2^{(i)}\right| \qquad (i=1,2,3).$$

Considering the above equations as a linear equation system in b_1,b_2 we conclude

$$H\le c_3\max_{1\le i\le 3}\left|log\left|\beta^{(i)}\right|\right|\le c_4logX. \tag{6}$$

The application of Baker's method is always based on an identity consisting of three terms, called Siegel's identity. In our case it has the form

$$(\alpha^{(1)}-\alpha^{(2)})\beta^{(3)}+(\alpha^{(2)}-\alpha^{(3)})\beta^{(1)}+(\alpha^{(3)}-\alpha^{(1)})\beta^{(2)}=0.$$

We can write this identity in the form

$$1-\frac{(\alpha^{(1)}-\alpha^{(3)})}{(\alpha^{(1)}-\alpha^{(2)})}\frac{\beta^{(2)}}{\beta^{(3)}}=\frac{(\alpha^{(3)}-\alpha^{(2)})}{(\alpha^{(1)}-\alpha^{(2)})}\frac{\beta^{(1)}}{\beta^{(3)}}. \tag{7}$$

Consider now the following sequence of inequalities:

$$\begin{aligned} exp(-c_5logH) &< \left|log\left|\frac{(\alpha^{(1)}-\alpha^{(3)})}{(\alpha^{(1)}-\alpha^{(2)})}\right|+b_1log\left|\frac{\eta_1^{(2)}}{\eta_1^{(3)}}\right|+b_2log\left|\frac{\eta_2^{(2)}}{\eta_2^{(3)}}\right|\right| \qquad (8)\\ &\le 2\left|1-\frac{(\alpha^{(1)}-\alpha^{(3)})}{(\alpha^{(1)}-\alpha^{(2)})}\left[\frac{\eta_1^{(2)}}{\eta_1^{(3)}}\right]^{b_1}\left[\frac{\eta_2^{(2)}}{\eta_2^{(3)}}\right]^{b_2}\right| \end{aligned}$$

$$\begin{aligned} &= 2\left|1-\frac{(\alpha^{(1)}-\alpha^{(3)})\,\beta^{(2)}}{(\alpha^{(1)}-\alpha^{(2)})\,\beta^{(3)}}\right| = 2\left|\frac{(\alpha^{(3)}-\alpha^{(2)})\,\beta^{(1)}}{(\alpha^{(1)}-\alpha^{(2)})\,\beta^{(3)}}\right| \\ &\le c_6\frac{c_2X^{-2}}{c_1X} = c_7X^{-3} = exp(c_8-3logX) \\ &\le exp(c_8-c_9H) \le exp(-c_{10}H). \end{aligned}$$

Here the first inequality is the application of Baker's estimates, the second follows from the well known inequality $|logz| < 2|z-1|$, holding for any complex number z satisfying $|z-1| \le \frac{1}{3}$ (otherwise we get a much better estimate for H). The following equation is obvious by (3) then we applied (7). In the next step we used (4) and (5), and then (6). The last estimate holds if H is sufficiently large, otherwise again we get a much better estimate for H.

If we compare the left and right sides of the above inequality sequence, then we conclude that H must be bounded by an effectively computable positive constant H_B, which, applying (3) enables us to give an upper bound X_B for X. We remark, that even in this very simple case H_B is about 10^{30}, whence X_B is about the exponential function of it.

Until here it was the usual way to apply Baker's method to Thue equations. The reason why we cited these arguments is that the reduction method of Baker and Davenport is based just for the inequality

$$\left|log\left|\frac{(\alpha^{(1)}-\alpha^{(3)})}{(\alpha^{(1)}-\alpha^{(2)})}\right| + b_1log\left|\frac{\eta_1^{(2)}}{\eta_1^{(3)}}\right| + b_2log\left|\frac{\eta_2^{(2)}}{\eta_2^{(3)}}\right|\right| < exp(-c_{10}H),$$

which is included in (8). Put

$$\theta = \frac{log\left|\frac{\eta_1^{(2)}}{\eta_1^{(3)}}\right|}{log\left|\frac{\eta_2^{(2)}}{\eta_2^{(3)}}\right|}; \qquad \gamma = -\frac{log\left|\frac{(\alpha^{(1)}-\alpha^{(3)})}{(\alpha^{(1)}-\alpha^{(2)})}\right|}{log\left|\frac{\eta_2^{(2)}}{\eta_2^{(3)}}\right|}.$$

Then the above inequality can be written as

$$|b_1\theta + b_2 - \gamma| < K^{-H}, \tag{9}$$

where K is an effectively computable positive constant. The heart of the reduction method is contained in the following lemma, usually cited as "Davenport's lemma":

Lemma 1.1 (Baker and Davenport [5], 1969) *Let θ, γ be given real numbers, M, B $(B > 6)$ given integers. If there exists an integer q with*

(a) $1 \le q \le MB$

(b) $\|q\theta\| < \dfrac{2}{MB}$

(c) $\|q\gamma\| \ge \dfrac{3}{B}$

then inequality (9) *has no solutions b_1, b_2 in the range*

$$\frac{log(MB^2)}{logK} \le H \le M, \tag{10}$$

where $H = \max(|b_1|, |b_2|)$ *and* $||.||$ *denotes the distance from the nearest integer.*

How to use this Lemma ? Take M to be H_B, that we got from Baker's method, let B be 100 or 1000. If we apply the continued fraction algorithm to the number θ, then we can compute a number q satisfying (a) and (b). Since we compute q to fit to θ, and γ has no influence for q, it is almost sure, that q will not be a good approximate of γ, and (c) will also be satisfied. But it can not be proved in advance, that means, in every case we have to calculate the actual value of q satisfying (a) and (b), and we have to test (c). It is necessary to remark that in all examples we considered until now, the number q satisfying (a) and (b) was suitable also for (c). If (c) would not hold, try other values of M and B, and see also the corresponding remarks of [10].

If we found the value of q then by (10) we get a much better bound for H. Since B and K usually have moderate values, hence this new bound is about the logarithm of the original bound. This reduction step can be repeated again, until the new bound is still less than the preceeding one. Usually, after three or four steps the new bound for H is about 10 and we can not reduce it further. A typical reduction sequence is shown in the following table (we took the value $log K = 0.8$ which is also typical):

Table 1

M	B	$q \leq$	$\|\|q\theta\|\| \leq$	$\|\|q\gamma\|\| \geq$	$newM$
10^{30}	100	10^{32}	$2 * 10^{-32}$	0.03	97
97	100	9700	$2.06 * 10^{-4}$	0.03	17
17	50	850	$2.3 * 10^{-3}$	0.06	13
13	50	650	$3.07 * 10^{-3}$	0.06	12

If we have a good bound for H, then testing all possibilities for b_1, b_2 in (3) we can easily find all solutions of our equation.

We have to add some *remarks* to the above algorithm.

A. In the first reduction step we have to use multiple precision arithmetics because of the large values of M and q. For this reason it is necessary to calculate the values of θ and γ with the accuracy of about 100 digits. This computation involves iteration procedures for determining the roots of the defining polynomial of α, and for calculating certain logarithms. These iterations with multiple precision numbers slow down the algorithm (to solve a cubic Thue equation on an IBM PC AT computer takes about 60 seconds), but even then this reduction algorithm is nearly the only way to solve completely such equations. Subroutines for the basic operations with multiple precision numbers can be written by using the classical algorithms of Knuth [25]. Such a subroutine package, including also the log, exp and some other useful procedures can be found in KANT [38].

B. If the unit rank of K is greater than 2, then we have to use a modified version of Lemma 1.1 (see e.g. Lemma 2.1). Extensions of Lemma 1.1 were given by Ellison [10], Ellison at al. [11], a general version was formulated by Pethő and Schulenberg [37]. If there are more than two variables in (9), then instead of the continued fraction algorithm it is necessary to use the LLL-reduction algorithm of Lenstra, Lenstra and Lovász [26] to solve a simultaneous approximation problem (cf.[37], [52]). General aspects of the reduction algorithm are detailed in [36].

Using the reduction algorithm complete solutions of cubic Thue equations were given by Baker and Davenport [5] and Ellison at al. [11] and Steiner [43]. Quartic Thue equations were completely solved by Pethő and Schulenberg [37], Tzanakis and de Weger [50], Zagier [53].

Example. Let us give an example for the complete resolution of a cubic Thue equation. Consider the equation

$$x^3 - 2x^2y - 5xy^2 - y^3 = \pm 1 \qquad \text{in } x, y \in \mathbb{Z}.$$

(It has discriminant 361.) From Baker's method we get $H_B = 10^{27}$, which can be reduced to 8 in three steps. All solutions of the equation are

$$(x, y) = (-7, -2), (-2, 9), (-1, 1), (0, -1), (1, 0), (9, -7)$$

and their negatives.

A fast algorithm for finding "small" solutions

As we mentioned above the reduction algorithm is rather slow, but it gives all solutions of the equations. On the other hand, from a practical point of view it is enough to know only the "small" solutions of the equation (that is the solutions less than some million in absolute value) and it is important to have a fast algorithm for finding these solutions. (We also have to remark, that until today nobody found Thue equations with really "large" solutions.) Such a fast algorithm for finding "small" solutions of Thue equations was worked out by Pethő [34].

Let α be an algebraic number of degree $n \geq 3$, let $K = \mathbb{Q}(\alpha)$ and $0 < m \in \mathbb{Z}$. The algorithm of Pethő in fact produces the solutions of *Thue inequalities* of the form

$$\left|N_{K/\mathbb{Q}}(x + \alpha y)\right| \leq m \qquad \text{in } x, y \in \mathbb{Z}, (x, y) = 1, |y| \leq y_0 \tag{11}$$

where y_0 is a prescribed bound.

The method is based on the continued fraction algorithm. If δ is any real number, then $[a_0; a_1, a_2, \ldots]$ will denote the convergents in its continued fraction expansion, and p_i/q_i $(i \geq 0)$ the partial quotients, where

$$\begin{aligned} p_0 &= 1, \quad p_1 = a_0, \quad p_n = a_n p_{n-1} + p_{n-2} \text{ for } n \geq 2, \\ q_0 &= 0, \quad q_1 = 1, \quad q_n = a_n q_{n-1} + q_{n-2} \text{ for } n \geq 2. \end{aligned}$$

The essence of the method is contained in the following theorem:

Theorem 1.2 (Pethő [34], 1987) *If $x, y \in \mathbb{Z}$ is a solution of* (11) *then either $|y| \leq d_1$, or there is a real conjugate $\alpha^{(i)}$ of α such that x/y is a partial quotient of $\alpha^{(i)}$ with*

$$|y| \leq d_2 \left[\frac{A+2}{2}\right]^{1/(n-2)},$$

where $A = \max_{1 \le j \le t} a_j$, a_j are the convergents in the continued fraction algorithm of $\alpha^{(i)}$, and the index t is determined by $q_{t-1} > y_0$, where q_j are the denominators of the partial quotients of $\alpha^{(i)}$.

In the theorem d_1 and d_2 are explicitely given positive constants of moderate size.

By Theorem 1.2 to find the solutions of (11) we have to test some partial quotients of the real conjugates of α, and to examine some small values of y.

Example. Let us give an example for finding small solutions of a Thue inequality with this fast algorithm. Pethő [34] considered the inequality

$$|f(x,y)| = |x^3 + x^2y - 2xy^2 - y^3| \le 200 \quad \text{in } x, y \in \mathbb{Z}, (x,y) = 1, |y| \le 10^{500}$$

and he found that all solutions are of the form

$$(x,y) = (u,v),\ (-u,-v),\ (-v,u+v),\ (v,-u-v),\ (-u-v,u),\ (u+v,-u)$$

where (u,v) runs through the following pairs:

Table 2

u	0	1	1	1	1	1	1	2	2	2	3
v	0	0	1	2	3	4	5	1	3	5	1
$f(u,v)$	0	1	−1	−13	−41	−91	−169	7	−43	−197	29

u	3	3	4	4	4	5	5	5	5	6	7	7
v	2	4	1	3	5	1	2	3	4	5	5	6
$f(u,v)$	13	−97	71	13	−181	139	27	83	1	−29	113	−83

u	8	9	11	14	16	101	207
v	7	7	9	11	13	81	166
$f(u,v)$	−167	71	−91	181	−181	−181	197

Experiences by using the fast algorithm

Applying the above fast algorithm Pethő [35] computed the solutions with $|y| \le 10^{41}$ of about 3000 Thue equations of the form

$$f(x,y) = 1 \qquad \text{in } x, y \in \mathbb{Z}, \tag{12}$$

where $f(x,y) = ax^3 + bx^2y + cxy^2 + dy^3$ is an irreducible form with rational integer coefficients, and he obtained interesting experiences for the *number of solutions* of (12). To formulate his conjecture let us introduce the following concepts. The form f is called *reversible*, if $a = d = 1$. Two cubic forms $f(x,y)$ and $g(x,y) \in \mathbb{Z}[x,y]$ are called *equivalent*, if f can be transformed into g with a unimodular transformation, that is if there exist integers a_1, a_2, a_3, a_4 with $a_1a_4 - a_2a_3 = \pm 1$ and

$$g(x,y) = f(a_1x + a_2y, a_3x + a_4y).$$

Obviously, if f is equivalent with g, then their discriminants are equal and the number of solutions of $f(x, y) = 1$ and $g(x, y) = 1$ are the same.

Denote by N the number of solutions of (12), and by D the discriminant of f. In the case of cubic forms with negative discriminants Delone [7] and Nagell [28] (see also Delone and Faddeev [8]) proved, that $N \leq 5$,

	N=	
if f is not equivalent to a revesible form (N = 0, 1, 2)	0	
	1	
	2	if f is eqivalent to a revesible form (N = 2, 3)
	3	
	4	if D=−44 and −31
	5	if D=−23

and there are infinitely many inequivalent forms with N=3.

In case of positive discriminants the corresponding result has not been verified until now. However, the above mentioned extensive computations of Pethő [35] led him to the following conjecture: $N \leq 9$ and

	N=	
if f is not equivalent to a revesible form (N = 0, 1, 2, 3)	0	
	1	
	2	
	3	if f is eqivalent to a reversible form (N = 2, 3, 4, 5)
	4	
	5	
	6	if D= 81, 229, 257 and 361
	7	no forms
	8	no forms
	9	if D= 49

Some more computational results on Thue equations

Let us close this part by mentioning some recent computational results connected with Thue equations.

We would like to recall, that there exists a p-adic analogue of the reduction algorithm. Combining the reduction method with its p-adic version, cubic Thue-Mahler equations were solved by Agrawal at al. [1] and by Tzanakis and de Weger [51], with one and three primes on the right side, respectively.

In a recent paper Thomas [47] considered a parametrized family of Thue equations of the form

$$x^3 - (n-1)x^2y - (n+2)xy^2 - y^3 = \pm 1 \qquad \text{in } x, y \in \mathbb{Z}, \tag{13}$$

with the parameter $0 \leq n \in \mathbb{Z}$. He called the solutions $(0, \pm 1)$, $(\pm 1, 0)$, $(\pm 1, \mp 1)$ trivial, and obtained the following result:

Theorem 1.3 (Thomas [47], 1989) *If* $n \geq 1.365 * 10^7$, *then equation* (13) *has only the trivial solutions. For* $0 \leq n \leq 10^3$ *all non-trivial solutions are given by*

$$n = 0 \quad (5,4), (-4,9), (-9,5), (-2,1), (1,1), (-1,2)$$
$$n = 1 \quad (2,1), (-3,2), (-1,3)$$
$$n = 3 \quad (7,2), (-2,9), (-9,7).$$

In his paper Thomas formulated also the conjecture, that $n = 0, 1, 3$ are all values of $n \geq 0$, when non-trivial solutions occur.[1]

Another attempt for solving several Thue equations instead of individual ones was made by Stroeker [44] (see also [45]). Let K be a quartic number field embedded into the real field, but with negative discriminant, and let $L = \mathbb{Q}(\sqrt{d})$ ($d \equiv 3 \ (mod\ 4)$) be its real quadratic subfield. Assume that $\xi > 1$ generates the subgroup of units relative to L in the group of positive units of K. Stroeker considered Thue equations of type

$$N_{K/\mathbb{Q}}(X(u,v) + Y(u,v)\xi^2) = 1 \qquad \text{in } u, v \in \mathbb{Z},$$

where $X(u,v)$ and $Y(u,v)$ are linear forms with coefficients in $\mathbb{Z}\left[\sqrt{d}\right]$. He proved, that under certain conditions, satisfied by several linear forms $X(u,v)$, $Y(u,v)$, the above equation has only the trivial solutions given by $XY = 0$.

In a paper Stroeker and Tzanakis [46] worked out an algorithm, based on Skolem's p-adic method, for the resolution of Thue equations. The algorithm is applicable for all cubic Thue equations, and for quartic Thue equations with negative discriminants. It yields somewhat less computation, than the reduction method, but it does not seem to be straitforward to generalize.

Our last remark, related to Thue equations is devoted to their so called "inhomogeneous" extension. Let α be an algebraic integer of degree ≥ 3, let $K = \mathbb{Q}(\alpha)$ and $0 \neq m \in \mathbb{Z}$. It was Sprindzuk [42] who first considered in 1974 *inhomogeneous Thue equations* of type

$$N_{K/\mathbb{Q}}(x + \alpha y + \lambda) = m \qquad \text{in } x, y \in \mathbb{Z},\ \lambda \in \mathbb{Z}_K. \tag{14}$$

Here λ is a new variable which takes algebraic integer values in K. Of course, to ensure the finiteness of the number of solutions (x, y, λ) of equation (14), we have to restrict the variables. Sprindzuk used the natural assumption, that the *non-dominating variable* λ is less than the *dominating variables* x, y, more precisely

$$\overline{|\lambda|} < X^{1-\xi}, \qquad X = \max(|x|, |y|), \tag{15}$$

where $\overline{|\lambda|}$ is the size of λ, that is the maximum absolute value of its conjugates, and ξ is a given small positive constant. Under these conditions Sprindzuk gave an effective upper bound for the sizes of all solutions (x, y, λ) of equation (14), which implies also that it has only finitely many solutions. In the special case $\lambda = 0$ his result includes Baker's Theorem 1.1 on Thue equations.

[1] The conjecture was recently proved by M. Mignotte

To illustrate the importance of this result, assume that λ is a fixed non-zero integer in K, and consider equation (14) as an equation in x, y. It is easy to see, that a special case of Sprindzuk's result gives an upper bound for all solutions x, y. But for a fixed $\lambda \neq 0$ equation (14) is an inhomogeneous polynomial equation in x, y, and formerly there were no effective theorems for such inhomogeneous equations.

Example. Let us give an example. Let α be defined by the equation $x^3 - 3x^2 - 4x - 1 = 0$. Let $K = \mathbb{Q}(\alpha)$ and consider the Thue equation

$$N_{K/\mathbb{Q}}(x + \alpha y) = 2 \qquad \text{in } x, y \in \mathbb{Z},$$

that is

$$x^3 + 3x^2y - 4xy^2 + y^3 = 2 \qquad \text{in } x, y \in \mathbb{Z}.$$

Put $\lambda = \alpha^2$. Then the corresponding inhomogeneous equation

$$N_{K/\mathbb{Q}}(x + \alpha y + \lambda) = 2$$

has the form

$$x^3 + y^3 + 3x^2y - 4xy^2 + 17x^2 + 3y^2 - 15xy + 10x - 4y - 1 = 0.$$

Combining Sprindzuk's method with the reduction algorithm, the author [15] gave a method for the complete resolution of inhomogeneous Thue equations of type (14). In the method the only restriction is that $\xi \geq 1/2$. Of course it is much more efficient if the degree and the unit rank of K is small and $m = 1$.

II. Index form equations

Let K be an algebraic number field of degree n with integral basis $1, \omega_2, \ldots, \omega_n$. As before, denote by $\gamma^{(i)}$ $(i = 1, \ldots, n)$ the conjugates of any $\gamma \in K$. Let

$$L^{(i)}(\underline{X}) = \omega_2^{(i)} X_2 + \ldots + \omega_n^{(i)} X_n \qquad (i = 1, \ldots, n).$$

The *discriminant form*

$$D_{K/\mathbb{Q}}(\omega_2 X_2 + \ldots + \omega_n X_n) = \prod_{1 \leq i < j \leq n} \left[L^{(i)}(\underline{X}) - L^{(j)}(\underline{X}) \right]^2$$

can be written in the form

$$D_{K/\mathbb{Q}}(\omega_2 X_2 + \ldots + \omega_n X_n) = [I(X_2, \ldots, X_n)]^2 \, D_{K/\mathbb{Q}},$$

where $I(X_2, \ldots, X_n)$ is a homogeneous polynomial on degree $n(n-1)/2$ with rational integer coefficients, called the *index form* corresponding to the basis $1, \omega_2, \ldots, \omega_n$ of K, and $D_{K/\mathbb{Q}}$ is the discriminant of the field K. An *index form equation* is a diophantine equation of the form

$$I(x_2, \ldots, x_n) = m \qquad \text{in } x_2, \ldots, x_n \in \mathbb{Z}, \tag{16}$$

with a non-zero rational integer m. The finiteness of the number of solutions of (16) was first proved by Győry [23], already in an effective form. His result was later extended by Trelina [49], Győry and Papp [24] and others.

The index form equations play an essential role in algebraic number theory. Let $x_1, x_2, \ldots, x_n$ be any rational integers and put $\alpha = x_1 + \omega_2 x_2 + \ldots + \omega_n x_n$. Then

$$|I(x_2, \ldots, x_n)| = \left[\mathbb{Z}_K^+ : \mathbb{Z}^+[\alpha]\right],$$

where $\mathbb{Z}_K^+$ and $\mathbb{Z}^+[\alpha]$ denote the additive groups of the ring of integers of K, and $\mathbb{Z}[\alpha]$, respectively. The above index is called the *index of* α and is denoted by $I(\alpha)$. It is easily seen that if α and α^* are integers in K with $\alpha - \alpha^* \in \mathbb{Z}$ (in which case they are called *equivalent*) then $I(\alpha) = I(\alpha^*)$. Moreover, $\left\{1, \alpha, \ldots, \alpha^{n-1}\right\}$ is an integral basis of K (a *power integral basis*) if and only if $I(\alpha) = 1$, that is if $x_2, \ldots, x_n$ is a solution of the equation

$$I(x_2, \ldots, x_n) = \pm 1 \qquad \text{in } x_2, \ldots, x_n \in \mathbb{Z}. \tag{17}$$

This shows why index form equations are so important: to determine the power integral bases of the field K, we have to find all solutions of (17). In connection with equation (17) the following problems arise naturally:

I. Solve equation (17), to find all power integral bases of K. If (17) is solvable, that is if there are power integral bases of the field, then K is called *monogene*.

II. If (17) is not solvable, then find the minimal m for which (16) is solvable, that is find the elements with *minimal index*.

The computational results obtained until now for index form equations correspond to cubic and quartic number fields. Thus, in the following we shall consider these types of fields, separately.

Cubic number fields

Authors studying the question of monogenity in cubic number fields often restrict themselves to cyclic cubic fields. M.N.Gras [20], [21], and Archinard [2] gave necessary and sufficient conditions for monogenity of cyclic cubic fields. They also decided in several numerical cases, whether or not there exist power integral bases. Dummit and Kisilevsky [9] proved that there exist infinitely many cyclic cubic fields which are monogene. On the other hand they showed that there are cyclic cubic fields in which the minimal index of the elements is arbitrary large.

If K is a cubic number field, then the index form $I(X, Y)$ corresponding to an integral basis of K is just an irreducible binary form, and the index form equation

$$I(x, y) = \pm 1 \qquad \text{in } x, y \in \mathbb{Z}$$

is a cubic Thue equation. This made possible to solve index form equations of cubic fields by using the reduction method as we described in part I. Together with Schulte [19] we solved completely index form equations corresponding to cubic number fields with discriminants $-300 \leq D \leq 3137$. It yields, that we determined all power integral bases in approximately 30 complex, and 100 totally real cubic fields with least discriminants. This was the first case, when Thue equations with leading coefficient not equal to 1, and Thue equations with negative discriminants were completely solved. Further, Schulte [39] computed the power integral bases, and in case of non-monogenity all elements with minimal index in all totally real cubic fields with discriminants ≤ 44521. This yields about 2000 fields. The new feature in this computation was, that she solved Thue equations with right hand side not equal to ± 1. These computations are very important also from the point of view of Thue equations, because formerly nobody solved Thue equations in such large number. The solutions of these index form equations can give important informations also for the solutions of Thue equations. In particular these examples make more probable the before mentioned conjecture of Pethő [35] on the number of solutions of Thue equations with positive discriminants.

Example. Let us give an example for index form equations of cubic number fields. Consider the cubic number field with discriminant $D = 2636$. This field can be generated by the element α with defining polynomial $x^3 - 14x + 4 = 0$. An integral basis in the field is $1, \omega_2 = \alpha, \omega_3 = \alpha^2/2$. The index form equation, corresponding to this integral basis is

$$2x^3 - 7xy^2 + y^3 = \pm 1 \qquad \text{in } x, y \in \mathbb{Z}.$$

All solutions of this equation are $(x, y) = (-2, 1), (0, 1)$ (and their negatives), which means that for any algebraic integer ρ of K, $\{1, \rho, \rho^2\}$ is a power integral basis of the field, if and only if

$$\rho = \pm(a + \omega_2 x + \omega_3 y),$$

where a is any rational integer and (x, y) is a solution of the index form equation.

Quartic number fields

Now let us turn to quartic number fields. Considering cyclic biquadratic fields, M.N.Gras [22] gave necessary and sufficient conditions for the monogenity of these fields. She also decided for several numerical examples, whether or not there exist power integral bases in the field. She proved, that imaginary cyclic biquadratic fields can have power integral bases only in two trivial cases. Nakahara showed, that although there exist cyclic biquadratic fields which have power integral bases [29], there are also cyclic biquadratic fields in which the minimal index of the integral elements is arbitrary large [31]. For non-cyclic, but Abelian biquadratic fields Nakahara [30] proved, that infinitely of them are monogene, but there are also fields of this type with arbitrary large minimal index.

Together with Pethő and Pohst [18] we gave a computational algorithm for the complete resolution of index form equations corresponding to totally real cyclic biquadratic number fields. Here the assumption that the field is cyclic is only technical, the purpose of which is to save computation time. The method for the resolution is to reduce the index form equation to weighted unit equations in two variables and to solve these unit equations by a variant of the reduction method. It turned out, that the algorithm does not utilize any special properties of index form equations, which means, that almost the same method can be applied also for the resolution of those other types of decomposable form equations, which can be reduced to finitely many unit equations in two variables. Wide classes of norm form equations, all the discriminant form equations and index form equations, and also some classes of general decomposable form equations have this property (cf. e.g. [13], [14]). This general feature makes it worthy to give a brief sketch of the method here (see also [16]).

Let K be a totally real cyclic biquadratic number field with fundamental units η_1, η_2, η_3. Take an integral basis of K, then the index form equation corresponding to it is

$$I(x_2, x_3, x_4) = \pm 1 \qquad \text{in } x_2, x_3, x_4 \in \mathbb{Z}. \tag{18}$$

The index form $I(x_2, x_3, x_4)$ can be factorized into linear factors $L_i(\underline{x})$ having algebraic integer coefficients in K (here we use that K is cyclic):

$$\prod_{i=1}^{6} L_i(\underline{x}) = \pm 1 \qquad \text{in } \underline{x} \in \mathbb{Z}^3. \tag{19}$$

There are three linear forms, say L_1, L_2, L_3 and algebraic integers τ_1, τ_2, τ_3 such that

$$\tau_1 L_1(\underline{X}) + \tau_2 L_2(\underline{X}) + \tau_3 L_3(\underline{X}) = 0 \tag{20}$$

is an identity. Take any fixed solution (x_2, x_3, x_4) of equation (18). Then, by (19) $L_i(\underline{x})$ are units in K, and hence there exist integers y_1, y_2, y_3 and z_1, z_2, z_3 such that (20) can be written as

$$\frac{\tau_1}{\tau_3}\eta_1^{y_1}\eta_2^{y_2}\eta_3^{y_3} + \frac{\tau_2}{\tau_3}\eta_1^{z_1}\eta_2^{z_2}\eta_3^{z_3} + 1 = 0. \tag{21}$$

This is just a unit equation in two varables. We may suppose that $Y = \max|y_i| > Z = \max|z_i|$ (the opposit case should be considered as well). Then we have the following sequence of inequalities (c_1, c_2 again denote effectively computable positive constants):

$$\begin{aligned} exp(-c_1 log Z) &< \left| z_1 log|\eta_1| + z_2 log|\eta_2| + log\left|\frac{\tau_2}{\tau_3}\right|\right| \qquad (22) \\ &= \left| log\left|\eta_1^{z_1}\eta_2^{z_2}\eta_3^{z_3}\frac{\tau_2}{\tau_3}\right|\right| \le 2\left|\eta_1^{z_1}\eta_2^{z_2}\eta_3^{z_3}\frac{\tau_2}{\tau_3} + 1\right| \\ &= 2\left|\eta_1^{y_1}\eta_2^{y_2}\eta_3^{y_3}\frac{\tau_1}{\tau_3}\right| < exp(-c_2 Y) < exp(-c_2 Z). \end{aligned}$$

Here the first is Baker's inequality, the next one is a trivial identity, then we applied $|log z| \le 2|z-1|$ (if it can not be used, that is if $|z-1| > 1/3$ then we get a better bound for Z). The next is equation (21), and the following estimate can be obtained by standard, elementary tools. Finally $Y > Z$ is applied. Comparing the beginning and the end of this inequality sequence, we get a bound Z_B for Z of magnitude 10^{35}. Dividing the linear form in the logarithms in the second line of (22) by the coefficient of z_3, and using the final estimate, we obtain an inequality of the form

$$|z_1\theta_1 + z_2\theta_2 + z_3 - \gamma| < K^{-Z}, \tag{23}$$

where θ_1, θ_2 and γ are real numbers (the quotients of two logarithms above) and K is a positive constant that can be explicitely determined. As before, the reduction algorithm is based on inequality (23). In this case we apply the following version of Lemma 1.1:

Lemma 2.1 (Gaál, Pethő and Pohst [18]) *Let* M, B *be integers with* $B > M^2$ *and let* $\theta_1, \theta_2, \gamma$ *be real numbers. If there exist an integer* q *with*

$$(a') \quad 1 \le q \le MB$$
$$(b') \quad \|q\theta_i\| < 2(MB)^{-1/2} \qquad (i = 1, 2)$$
$$(c') \quad \|q\gamma\| \ge 5B^{-1/4},$$

then inequality (23) *has no integer solutions* z_1, z_2, z_3 *with*

$$\frac{log\left[MB^{5/4}\right]}{logK} \le Z \le M,$$

where $Z = max\,|z_i|$, *and* $\|.\|$ *denotes the distance from the nearest integer.*

The above lemma makes possible to reduce the original bound 10^{35} to about 100. First take M to be the Baker's bound and $B = M^2 + 1$. Applying the LLL-reduction algorithm [26] we can find a number q satisfying (a') and (b'). In all cases we considered, this number satisfied also (c') and made possible the reduction. In the next step we take M to be the new bound for Z, and B as before. Within 4 or 5 reduction steps we get a bound that we can not reduce further (≈ 100). Here we also have to remark, that in the first reduction step we have to use multiple precision arithmetics and we have to calculate the values of θ_1, θ_2 and γ with the accuracy of about 170 digits.

Using easy tricks (cf. [18]), based on inequality (23), we can drop out the majority of triples (z_1, z_2, z_3), which are under the reduced bound in absolute value, but in fact do not satisfy (23). For the remaining few triples we determine the corresponding (y_1, y_2, y_3) from (21), and finally the solutions (x_2, x_3, x_4) of the original equation (18). The whole computation took about $3.5 - 4$ hours on an IBM PC AT computer.

The algorithm is applicable also if K is not cyclic, but then we have to work in the smallest normal extension field of K, in which the number of fundamental units is larger, which would make the algorithm much slower. As it is easy to see, the algorithm works without any principal change also for any decomposable form equation, which can be reduced to unit equations in two variables of type (21). It is interesting to remark, that this was the first case, when decomposable form equations in more than two variables were completely solved.

Example. To give an idea about the difficulties appearing here let us see an example. Consider the totally real cyclic biquadratic field $K = \mathbb{Q}(\sqrt{\mu})$, where $\mu = 55 + 22\sqrt{5}$. The discriminant of this field is $D = 15125$, an integral basis is $1, \omega, \psi, \omega\psi$, where $\omega = (1+\sqrt{5})/2$ and $\psi = (1+\sqrt{\mu})/2$. The index form corresponding to this integral basis is

$$\begin{aligned} I(x,y,z) = \; & - \; 5x^4y^2 - 5x^4yz + 5x^4y^2 - 10x^3y^2z - 10x^3yz^2 + 10x^3z^3 \\ & + \; 55x^2y^4 + 220x^2y^3z + 240x^2y^2z^2 - 35x^2yz^3 - 130x^2z^4 \\ & + \; 55xy^4z + 220xy^3z^2 + 245xy^2z^3 - 30xyz^4 - 135xz^5 \\ & - \; 121y^6 - 968y^5z - 2860y^4z^2 - 3575y^3z^3 - 1035y^2z^4 \\ & + \; 1407yz^5 + 881z^6. \end{aligned}$$

This is quite complicated (even for this field with rather small discriminant), although the index form (for biquadratic fields) can always be factorized into the product of a quadratic and a quartic form:

$$\begin{aligned} I(x,y,z) = \left[y^2 + yz - z^2\right] \cdot [\; & - \; 5x^4 - 10x^3z + 55x^2y^2 + 165x^2yz + 130x^2z^2 \\ & + \; 55xy^2z + 165xyz^2 + 135xz^3 - 121y^4 \\ & - \; 847y^3z - 2134y^2z^2 - 2288yz^3 - 881z^4 \;]. \end{aligned}$$

Of course, taking the above forms, the original index form equation

$$I(x,y,z) = \pm 1 \qquad \text{in } x,y,z \in \mathbb{Z} \tag{24}$$

can be replaced with the system of a quadratic and a quartic equation, but unfortunately both of them have infinitely many solutions.

Using our algorithm we could determine all solutions of (24), which are $(x,y,z) = (-1,-2,1), (-1,-1,1), (0,-2,1), (0,-1,1)$ (and their negatives). It means that for any algebraic integer ρ of K, $\{1, \rho, \rho^2, \rho^3\}$ is a power integral basis of K, if and only if

$$\rho = \pm(a + \omega x + \psi y + \omega\psi z),$$

where a is any rational integer and (x,y,z) is a solution of (24).

As we mentioned above, the reduction algorithm for the resolution of index form equations corresponding to cyclic biquadratic number fields is rather time consuming. At this point again arises the need of a fast algorithm producing only the "small" solutions (that is the solutions with absolute values less than some million). In practice only those power integral bases are usable, which are generated by elements having moderate coefficients in the integral basis.

Together with Pethő and Pohst [17] we worked out such a fast algorithm, applicable for finding small solutions of index form equations corresponding to any kind of biquadratic number fields. The algorithm reduces the problem to searching for squares (more precisely squares translated with a constant) in second order linear recurrence sequences. We remark, that the problem of finding polynomial values in linear recurrence sequences was considered formerly by Pethő [33], Nemes and Pethő [32], and others.

To give numerical examples of this algorithm we considered index form equations of biquadratic number fields of the form

$$K = \mathbb{Q}(\sqrt{\mu}), \qquad \mu = \frac{e + f\sqrt{m}}{2},$$

where e, f and m are certain rational integers. We computed the small solutions of index form equations of all those totally real biquadratic fields of the above type for which $m = 2$ or 5, the discriminant is ≤ 1000000 and the Galois group is C4 or D8. The numerical data of the fields were taken from the table of Buchmann and Ford [6]. It yields 34 fields with Galois group C4 and 2636 fields with Galois group D8.[1]

Example. To see an example from this computation consider the field $K = \mathbb{Q}(\sqrt{\mu})$ where $\mu = 43 + 26\sqrt{2}$. This field has discriminant $D = 31808$ and integral basis $1, \omega, \psi, \omega\psi$, where $\omega = \sqrt{2}$ and $\psi = (1 + \sqrt{2} + \sqrt{\mu})/2$. The solutions with $\max(|x_2|, |x_3|, |x_4|) < 10^7$ of the index form equation

$$I(x, y, z) = \pm 1 \qquad \text{in } x, y, z \in \mathbb{Z}$$

are $(x, y, z) = (-5, 1, 1)$ and $(3, 1, 1)$ (and their negatives), which means that very probably for any algebraic integer ρ, $\{1, \rho, \rho^2, \rho^3\}$ is a power integral basis of the field K, if and only if

$$\rho = \pm(a + \omega x + \psi y + \omega\psi z),$$

where a is any rational integer and (x, y, z) is either

$$(-5, 1, 1) \quad \text{or} \quad (3, 1, 1).$$

Extending the above mentioned algorithm we obtained a fast computational method for determining the elements with *minimal index* among elements having "small" coefficients in the integral basis of biquadratic fields. This description seems to be a little artificial at the first look, but let us take into consideration two simple facts. First, our experience shows, that such equations never have really "large" solutions. That is, if for a certian index m, there are no elements with index m and "small" coefficients in the integral basis, then almost surely there are not such elements at all. On the other hand, from a practical point of view, it is much better to know in a field an element with somewhat larger index and small coefficients in the integral basis (which yields a more simple form of the element), than another element with somewhat smaller index, but very large coefficients in the integral basis (which can make the element unusable for practical calculations).

We executed this algorithm for all those fields we considered in the preceeding method.[1] Let us close with an example of this kind.

Example. Let $K = \mathbb{Q}(\sqrt{\mu})$ with $\mu = 39 + 10\sqrt{5}$. The discriminant is $D = 25525$, the integral basis is $1, \omega, \psi, \omega\psi$, where $\omega = (1+\sqrt{5})/2, \psi = (1+\sqrt{\mu})/2$.

[1] Recently we extended this computation to all totally real biquadratic fields with Galois group C4 or D8 and discriminant ≤ 1000000.

Denote by $I(x,y,z)$ the index form corresponding to this integral basis. The least m, for which the equation

$$I(x,y,z) = \pm m \qquad \text{in } x,y,z \in \mathbb{Z}$$

has solutions with $\max(|x|,|y|,|z|) < 10^7$ is $m = 3$, and these are

$$(x,y,z) = (-5,5,-3),(-1,-1,1),(0,-1,1),(8,5,-3).$$

Hence, very probably in this field the elements with least index have index 3, and all these elements are of the form

$$\rho = \pm(a + \omega x + \psi y + \omega\psi z),$$

where a is any rational integer and (x,y,z) is one of the above triples.

References

[1] A.K. Agrawal, J.H. Coates, D.C. Hunt and A.J. van der Poorten, *Elliptic curves of conductor 11*. Math. Comp **35** (1980), 991-1002.

[2] G. Archinard, *Extensions cubiques cycliques de* $\mathbb{Q}$ *dont l'anneau des entiers est monogéne*, Enseignement Math. **20** (1974), 179-203.

[3] A. Baker, *Linear forms in the logarithms of algebraic numbers*, Mathematika **13** (1966), 204-216.

[4] A. Baker, *Contributions to the theory of Diophantine equations*, Philos. Trans. Roy. Soc. London **263** (1968), 173-208.

[5] A. Baker and H. Davenport, *The equations* $3x^2 - 2 = y^2$ *and* $8x^2 - 7 = z^2$. Quart. J. Math. Oxford **20** (1969), 129-137.

[6] J. Buchmann and D. Ford, *On the computation of totally real quartic fields of small discriminant*, Math. Comp. **52** (1989), 161-174.

[7] B.N. Delone, *Über die Darstellung der Zahlen durch die binare kubischen Formen von negativer Diskriminante*, Math. Zeitschr. **31** (1930), 1-26.

[8] B.N. Delone and D.K. Faddeev, *The theory of irrationalities of the third degree*, Am. Math. Soc. Transl. of Math. Monographs 10 Providence, 1964.

[9] D.S. Dummit and H. Kisilevsky, *Indices in cyclic cubic fields* . "Number Theory and Algebra", Academic Press 1977, 29-42.

[10] W.J. Ellison, *Recipes for solving diophantine problems by Baker's method*. Sém. Théorie Nombres, Année 1970-1971, **exp. no.11.**

[11] W.J. Ellison, F. Ellison, J. Pesek, C.E. Stahl and D.S. Stall, *The diophantine equation* $y^2 + k = x^3$, J. Number Theory **4** (1972), 107-117.

[12] J.H. Evertse, *On the representation of integers by binary cubic forms of positive discriminant*, Invent. Math. **73** (1983), 117-138.

[13] J.H. Evertse and K. Győry, *Decomposable form equations*, "New Advances in Transcendence Theory", ed. by A.Baker, Cambridge University Press 1988, 175-202.

[14] J.H. Evertse, K. Győry, C.L. Stewart and R. Tijdeman, *S-unit equations and their applications*, "New Advances in Transcendence Theory", ed. by A.Baker, Cambridge University Press 1988, 110-174.

[15] I. Gaál, *On the resolution of inhomogeneous norm form equations in two dominating variables*, Math. Comp. **51** (1988), 359-373.

[16] I. Gaál, *On the computer resolution of index form equations*, "Algebra and Number Theory", ed. by A. Grytczuk, Pedagogical University, Zielona Gora, 1990, 21-27.

[17] I. Gaál, A. Pethő and M. Pohst, *On the resolution of index form equations in biquadratic number fields, I.*, J. Number Theory, to appear.

[18] I. Gaál, A. Pethő and M. Pohst, *On the resolution of index form equations in biquadratic number fields, II.*, J. Number Theory, to appear.

[19] I. Gaál and N. Schulte, *Computing all power integral bases of cubic fields*, Math. Comp. **53** (1989), 689-696.

[20] M.N. Gras, *Sur les corps cubiques cycliques dont l'anneau des entiers monogéne*, Publ. Math. Fac. Sci. Besancon, (1973).

[21] M.N. Gras, *Lien entre le groupe des unites et la monogéneite des corps cubiques cycliques*, Sém. Théorie Nombr. Besancon, Année 1975-1976.

[22] M.N. Gras, *$\mathbb{Z}$-bases d'entiers $1, \theta, \theta^2, \theta^3$ dans les extensions cycliques de degre 4 de $\mathbb{Q}$*, Theorie des Nombres Besancon, Années 1979-1980 et 1980-1981.

[23] K. Győry, *Sur les polynomes á coefficients entiers et de discriminant donné, III.*, Publ. Math. Debrecen **23** (1976), 141-165.

[24] K. Győry and Z.Z. Papp, *Effective estimates for the integer solutions of norm form and discriminant form equations*, Publ. Math. (Debrecen) **25** (1978), 311-325.

[25] D.E. Knuth, "The art of computer programming", Vol.2. (Seminumerical algorithms), 2nd.ed., Addison Wesley Publ. Comp., Reading, Mass. 1981.

[26] A.K. Lenstra, H.W. Lenstra Jr., and L. Lovász, *Factoring polynomials with rational integer coefficients*, Math. Ann. **261** (1982), 515-534.

[27] K. Mahler, *Zur Approximation algebraischer Zahlen I: Über den grössten Primteiler binärer Formen*, Math. Ann. **107** (1933), 691-730.

[28] T. Nagell, *Darstellung ganzer Zahlen durch binare kubischen Formen mit negativer Diskriminante*, Math. Zeitschr. **28** (1928), 10-29.

[29] T. Nakahara, *On cyclic biquadratic fields related to a problem of Hasse*, Mon. Math. **94** (1982), 125-132.

[30] T. Nakahara, *On the indices and integral bases of non-cyclic but abelian biquadratic fields*, Archiv der Math. **41** (1983), 504-508.

[31] T. Nakahara. *On the minimum index of a cyclic quartic field* Arch. Math. **48** (1987), 322-325.

[32] I. Nemes and A. Pethő, *Polynomial values in linear recurrences*, Publ. Math. (Debrecen) **31** (1984), 229-233.

[33] A. Pethő, *Perfect powers in second order linear recurrences*, J. Number Theory **15** (1982), 5-13.

[34] A. Pethő, *On the resolution of Thue inequalities*, J. Symbolic Comp. **4** (1987), 103-109.

[35] A. Pethő, *On the representation of 1 by binary cubic forms with positive discriminant*, Number Theory, Ulm (1987), eds. H.P.Schlikewei and E.Wirsing, Lecture Notes in Math. No. 1380, Springer Verlag 1989, 185-196.

[36] A. Pethő, *Computational methods for the resolution of diophantine equations*, Number theory, ed. R.Mollin, de Gruyter , Berlin-New York 1990, 477-492.

[37] A. Pethő and R.Schulenberg, *Effektives Lösen von Thue Gleichungen*, Publ. Math. (Debrecen) **34** (1987), 189-196.

[38] J. Graf v. Schmettow, *KANT - a tool for computations in algebraic number fields, these proceedings,*

[39] N. Schulte, *Indexgleichungen in kubischen Zahlkörpern*, Diplomarbeit, Düsseldorf 1989.

[40] T.N. Shorey and R. Tijdeman, *Exponential diophantine equations*, Cambridge University Press 1986.

[41] C.L. Siegel, *Approximation algebraischer Zahlen*, Math. Z. **10** (1921), 173-213.

[42] V.G. Sprindzuk, *Representation of numbers by the norm forms with two dominating variables*, J. Number Theory **6** (1974), 481-486.

[43] R.P. Steiner, *On Mordell's equation $y^2 - k = x^3$: A problem of Stolarsky*, Math. Comp. **46** (1986), 703-714.

[44] R.J. Stroeker, *On quartic Thue equations with trivial solutions*, Math. Comp. **52** (1989), 175-187.

[45] R.J. Stroeker, *On Thue equations associated with certian quartic number fields, these proceedings.*

[46] R.J. Stroeker and N.Tzanakis, *On the application of Skolem's p-adic method to the solution of Thue equations*, J. Number Theory **29** (1988), 166-195.

[47] E. Thomas, *Complete solutions to a family of cubic diophantine equations* , J. Number Theory **34** (1990), 235-250.

[48] A. Thue, *Über Annäherungswerte algebraischer Zahlen*, J. Reine Angew. Math. **135** (1909), 284-305.

[49] L.A. Trelina, *On the greatest prime factor of an index form* , Dokl. Akad. Nauk BSSR **21** (1977), 975-976.

[50] N. Tzanakis and B.M.M. de Weger, *On the practical solution of the Thue equation*, University of Twente, Memorandum **668** (1987),

[51] N. Tzanakis and B.M.M. de Weger, *Solving a specific Thue-Mahler equation*, University of Twente, Memorandum **793** (1989).

[52] B.M.M. de Weger, *Algorithms for diophantine equations*, Ph.D.Thesis, Amsterdam, (1987).

[53] D. Zagier, *Large integral points on elliptic curves*, Math. Comp. **48** (1987), 425-436.

Index form equations in cubic number fields

N. Schulte

Abstract. Let F be an algebraic number field and o_F the ring of algebraic integers of F. let $m(F) = \min_{\omega \in o_F}(o_F : \mathbb{Z}[\omega])$. Applying Baker's effective method and the reduction procedure of Baker and Davenport we present a method for computing the minimal indices $m(F)$ in cubic number fields and also all solutions $\theta \in o_F$ of $(o_F : \mathbb{Z}[\theta]) = m(F)$.

1 Introduction

Let F be a number field of degree n and denote by o_F the ring of algebraic integers of F. If $\omega_1 = 1, ..., \omega_n$ is an integral basis of F the index $(o_F : \mathbb{Z}[\alpha])$ $(\alpha \in o_F \backslash \mathbb{Z})$ is the absolute value of the determinant of the matrix transforming the basis $\omega_1, ..., \omega_n$ to $1, \alpha, \alpha^2, ..., \alpha^{n-1}$.

From now on let F be a cubic number field. Then each element of o_F has the form:

$$\alpha = z + x\omega_2 + y\omega_3 \ (z, x, y \in \mathbb{Z}).$$

Hence, we want to compute

$$m(F) = \min_{x,y,z \in \mathbb{Z}} (o_F : \mathbb{Z}[z + x\omega_2 + y\omega_3]).$$

Because of $\mathbb{Z}[z + x\omega_2 + y\omega_3] = \mathbb{Z}[x\omega_2 + y\omega_3]$ we need to consider only elements of the form: $x\omega_2 + y\omega_3$. Let $M_{xy} \in \mathbb{Z}[x, y]^{3\times3}$ be the matrix transforming $1, \omega_2, \omega_3$ to $1, x\omega_2 + y\omega_3, (x\omega_2 + y\omega_3)^2$. Then the determinant of M_{xy} is a homogeneous cubic polynomial with rational integral coefficients. Then $I(x, y) := \det(M_{xy})$ is called the *index form of F*. There exist $I_0, I_1, I_2, I_3 \in \mathbb{Z}$ with:

$$I(x, y) = I_0x^3 + I_1x^2\,y + I_2x\,y^2 + I_3y^3.$$

Our problem is to find solutions $x, y \in \mathbb{Z}$ of

$$I(x, y) = \pm K \tag{1}$$

for given $K \in \mathbb{N}$. If $x, y \in \mathbb{Z}$ is a solution of (1) for minimal K we know: $m(F) = K = (o_F : \mathbb{Z}[x\omega_2 + y\omega_3])$. Obviously, o_F has a power integral basis if and only if there exist $x, y \in \mathbb{Z}$ with $|I(x, y)| = 1$.

Computational Number Theory

A. Thue ([12]) proved in 1909, that each equation of the form: $I(x,y) = K$ with a homogeneous form $I(x,y)$ has only finitely many solutions. In 1968 A. Baker ([1]) gave an upper bound for the solutions x, y which depends only on the coefficients of $I(x,y)$ and on K. Unfortunately this constant is too large for practical applications.

In the next section we will present a method for obtaining an upper bound for possible solutions which bases on Baker's method and the reduction procedure of Baker and Davenport. All solutions will then be computed with a procedure of A. Pethö ([10]). Because the method for solving cubic Thue equations was already presented by I. Gaál and N. Schulte ([7]) there is no need to go into details.

2 Brief sketch of the algorithm

Let the index form equation (1) (corresponding to the integral basis $\{1, \omega_2, \omega_3\}$) of F be

$$I(x,y) = I_0 x^3 + I_1 x^2\, y + I_2 x\, y^2 + I_3 y^3 = \pm K \qquad (x, y \in \mathbb{Z}) \tag{2}$$

Denote by α a root of $I(x,1) = 0$. We note that α can be chosen as a generator of F over $\mathbb{Q}$. Let $x, y \in \mathbb{Z}$ be a solution of (2) and set

$$\beta = x - \alpha y.$$

Then β is an algebraic integer and (2) can be written in the form

$$\beta^{(1)}\, \beta^{(2)}\, \beta^{(3)} = \pm I_0^2\, K,$$

where $\beta^{(i)}$ $(1 \le i \le 3)$ denote the conjugates of $\beta = \beta^{(1)}$. If η_1, η_2 denote a set of fundamental units in the real case, η_1 in the complex case resp., then we obtain

$$\beta = \gamma \cdot \eta_1^{b_1} \cdot \eta_2^{b_2} \qquad (\beta = \gamma \cdot \eta_1^{b_1} \text{ resp.}), \tag{3}$$

where $b_1, b_2 \in \mathbb{Z}$ and γ is some algebraic integer of F with norm equal to $I_0^2{\cdot}K$. If $|I_0| \neq 1$ and $K \neq 1$ then in the sequel we have to consider seperately each element of a full set of non-associate elements γ of norm $I_0^2{\cdot}K$. Such a set can be computed with the method of Fincke and Pohst ([5]).

Applying Siegel's identity and some standard estimates, we obtain

$$|b_1 \log|\tau_1| + b_2 \log|\tau_2| - \log|\tau_3|| \le \exp(c_1 - c_2 B) \tag{4}$$

in the real case, and in the complex case

$$|b_1 \log|\tau_1| - c \log(-1) - \log \tau_3| \le \exp(c_1 - c_2 B) \tag{5}$$

where τ_1, τ_2 depend on the fundamental unit(s), τ_3 depends on the fundamental unit(s) and on γ and log denotes the principal value. The constants c_1, c_2, c are easily computable.

Applying Baker's method with the better estimates of Waldschmidt ([13]), in both cases the linear forms in logarithms (4),(5) can be estimated from below by a constant of the form

$$\exp(-c_3(\log B + c_4))$$

with $B = \max\{|b_1|, |b_2|\}$ in the real and $B = \max\{|b_1|, |c|\}$ in the complex case.

Comparing the lower and upper estimates obtained for the linear forms (4) and (5) we get an upper bound B_U for B, which is about 10^{80} in the cases we computed. Dividing (4) by $\log|\tau_2|$, (5) by $\log(-1)$ in the complex case, we obtain

$$|b_1\theta_1 + b_2 - \theta_2| \leq c_5 \cdot c_6^{-B} \tag{6}$$

in the real case and in the complex case

$$|b_1\theta_1 + c - \theta_2| \leq c_5 \cdot c_6^{-B} \tag{7}$$

where θ_1, θ_2 are quotients of two logarithms. We can now apply the Baker-Davenport ([2]) reduction method to these inequalities. We must find a good approximation of θ_1, which is not a good approximation to θ_2. Then each solution b_1, b_2 of (6), b_1, c of (7) resp., satisfies

$$B \leq \frac{c_7 + \log B_U}{c_8}.$$

Here c_7, c_8 are computable constants depending on B_U. Thus we can reduce B_U almost to its logarithm and get a much better bound for B. Repeating the reduction as long as the new bound is still less than the original one (4 or 5 times in the computed cases) we get a quite small bound for B which is usually ≤ 100.

Using the reduced bound for B from (3) it is easy to derive a bound Y_0 for $|y|$, and the procedure is completed by applying a fast algorithm of A. Pethö ([10]).

3 The minimal index K

The minimal index K is computed as follows. Define

$$m(\alpha) := (o_F : \mathbb{Z}[\alpha]) \quad (\alpha \in F)$$

and

$$M(F) := \gcd(\{m(\alpha) \mid \alpha \in F\}),$$

which we call field index. It is known ([9]) that 2, 3 are the only prime numbers which can divide $M(F)$.

Hence, for all x, y modulo $2^3 \cdot 3^2$ we calculate $K_{xy} \in \{0, ..., 71\}$ subject to

$$I(x, y) \equiv K_{xy} \bmod 72.$$

Then we know that $K_{XY} := \gcd\{K_{xy} \mid x, y \in \{0, .., 71\}\}$ is a divisor of $m(F)$ if K_{XY} is a divisor of 72, otherwise $M(F) = 1$. For $\nu = 1, 2, 3, ...$ and $K = \nu \cdot K_{XY}$ we test (1) for solvability until a solution (with ν minimal) is found.

Calculating the index form for the first 1831 totally real cubic fields we found $K_{XY} = 2$ in 73 cases. In all other cases we obtained $K_{XY} = 1$. For 79 cyclic cubic fields we got $K_{XY} = 2$ in 24 cases and $K_{XY} = 1$ otherwise.

4 Computational results

The program has been implemented in ANSI-Fortran-77 on an Apollo workstation. Most of the subroutines belong to the library KANT ([11]).

Using the method above we computed for all cubic number fields with discriminant between 49 and 44521 and some other cyclic cubic fields the minimal index $m(F)$ and all solutions satisfying the index form equation for the minimal index. The largest solution we found, was

$$x = 508, \quad y = 273$$

for discriminant 229. The maximal number of solutions occured was 9 only for discriminant 49.

We conclude with some results for cyclic fields of discriminant between 7^2 and 997^2 which were also considered by M.N. Gras ([8]) and D.S. Dummit and H. Kisilevsky ([4]).

The first column contains the discriminant d_F of the field. In the second column a, b are listed with the property that the generating polynomial f has the form: $f(t) = t^3 + t^2 + \frac{1-a}{3}\, t - \frac{a\,(3+b)-1}{27}$. The third column contains the integral basis $1, \omega_2, \omega_3$, where ρ denotes a root of the generating polynomial, the fourth the coefficients of $I(x, y)$, the fifth the minimal index K and the sixth column contains one solution of $|I(x, y)| = K$ and in the last column the number of solutions is listed. The results have been compared with those from ([8]) and ([4]). In these papers the minimal index could not be determined for every field, but possible candidates for the minimal index were given. These candidates are listed in parentheses in front of the minimal index in column 5.

Table 1

d_F	a, b	$1, \omega_2, \omega_3$	$I(x, y)$	K	(x, y)	#
7^2	$7, 1$	$1, \rho, \rho^2$	$1, -2, -1, 1$	1	$(1, 0)$	9
13^2	$13, 5$	$1, \rho, \rho^2$	$1, -2, -3, 5$	1	$(1, 0)$	3
19^2	$19, 7$	$1, \rho, \rho^2$	$1, -2, -5, -1$	1	$(1, 0)$	6
31^2	$31, 4$	$1, \rho, (\rho + \rho^2)/2$	$2, 1, -5, -2$	2	$(1, 0)$	3
37^2	$37, 11$	$1, \rho, \rho^2$	$1, -2, -11, 23$	1	$(1, 0)$	3
43^2	$43, 8$	$1, \rho, (\rho + \rho^2)/2$	$2, 1, -7, 2$	2	$(1, 0)$	6
61^2	$61, 1$	$1, \rho, (-\rho + \rho^2)/3$	$3, -5, -4, 3$	1	$(-1, 1)$	3
67^2	$67, 5$	$1, \rho, (-1 + \rho^2)/3$	$3, -2, -7, 3$	3	$(1, 0)$	3
73^2	$73, 7$	$1, \rho, (\rho + \rho^2)/3$	$3, 1, -8, -3$	3	$(1, 0)$	3
79^2	$79, 17$	$1, \rho, \rho^2$	$1, -2, -25, 67$	1	$(1, 0)$	3
97^2	$97, 19$	$1, \rho, \rho^2$	$1, -2, -31, -47$	1	$(1, 0)$	3
103^2	$103, 13$	$1, \rho, (-1 + \rho^2)/3$	$3, -2, -11, -3$	3	$(1, 0)$	3
109^2	$109, 2$	$1, \rho, (2 - \rho + \rho^2)/4$	$4, -5, -7, 4$	2	$(-1, 1)$	3
127^2	$127, 20$	$1, \rho, (\rho + \rho^2)/2$	$2, 1, -21, 20$	2	$(1, 0)$	3
139^2	$139, 23$	$1, \rho, \rho^2$	$1, -2, -45, 149$	1	$(1, 0)$	3

Table 1 (continued)

d_F	a, b	$1, \omega_2, \omega_3$	$I(x,y)$	K	(x,y)	#
151^2	$151, 19$	$1, \rho, (-\rho + \rho^2)/3$	$3, -5, -14, -3$	3	$(1,0)$	3
157^2	$157, 14$	$1, \rho, (\rho + \rho^2)/4$	$4, 1, -13, 4$	2	$(-2,1)$	3
163^2	$163, 25$	$1, \rho, \rho^2$	$1, -2, -53, -115$	1	$(1,0)$	3
181^2	$181, 7$	$1, \rho, (2 + 2\rho + \rho^2)/5$	$5, 4, -11, -5$	5	$(1,0)$	3
193^2	$193, 23$	$1, \rho, (-1 + \rho^2)/3$	$3, -2, -21, 23$	3	$(1,0)$	3
199^2	$199, 11$	$1, \rho, (-1 + \rho^2)/5$	$5, -2, -13, 5$	$(1,5)\ 5$	$(1,0)$	3
211^2	$211, 13$	$1, \rho, (\rho + \rho^2)/5$	$5, 1, -14, -5$	$(1,5)\ 5$	$(1,0)$	3
223^2	$223, 28$	$1, \rho, (\rho + \rho^2)/2$	$2, 1, -37, -64$	2	$(1,0)$	3
229^2	$229, 22$	$1, \rho, (2 - \rho + \rho^2)/4$	$4, -5, -17, -4$	$(2,4)\ 4$	$(1,0)$	3
241^2	$241, 17$	$1, \rho, (\rho + \rho^2)/5$	$5, 1, -16, 5$	1	$(1,3)$	3
271^2	$271, 26$	$1, \rho, (\rho + \rho^2)/3$	$3, 1, -30, 29$	3	$(1,0)$	3
277^2	$277, 26$	$1, \rho, (2 - \rho + \rho^2)/4$	$4, -5, -21, 26$	4	$(1,0)$	3
283^2	$283, 32$	$1, \rho, (\rho + \rho^2)/2$	$2, 1, -47, 76$	2	$(1,0)$	3
307^2	$307, 16$	$1, \rho, (\rho + \rho^2)/6$	$6, 1, -17, -6$	$(2,6)\ 6$	$(1,0)$	3
313^2	$313, 35$	$1, \rho, \rho^2$	$1, -2, -103, 475$	1	$(1,0)$	3
331^2	$331, 1$	$1, \rho, (-3\rho + \rho^2)/7$	$7, -11, -10, 7$	1	$(1,-1)$	3
337^2	$337, 5$	$1, \rho, (-1 - 2\rho + \rho^2)/7$	$7, -8, -13, 7$	$(1,5)\ 5$	$(1,2)$	3
349^2	$349, 37$	$1, \rho, \rho^2$	$1, -2, -115, -401$	1	$(1,0)$	3
367^2	$367, 35$	$1, \rho, (-\rho + \rho^2)/3$	$3, -5, -38, 75$	1	$(1,0)$	3
373^2	$373, 13$	$1, \rho, (-3 + 2\rho + \rho^2)/7$	$7, 4, -17, -7$	1	$(2,5)$	3
379^2	$379, 29$	$1, \rho, (-2\rho + \rho^2)/5$	$5, -8, -21, 29$	1	$(2,-1)$	3
397^2	$397, 34$	$1, \rho, (\rho + \rho^2)/4$	$4, 1, -33, -34$	4	$(1,0)$	3
409^2	$409, 31$	$1, \rho, (-2\rho + \rho^2)/5$	$5, -8, -23, -5$	$(1,5)\ 5$	$(1,0)$	3
421^2	$421, 19$	$1, \rho, (\rho + \rho^2)/7$	$7, 1, -20, -7$	$(1,7)\ 7$	$(1,0)$	3
433^2	$433, 2$	$1, \rho, (4 - 3\rho + \rho^2)/8$	$8, -11, -13, 8$	2	$(-1,1)$	3
439^2	$439, 28$	$1, \rho, (-\rho + \rho^2)/6$	$6, -5, -23, -6$	6	$(1,0)$	3
457^2	$457, 10$	$1, \rho, (-2 + 3\rho + \rho^2)/8$	$8, 7, -17, -8$	$(2,4,8)\ 8$	$(1,0)$	3
463^2	$463, 23$	$1, \rho, (\rho + \rho^2)/7$	$7, 1, -22, 7$	1	$(2,-1)$	3
487^2	$487, 25$	$1, \rho, (-1 + \rho^2)/7$	$7, -2, -23, -7$	$(5,7)\ 5$	$(2,1)$	3
499^2	$499, 32$	$1, \rho, (2 + 3\rho + \rho^2)/6$	$6, 7, -25, 6$	6	$(1,0)$	3
523^2	$523, 43$	$1, \rho, (\rho + \rho^2)/3$	$3, 1, -58, -99$	3	$(1,0)$	3
541^2	$541, 29$	$1, \rho, (-3 + 2\rho + \rho^2)/7$	$7, 4, -25, 7$	7	$(1,0)$	3
547^2	$547, 1$	$1, \rho, (-4\rho + \rho^2)/9$	$9, -14, -13, 9$	1	$(-1,1)$	3
571^2	$571, 31$	$1, \rho, (1 - \rho + \rho^2)/7$	$7, -5, -26, -7$	7	$(1,0)$	3
577^2	$577, 11$	$1, \rho, (3 - 2\rho + \rho^2)/9$	$9, -8, -19, 9$	9	$(1,0)$	3
607^2	$607, 49$	$1, \rho, \rho^2$	$1, -2, -201, -967$	1	$(1,0)$	3
613^2	$613, 47$	$1, \rho, (\rho + \rho^2)/3$	$3, 1, -68, 111$	3	$(1,0)$	3
619^2	$619, 17$	$1, \rho, (3 - \rho + \rho^2)/9$	$9, -5, -22, 9$	9	$(1,0)$	3

Table 1 (continued)

d_F	a, b	$1, \omega_2, \omega_3$	$I(x, y)$	K	(x, y)	#
631^2	631, 43	$1, \rho, (\rho+\rho^2)/5$	5, 1, −42, −43	5	(1, 0)	3
643^2	643, 40	$1, \rho, (2+3\rho+\rho^2)/6$	6, 7, −33, −40	6	(1, 0)	3
661^2	661, 49	$1, \rho, (-1+\rho^2)/3$	3, −2, −73, −117	3	(1, 0)	3
673^2	673, 37	$1, \rho, (-1-2\rho+\rho^2)/7$	7, −8, −29, −7	(1, 7)7	(1, 0)	6
691^2	691, 8	$1, \rho, (2-3\rho+\rho^2)/10$	10, −11, −19, 10	(2, 4, 8)8	(2, 1)	3
709^2	709, 53	$1, \rho, \rho^2$	1, −2, −235, 1549	1	(1, 0)	3
727^2	727, 44	$1, \rho, (-\rho+\rho^2)/6$	6, −5, −39, 44	6	(1, 0)	3
733^2	733, 50	$1, \rho, (2-\rho+\rho^2)/4$	4, −5, −59, 110	(2, 4)4	(1, 0)	3
739^2	739, 16	$1, \rho, (3\rho+\rho^2)/10$	10, 7, −23, −10	(4, 6, 8, 10)10	(1, 0)	3
751^2	751, 41	$1, \rho, (-3\rho+\rho^2)/7$	7, −11, −30, 41	1	(2, −1)	3
757^2	757, 29	$1, \rho, (\rho+\rho^2)/9$	9, 1, −28, 9	3	(2, −1)	3
769^2	769, 49	$1, \rho, (-1+\rho^2)/5$	5, −2, −51, −49	(1, 5)5	(1, 0)	3
787^2	787, 31	$1, \rho, (-1+\rho^2)/9$	9, −2, −29, −9	3	(2, 1)	3
811^2	811, 56	$1, \rho, (\rho+\rho^2)/2$	2, 1, −135, 398	2	(1, 0)	3
823^2	823, 5	$1, \rho, (-1-4\rho+\rho^2)/11$	11, −14, −19, 11	(1, 5)5	(−1, 1)	3
829^2	829, 7	$1, \rho, (-3+5\rho+\rho^2)/11$	11, 13, −20, −11	7	(1, 1)	3
853^2	853, 35	$1, \rho, (-3+2\rho+\rho^2)/9$	9, 4, −31, 9	(1, 9)9	(1, 0)	3
877^2	877, 59	$1, \rho, \rho^2$	1, −2, −291, 2111	1	(1, 0)	3
883^2	883, 47	$1, \rho, (-1-2\rho+\rho^2)/7$	7, −8, −39, 47	(1, 7)7	(1, 0)	3
907^2	907, 19	$1, \rho, (1+3\rho+\rho^2)/11$	11, 7, −26, −11	(1, 11)11	(1, 0)	3
919^2	919, 52	$1, \rho, (\rho+\rho^2)/6$	6, 1, −51, −52	(2, 6)6	(1, 0)	3
937^2	937, 61	$1, \rho, \rho^2$	1, −2, −311, −1909	1	(1, 0)	3
967^2	967, 41	$1, \rho, (-1+3\rho+\rho^2)/9$	9, 7, −34, 9	(3, 9)9	(1, 0)	3
991^2	991, 61	$1, \rho, (\rho+\rho^2)/3$	3, 1, −110, −261	3	(1, 0)	3
997^2	997, 10	$1, \rho, (5\rho+\rho^2)/12$	12, 13, −23, −12	(6, 8, 10)10	(1, 1)	3

References

[1] A. Baker, *Contributions to the theory of diophantine equations I, On the representation of integers by binary forms II, The diophantine equation* $y^2 = x^3 + k$, Phil. Trans. R. Soc. London **263** (1968), 173-208.

[2] A. Baker, H. Davenport, *The equation* $3x^2 - 2 = y^2$ *and* $8x^2 - 7 = z^2$,Quart. J. Math., Oxford **20** (1969),129-137.

[3] J. Buchmann, *A Generalization of Voronoi's Unit Algorithm II*, J. Number Theory **20** (1983), 192-209.

[4] D.S. Dummit and H. Kisilevsky, *Indices in Cyclic Cubic Fields*, Number Theory and Algebra, Academic Press, New York 1977, 29-42.

[5] U. Fincke and M. Pohst, *A procedure for determing algebraic integers of given norm*, Proc. EUROCAL'83, Springer Lecture Notes in Comp.Sc. **162** (1983), 194-202.

[6] D.J. Ford, *The construction of maximal orders over a Dedekind domain*, J. Symbolic Computations **4** (1987), 69-75.

[7] I. Gaál and N. Schulte, *Computing all Power Integral Bases of Cubic Fields*, erscheint in Math. Comp. **53** (Oktober 1989).

[8] M.N. Gras, *Lien entre le Group des Unités et la Monogéneité der Corps Cycliques*, Publ. Math. Fac. Besançon 1975-1976.

[9] K. Hensel, "Theorie der algebraischen Zahlen", Teubner Verlag, Leipzig und Berlin 1908.

[10] A. Pethö, *On the resolution of Thue inequalities*, J. Symbolic Computations **4** (1987), 103-109.

[11] J. von Schmettow, *KANT - a tool for computations in algebraic number fields.*

[12] A. Thue, *Über Annäherungswerte algebraischer Zahlen*, J. reine Math. **135** (1909), 284-305.

[13] M. Waldschmidt, *A lower bound for linear forms in logarithms*, Acta Arith. **37** (1980) , 257-283.

On the practical solution of the Thue-Mahler equation *

N. Tzanakis and B.M.M. de Weger

1 Introduction

We are interested in the following problem: Given any Thue-Mahler equation, to find *explicitly* all its solutions. Although there is an effective way to solve any Thue-Mahler equation (see, for example, Chapter 7 of [5]), in practice one faces a variety of computational problems. To the best of our knowledge, only one or two Thue-Mahler equations have been solved so far (see [1] and [7]) and in these, only one prime power and one fundamental unit (thus, two unknown exponents) are involved.

We have developed a general method, which we now want to illustrate by solving a specific equation. In this example four unknown exponents are involved.

A famous Thue equation is Ljunggren's $x^3 - 3xy^2 - y^3 = 1$ (see [2]; Chapter 23 of Mordell's book [4] and [6]). We decided to solve

$$x^3 - 3xy^2 - y^3 = \pm 3^{n_0} 17^{n_1} 19^{n_2}, \tag{1}$$

which "includes" Ljunggren's equation. The primes 3, 17, 19 have not been arbitrarily chosen; in fact, these are the only primes $p < 20$ for which $x^3 - 3xy^2 - y^3 \equiv 0 \pmod{p}$ is solvable.

2 Solution of (1)

We worked in $\mathbb{Q}(\theta)$, where $\theta^3 - 3\theta - 1 = 0$. This is a Galois field. This fact is not essential in our method, but surely simplified our computations and clarified our exposition.

To every solution (x, y), a class of six solutions is associated:

$$\{\pm(x,y), \pm(y, -x-y), \pm(-x-y, x)\}.$$

*Lecture given by N.Tzanakis

Computational Number Theory

Using this observation, it was easy to show that, in order to solve (1), it suffices to solve

$$\beta := x - y\theta = \pm\psi^{n_0}\pi^{n_1}\rho^{n_2}\theta^{a_1}\eta^{a_2}. \tag{2}$$

Here

$$\psi = -1 + \theta \text{ is the prime divisor of 3,}$$

$$\pi = 3 - \theta \text{ is a prime divisor of 17,}$$

$$\rho \in \{3+\theta, 5-\theta^2, 1-\theta+\theta^2\}, \text{ the set of prime divisors of 19,}$$

$$\theta, \eta = 1 + \theta \text{ is a pair of fundamental units,}$$

$$n_0 \in \{0, 1\} \text{ (an obvious fact),}$$

$$n_1, n_2 \text{ are unknown nonnegative integers,}$$

$$a_1, a_2 \text{ are unknown integers.}$$

Of course, solving (2) means to find all $(n_0, n_1, n_2, a_1, a_2)$ for which the coefficient of θ^2 is zero when the right-hand side of (2) is expressed in terms of the basis $1, \theta, \theta^2$.

Using (2) and its other two conjugate relations we eliminated x, y and got the relation

$$(-1)^{a_2}\xi^{n_1}\chi^{n_2}\theta^{a_1'}\eta^{a_2'} - 1 = (-1)^{a_1+a_2+n_0}\xi^{*n_1}\chi^{*n_2}\theta^{2a_1+a_2-1}\eta^{-a_1+a_2-n_0}, \tag{3}$$

where ξ, χ, ξ^*, χ^* are explicitly known elements of $\mathbb{Q}(\theta)$ and $a_1' = a_1 + 2a_2 - n_0 - 1$, $a_2' = -2a_1 - a_2 + 1$.

In the right-hand side of (3) all elements apart from ξ^* have π-adic order zero, while $\operatorname{ord}_\pi(\xi^*) = 1$. Then (3) implies that

$$\operatorname{ord}_\pi\left((-1)^{a_2}\xi^{n_1}\chi^{n_2}\theta^{a_1'}\eta^{a_2'} - 1\right) = n_1. \tag{4}$$

Put $N = \max\{n_1, n_2\}$, $A = \max\{|a_1'|, |a_2'|\}$, $H = \max\{N, A\}$. Since ξ, χ, θ, η are explicitly known, we could apply the p-adic theory of linear forms in logarithms of algebraic numbers to the left-hand side of (4), in order to find an upper bound of n_1 in terms of H. This was accomplished thanks to a recent theorem of Kunrui Yu ([12]). Analogously, working ρ-adically, we found an upper bound of n_2. Thus we have an upper bound of N, which, by Yu's theorem, is of the form

$$N \leq C_{13}(\log H + C_{14}). \tag{5}$$

In our case we computed $C_{13} = 6.190047 \cdot 10^{24}$ and $C_{14} = 4.28$.

Our next task was to find an upper bound of A similar to that of N. Using elementary arguments we did so, under the assumption that

$$\min_{1 \leq i \leq 3} |\beta^{(i)}| > e^{-C_{16}A}, \tag{6}$$

for a suitable C_{16}. Then A is at most const. + const. $\log H$, which in combination with (5) gives

$$H < C_{19} + C_{20}\log H.$$

In our case we derived in this way

$$H < 4.38 \cdot 10^{27}, \tag{7}$$

under the restriction (6), where $C_{16} = 0.085$.

The case opposite to (6) remained. Here the philosophy is to work as in the case of a Thue equation, where also one of the conjugates of $\beta = x - y\theta$ is necessarily "very small" (see e.g. [8]). We proved by elementary arguments

$$0 < |\Lambda_0| < \text{const.} e^{-C_{16}H}, \tag{8}$$

where

$$\Lambda_0 = n_1 \log|\xi^{(i_0)}| + n_2 \log|\chi^{(i_0)}| + a_1' \log|\theta^{(i_0)}| + a_2' \log|\eta^{(i_0)}|$$

($i_0 \in \{1,2,3\}$, so we had to consider three cases), and the constant in (8) is explicitly calculable. On the other hand, we applied the (real-complex) theory of linear forms in logarithms of algebraic numbers (in our case we applied Waldschmidt's theorem [10]) to conclude that

$$|\Lambda_0| > e^{-C_7(\log H + C_8)},$$

where $C_7 = 2.6467 \cdot 10^{29}$ and $C_8 = 2.442325$. This relation, combined with (8), leads to a relation of the form

$$H < \text{const.} + \text{const.} \log H,$$

from which we have found

$$H < K_0 := 5.76 \cdot 10^{32}. \tag{9}$$

Compare this with (7) to see that in all cases (9) is true. In view of (9), the solution of (2) and, hence, of (1), is effective (by enumeration), but clearly such a task is thoroughly unrealistic.

3 Reduction of H

We found a reduced upper bound of n_1 as follows: By some elementary arguments (4) can be written as

$$n_1 = \text{ord}_{17}(\Lambda_1),$$

where Λ_1 is a linear form in 17-adic logarithms:

$$\Lambda_1 = n_1\lambda_1 + n_2\lambda_2 + a_1'\mu_1 + a_2'\mu_2.$$

Here, $\lambda_1, \lambda_2, \mu_1, \mu_2$ are explicitly known 17-adic logarithms of algebraic numbers. Next, put

$$\beta_1 = -\frac{\lambda_1}{\lambda_2}, \quad \alpha_1 = -\frac{\mu_1}{\lambda_2}, \quad \alpha_2 = -\frac{\mu_2}{\lambda_2} \quad \text{(17-adic integers)}.$$

For a positive integer m, to be specified later, we computed rational approximations $\beta_1^{(m)}$, $\alpha_1^{(m)}$, $\alpha_2^{(m)}$ of β_1, α_1, α_2 respectively, and considered the lattice Γ_m with basis vectors the columns of the matrix

$$\begin{pmatrix} 1 & 0 & 0 & 0 \\ 0 & 1 & 0 & 0 \\ 0 & 0 & 1 & 0 \\ \alpha_1^{(m)} & \alpha_2^{(m)} & \beta_1^{(m)} & 17^m \end{pmatrix}.$$

This lattice is just the set of vectors $(n_1, n_2, a_1', a_2')^t \in \mathbb{Z}^4$ for which $\mathrm{ord}_{17}(\Lambda_1) \geq 17^{m-1}$ (this shows the meaning of m).

By the L^3-algorithm ([3], see also [11] for various applications of this algorithm to diophantine problems) we computed a reduced basis $\mathbf{b}_1, \ldots, \mathbf{b}_4$ of Γ_m. We proved that

$$\text{if } |\mathbf{b}_1| > 2^{5/2} K_0 \text{ then } n_1 \leq m.$$

In practice, the condition $|\mathbf{b}_1| > 2^{5/2} K_0$ was satisfied, when we took m "somewhat larger" than $4 \log K_0 / \log 17$. To give an idea, in our example $4 \log K_0 / \log 17 \simeq 106.5$ and it turned out that $m = 110$ was sufficient; hence $n_1 \leq 110$. Working analogously with n_2 we came to the conclusion that $n_2 \leq 111$. This completes the *p-adic reduction step.*

Then we proceeded with the *real reduction step*: From (9) we have $A < K_0$, while from the p-adic reduction step we know that $N \leq 111$. We chose an integer C of the size of $K_0^2 \cdot 111^2$ (in our case $C = 10^{75}$ was good) and we put

$$\beta_1 = [C \log |\xi^{(i_0)}|], \quad \beta_2 = [C \log |\chi^{(i_0)}|],$$
$$\alpha_1 = [C \log |\theta^{(i_0)}|], \quad \alpha_2 = [C \log |\eta^{(i_0)}|].$$

Then we considered the lattice Γ with basis vectors the columns of the matrix

$$\begin{pmatrix} [K_0/111] & 0 & 0 & 0 \\ 0 & [K_0/111] & 0 & 0 \\ 0 & 0 & 1 & 0 \\ \beta_1 & \beta_2 & \alpha_1 & \alpha_2 \end{pmatrix},$$

and we calculated a reduced basis $\mathbf{b}_1, \ldots, \mathbf{b}_4$ of Γ by the L^3-algorithm. We have proved (using a.o. the fact that $N \leq 111$) that

$$\text{if } |\mathbf{b}_1| > 2^{3/2} \sqrt{4(K_0 + 111)^2 + 3K_0^2} \text{ then}$$
$$H < \text{some constant of the size of } \log K_0.$$

In our case this constant turned out to be equal to 1169.

Thus, the p-adic reduction step applied twice (with $p = 17$ and $p = 19$), combined with the real reduction step, has given the new upper bound

$$H < K_0' := 1169,$$

which is a big progress compared to (9). It is still large, however, therefore we have repeated the same process, with K_0' in place of K_0, and found a new upper bound

$$H < K_0'' := 296, \quad N \leq 14.$$

Then, $n_0 \in \{0, 1\}$, $\max\{|a_1|, |a_2|\} \leq 297$, $\max\{n_1, n_2\} \leq 14$, and we had to decide which tuples $(n_0, n_1, n_2, a_1, a_2)$ satisfy (2). It is not realistic to check all tuples in the above ranges separately. Working modulo conveniently chosen primes, we were able to exclude most of them, and the few that remained we checked by straightforward computations. The final list of solutions of (1) is given in the table below.

Final remarks: 1. A detailed exposition of the content of this talk can be found in our paper [9].

2. All the computations needed in this paper have been performed on an AT personal computer. Most of the computer output is included in an appendix of [9].

Table 1. Solutions of (1)

n_0	n_1	n_2	$\pm(x, y)$		
0	0	0	(1, 0)	(0, −1)	(1, −1)
0	0	0	(2, 1)	(1, −3)	(3, −2)
0	0	1	(1, 2)	(2, −3)	(3, −1)
0	0	2	(4, 5)	(5, −9)	(9, −4)
0	1	0	(3, 1)	(1, −4)	(4, −3)
0	1	0	(3, 2)	(2, −5)	(5, −3)
0	1	0	(15, 8)	(8, −23)	(23, −15)
0	1	1	(1, 6)	(6, −7)	(7, −1)
0	1	1	(3, 5)	(5, −8)	(8, −3)
0	1	1	(28, 15)	(15, −43)	(43, −28)
0	2	1	(59, 31)	(31, −90)	(90, −59)
0	2	1	(31, 15)	(15, −46)	(46, −31)
0	2	1	(18, 13)	(13, −31)	(31, −18)
0	2	2	(206, 109)	(109, −315)	(315, −206)
0	2	5	(896, 37)	(37, −933)	(933, −896)
0	3	1	(97, 54)	(54, −151)	(151, −97)
1	0	0	(1, 1)	(1, −2)	(2, −1)
1	0	1	(7, 4)	(4, −11)	(11, −7)
1	0	1	(13, 7)	(7, −20)	(20, −13)
1	0	1	(5, 2)	(2, −7)	(7, −5)
1	1	0	(4, 1)	(1, −5)	(5, −4)
1	1	1	(10, 1)	(1, −11)	(11, −10)
1	1	2	(29, 20)	(20, −49)	(49, −29)
1	1	3	(73, 13)	(13, −86)	(86, −73)
1	2	0	(4, 7)	(7, −11)	(11, −4)
1	2	2	(712, 379)	(379, −1091)	(1091, −712)

References

[1] A.K. Agrawal, J.H. Coates, D.C. Hunt and A.J. van der Poorten, *Elliptic curves of conductor 11*, Math. Comp. **35** (1980), 991-1002.

[2] W. Ljunggren, *Einige Bemerkungen über die Darstellung ganzer Zahlen durch binäre kubische Formen mit positiver Diskriminante*, Acta Math. **75** (1942), 1-21.

[3] A.K. Lenstra, H.W. Lenstra jr. and L. Lovász, *Factoring polynomials with rational coefficients*, Math. Ann. **261** (1982), 515-534.

[4] L.J. Mordell, *Diophantine equations*, Pure and Applied Mathematics Vol. 30, Academic Press, London & New York, 1969.

[5] T.N. Shorey and R. Tijdeman, *Exponential diophantine equations*, Cambridge Tracts in Mathematics Vol. 87, Cambridge University Press, Cambridge, 1986.

[6] N. Tzanakis, *The diophantine equation* $x^3 - 3xy^2 - y^3 = 1$ *and related equations*, J. Number Th. **18** (1984), 192-205.

[7] N. Tzanakis, *The complete solution in integers of* $x^3 + 2y^3 = 2^n$, J. Number Th. **19** (1984), 203-208.

[8] N. Tzanakis and B.M.M. de Weger, *On the practical solution of the Thue equation*, J. Number Th. **31** (1989), 99-132.

[9] N. Tzanakis and B.M.M. de Weger, *Solving a specific Thue-Mahler equation*, Memorandum No. 793, Faculty of Applied Mathematics, University of Twente, 1989, to appear in Math. Comp.

[10] M. Waldschmidt, *A lower bound for linear forms in logarithms*, Acta Arith. **37** (1980), 257-283.

[11] B.M.M. de Weger, *Algorithms for diophantine equations*, CWI-Tract No. 65, Centre for Mathematics and Computer Science, Amsterdam, 1989.

[12] K.R. Yu, *Linear forms in p-adic logarithms II*, Compositio Math. **74** (1990), 15-113.

Some results on Thue equations and Thue-Mahler equations

J.H. Evertse *and *K.Győry* †

Abstract. In this survey we present some recent results on Thue equations, Thue inequalities and Thue-Mahler equations which were obtained by Brindza and the present authors [4] and the authors [10], [11], respectively. In Section 1, some effective upper bounds from [4] are formulated for the solutions which are considerable improvements of the earlier ones in terms of the height of the binary form involved in the equation. These results show that very good (and, in certain important particular cases, best possible) upper bounds can be established in terms of the height, if one takes into consideration also the discriminant of the binary form. We hope that these new bounds and their proofs (see Section 2) will prove useful in practical solutions of the equations in question. In Section 3, some good (and, under certain additional assumption, sharp) upper bounds from [10] and [11] are presented for the number of solutions of Thue inequalities. For given $n \geq 3$ and given non-zero constant term, these bounds are valid for all but finitely many equivalence classes of binary forms of degree n and non-zero discriminant.

1 Bounds for the solutions of Thue equations and Thue-Mahler equations

Let $F(X,Y) \in \mathbb{Z}[X,Y]$ be a binary form of degree $n \geq 3$ with discriminant $D(F) \neq 0$, and let m be a non-zero integer. By using his fundamental results on linear forms in logarithms, Baker [1] derived the first explicit upper bound for the absolute values of the solutions of the Thue equation

$$F(x,y) = m \qquad \text{in } x, y \in \mathbb{Z}. \tag{1}$$

This bound depends only on $n, |m|$ and the height $H(F)$ of F, that is the maximum of the absolute values of the coefficients of F. For a finite set of primes $S = \{p_1, \ldots, p_s\}$ $(s \geq 0)$ and for any non-zero rational integer a, we denote by $|a|_S$

*The research of the first author has been made possible by a fellowship of the Royal Netherlands Academy of Arts and Sciences

†The research of the second author has been supported in part by Grant 273 from the Hungarian National Foundation for Scientific Research

Computational Number Theory

the maximal positive divisor of a which is relatively prime to $p_1, \ldots, p_s$. Coates [6] extended Baker's result to the equation

$$|F(x,y)|_S = m \qquad \text{in } x, y \in \mathbb{Z} \qquad \text{with } (x, y, p_1, \ldots, p_s) = 1; \tag{2}$$

he derived an explicit upper bound for the solutions of (2). Equation (2) is called a Thue-Mahler equation. Baker and Coates made the assumption that F is irreducible. Further, their upper bounds were exponential in terms of each parameter. Their results were later considerably improved and generalized by several authors; for references see, for example, [2],[23],[22] and [9].

In terms of $H(F)$ and $|m|$, much better upper bounds can be obtained if one takes into consideration the splitting field of F. Denote by G the splitting field of F over $\mathbb{Q}$, and let g and D_G denote the degree and discriminant of G over $\mathbb{Q}$. It was proved in [17] that all solutions of (2) satisfy

$$max(|x|, |y|) < c_1(H(F) \cdot |m|_S)^{c_2} \tag{3}$$

where c_1, c_2 are effectively computable numbers depending only on $n, g, |D_G|, s$ and the maximum P of $p_1, \ldots, p_s$. By substituting $S = \emptyset$, $s = 0$, $P = 1$, $|m|_S = |m|$ in (3) we get an upper bound for the solutions of (1); in that case, c_1 and c_2 depend only on n, g and $|D_G|$.

The height $H(F)$ of F can be arbitrarily large with respect to $|D(F)|$. For large $H(F)$, considerable improvements of (3) can be obtained in terms of $H(F)$. The binary form $F(X,Y)$ is called *monic* if $F(1,0) = 1$. We proved in our joint paper [4] with Brindza the following results.

Theorem 1.1 ([4]) *All solutions of* (2) *satisfy*

$$\max(|x|, |y|) < c_3 H(F)^{\frac{3}{n}} |D(F) \cdot m|_S^{c_4} \tag{4}$$

where c_3 and c_4 are effectively computable positive numbers which depend only on n, g, s, P and $|D_G|$.

Theorem 1.2 ([4]) *Suppose that F is monic. Then all solutions of* (2) *satisfy*

$$|y| < c_5 |D(F) \cdot m|_S^{c_6}, \tag{5}$$

$$|x| < c_5 \left(|D(F)|^{\frac{1}{n(n-1)}} + |F(0,1)|^{\frac{1}{n}}\right) |D(F) \cdot m|_S^{c_6} \tag{6}$$

and all solutions (x,y) of (1) *satisfy*

$$|y| < c_5 |m|^{\frac{1}{n}} |D(F) \cdot m|_S^{c_6}, \tag{5'}$$

$$|x| < c_5 |m|^{\frac{1}{n}} \left(1 + |F(0,1)|^{\frac{1}{n}}\right) |D(F) \cdot m|_S^{c_6} \tag{6'}$$

where c_5, c_6 are effectively computable positive numbers depending only on n, g, s, P and $|D_G|$.

It should be noticed that the bounds in (5), (5′) are independent of $H(F)$. Further, it is easy to see (cf. [4]) that the exponent $\frac{1}{n}$ of $|F(0,1)|$ in (6) and (6′) is already best possible.

It is easy to prove that $g \leq n!$ and that

$$|D_G| \leq (nP)^{sng} |D(F)|_S^g$$

(cf. [4]). Further, by prime number theory we have $s \leq \frac{2P}{logP}$. Hence Theorems 1.1 and 1.2 imply upper bounds for $|x|$ and $|y|$ in which g, s and $|D_G|$ do not occur. For example, Theorem 1.1 implies that all solutions of (2) satisfy

$$\max(|x|, |y|) < c_7 H(F)^{\frac{3}{n}} |m|_S^{c_8} \tag{7}$$

where c_7, c_8 are effectively computable numbers which depend only on n, P and $|D(F)|_S$. If in particular $S = \emptyset$ then c_7 and c_8 depend only on n and $|D(F)|$. Moreover, by the inequality of Győry [14]

$$n = \deg(F) \leq 3 + \frac{2}{\log 3} \log |D(F)|,$$

c_7 and c_8 can be replaced by numbers depending only on $|D(F)|$.

Of particular interest is the special case when, in (4), (5), (6), (5′), (6′) and (7), S consists exactly of the prime factors of $D(F)$ and m. Then $|D(F)|_S = |m|_S = 1$ and for instance (7) implies

$$\max(|x|, |y|) < c_9 H(F)^{\frac{3}{n}}$$

for each solution of (2), where c_9 is an effectively computable number depending only on n and the greatest prime factor of m and $D(F)$.

For practical solutions of Thue equations and Thue-Mahler equations, we refer to the papers of Gaál, Stroeker, Tzanakis and de Weger, respectively, in these Proceedings and to the references given there. We are hopeful that our above results and their proofs will be applicable in practical solutions of the equations under consideration.

2 Brief sketch of the proof of (5′) and (6′)

Theorems 1.1 and 1.2 can be proven by transforming first $F(X, Y)$ by an appropriate element of $SL(2, \mathbb{Z})$ into a binary form having the same discriminant and a "small" height with respect to $|D(F)|$, and then applying (3) to the solutions of the transformed equation. In the general case, such a transformation can be carried out by using a recent effective theorem of Evertse and Győry [11] on binary forms of given discriminant; in the monic case, an effective result of Győry [18] can be applied on monic polynomials of given discriminant. We note that in the proofs of (3) and the results in question of [12] and [18], the theory of linear forms in logarithms and its p-adic analogue were involved.

We outline now another, more direct proof for (5′) and (6′) of Theorem 1.2 which yields better dependence in n and the absolute constant and which seems to be more appropriate for practical solutions. For simplicity, we restrict ourselves to the case $S = \emptyset$. The next proof is in fact a combination of an argument of Győry and Papp [19] for deriving bounds for the solutions of Thue equations (and, more generally, for decomposable form equations) with that of Győry [14] for proving a quantitative and effective finiteness result for polynomials of given discriminant.

The binary form $F(\underline{X}) = F(X,Y)$ can be factorized in the form $F(\underline{X}) = l_1(\underline{X}) \dots l_n(\underline{X})$ where $l_i(\underline{X}) = X - \alpha_i Y$ with $\alpha_i \in O_G$. Here O_G denotes the ring of integers of G. The numbers $\Delta_{jk} := \alpha_j - \alpha_k$ divide $D(F)$ in O_G for each distinct $j,k \in \{1,\dots,n\}$. Hence it follows (see e.g. [16]) that

$$\Delta_{jk} = \delta_{jk} \eta_1^{a_{1,jk}} \dots \eta_r^{a_{r,jk}} \tag{8}$$

where $\delta_{jk} \in O_G$ with height $H(\delta_{jk}) < c_{10} |D(F)|^{c_{11}}$, $\eta_1, \dots, \eta_r$ are units in O_G with heights at most c_{12} and $a_{1,jk}, \dots, a_{r,jk}$ are rational integers. Here c_{10}, c_{11}, c_{12} and $c_{13}, \dots$ below are positive numbers depending only on g and $|D_G|$ which can be given explicitly. We have now

$$\Delta_{1i} + \Delta_{i2} + \Delta_{21} = 0 \qquad \text{for } i = 3, \dots, n. \tag{9}$$

Substituting the expression (8) into (9), we get a homogeneous unit equation. Now, by using the theory of linear forms in logarithms one can show (see e.g. [16]) that

$$\max\left\{ H\left(\frac{\Delta_{1i}}{\Delta_{12}}\right), H\left(\frac{\Delta_{i2}}{\Delta_{12}}\right)\right\} < c_{13} |D(F)|^{c_{14}}.$$

For each $j \geq 3$ which is different from i, one can get a similar upper bound for the heights of $\frac{\Delta_{1j}}{\Delta_{12}}$ and $\frac{\Delta_{j2}}{\Delta_{12}}$. But $\Delta_{ij} = \Delta_{1j} - \Delta_{1i}$, hence $H\left(\frac{\Delta_{ij}}{\Delta_{12}}\right)$ also has an upper bound of the same form for each distinct $i,j \in \{1,\dots,n\}$. Together with

$$\Delta_{12}^{n(n-1)} \prod_{\substack{1 \leq i,j \leq n \\ i \neq j}} \left(\frac{\Delta_{ij}}{\Delta_{12}}\right) = D(F),$$

this implies a similar upper bound for $H(\Delta_{12})$, whence

$$H(\Delta_{ij}) < c_{15} |D(F)|^{c_{16}} \qquad \text{for each } i,j \in \{1,\dots,n\}. \tag{10}$$

Let now $\underline{x} = (x,y) \in \mathbb{Z}^2$ be an arbitrary but fixed solution of (1). Then $l_i(\underline{x})$ divides m in O_G, hence

$$l_i(\underline{x}) = \tau_i \eta_1^{b_{1i}} \dots \eta_r^{b_{ri}} \tag{11}$$

with some $\tau_i \in O_G$ of height at most $c_{17} |m|^{c_{18}}$ and with rational integers $b_{1i}, \dots, b_{ri}$. Further, we have

$$\Delta_{1i} l_2(\underline{X}) + \Delta_{i2} l_1(\underline{X}) + \Delta_{21} l_i(\underline{X}) = 0 \tag{12}$$

identically in $\underline{X}$. Substituting the expression (11) into (12) and using (10), one can apply again an estimate for linear forms in logarithms to derive

$$H\left(\frac{l_i(\underline{x})}{l_1(\underline{x})}\right) < c_{19} |D(F)m|^{c_{20}}. \tag{13}$$

But (1) implies

$$(l_1(\underline{x}))^n \prod_{i=1}^{n} \left(\frac{l_i(\underline{x})}{l_1(\underline{x})}\right) = m. \tag{14}$$

Hence, from (13) and (14), one can derive an upper bound of the form $c_{21}\,|D(F)m|^{c_{22}}$ first for $H\,(l_1(\underline{x}))$ and then for $H\,(l_i(\underline{x}))$, $i = 1, \dots, n$. Now (5′) follows from this, (10) and $l_1(\underline{x}) = x - \alpha_1 y$, $\ l_2(\underline{x}) = x - \alpha_2 y$ by means of Cramer's rule. If $\alpha_i = 0$ for some i then (6′) follows immediately. Further, if $\alpha_i \neq 0$ for each i then there is an i with $|\alpha_i| \leq |F(0,1)|^{\frac{1}{n}}$ and then (5′) and $x = l_i(\underline{x}) + \alpha_i y$ imply (6′).

We remark that the best known and completely explicit estimates for linear forms in logarithms can be found in Philippon and Waldschmidt [21]. In the paper [16] quoted above, earlier and slightly weaker estimates were utilized from the theory of linear forms to bound the solutions of unit equations.

3 Bounds for the number of solutions of Thue equations and Thue inequalities

Several upper bounds are known for the number of solutions of (1) which depend on the height of F; for references we refer to [7]. In [8] it was proved for the first time that every equation (1) has an upper bound for the number of solutions that is independent of the coefficients of F; from the results in that paper it follows that (1) has at most $n \cdot 7^{n^3(2\omega(m)+3)}$ solutions where $\omega(m)$ denotes the number of distinct primes dividing m. Under the assumption that F is irreducible, Bombieri and Schmidt [3] estabilished later the bound $cn^{\omega(m)+1}$ for the number of solutions of (1) with $(x,y) = 1$, where c is some absolute constant which can be taken smaller than 215 if n is sufficiently large. Recently, Mueller and Schmidt [20] proved that for given $m \geq 1$, $|F(x,y)| \leq m$ has at most $c'u^2 m^{\frac{2}{n}}\left(1 + \log m^{\frac{1}{n}}\right)$ solutions $(x,y) \in \mathbb{Z}^2$, where c' is some absolute constant and u is the number of non-zero coefficients of the form F.

Next we present some new results which, for large n and apart from certain exceptional binary forms, considerably improve the estimate of Bombieri and Schmidt. For given $m \geq 1$, we consider more generally the number of solutions of the Thue inequality

$$0 < |F(x,y)| \leq m \quad \text{in } x, y \in \mathbb{Z} \quad \text{with } (x,y) = 1,\ y > 0 \text{ or } (x,y) = (1,0), \tag{15}$$

where $F(X,Y) \in \mathbb{Z}[X,Y]$ is again a binary form of degree $n \geq 3$ with non-zero discriminant. Put

$$N(n) = \begin{cases} 6n \cdot 7^{2\binom{n}{3}} & \text{for } n < 400, \\ 6n & \text{for } n \geq 400. \end{cases}$$

By using the arguments from [3] and the result from [8] quoted above, we proved in [11] the following

Theorem 3.1 ([11]) *Put* $\delta(n) = \frac{5}{6}n(n-1)$ *if* $n < 400$ *and* $\delta(n) = 120(n-1)$ *if* $n \geq 400$ *and assume that*

$$|D(F)| \geq \exp\{80n(n-1)\}\, m^{\delta(n)}.$$

Then (15) *has at most* $N(n)$ *solutions.*

If F(X,Y) is a binary form and $A = \begin{pmatrix} a & b \\ c & d \end{pmatrix}$ is a matrix, then we put $F_A(X,Y) := F(aX + bY, cX + dY)$. Two binary forms $F, G \in \mathbb{Z}[X,Y]$ are called *equivalent* if $G = F_U$ for some U in $SL(2,\mathbb{Z})$. It is obvious that the number of solutions of (15) does not change when F is replaced by an equivalent form. From Theorem 3.1 above and from an effective result of Győry [15] on equivalence of polynomials, we deduced in [11] the following theorem.

Theorem 3.2 ([11]) *If* (15) *has more than* $N(n)$ *solutions then there is a matrix* U *in* $SL(2,\mathbb{Z})$ *such that*

$$H(F_U) < \exp\left\{c(n)m^{10(n-1)^4}\right\}, \tag{16}$$

where $c(n)$ *is an effectively computable number depending only on* n. *Further, if* $0 < |F(1,0)| \leq m$, *then* (16) *holds for some* U *of the form* $\begin{pmatrix} 1 & b \\ 0 & 1 \end{pmatrix}$ *with* $b \in \mathbb{Z}$.

This shows that, for given $n \geq 400$, (15) has at most $6n$ solutions for all but finitely many equivalence classes of binary forms F of degree n with non-zero discriminant, and that a representative for each exceptional class can be determined effectively. We note that for given $n < 400$, for $m = 1$ and for irreducible binary forms F, Brindza, van der Poorten and Waldschmidt [5] have recently proved this result with the bound $2n^2 + 10n$ instead of $N(n)$. These results are not far from the best possible, as is shown by the binary forms

$$F(X,Y) = X^n + (a_1X - Y)\ldots(a_nX - Y) \tag{17}$$

with distinct $a_1, \ldots, a_n \in \mathbb{N}$. Then the inequality (15) with $m = 1$ has $n+1$ solutions, namely $(1,a_1), \ldots, (1,a_n), (0,1)$, and among the forms of type (17) there are infinitely many pairwise inequivalent ones.

Even better upper bound can be obtained for the number of solutions of (15) if we fix the splitting field of the binary forms F. Let $n \geq 3$ be an integer, and let G be a finite, normal extension of $\mathbb{Q}$. In [10], we proved the following

Theorem 3.3 ([10]) *There are only finitely many equivalence classes of binary forms* $F(X,Y) \in \mathbb{Z}[X,Y]$ *of degree* n *with non-zero discriminant and with splitting field* G *over* $\mathbb{Q}$ *for which* (15) *has more than two solutions.*

Here the bound "two" is already best possible. Theorem 3.3 is ineffective; an effective but weaker version of Theorem 3.3 was also established in [10] with $\min\{n(n-1)(n-2), [G:\mathbb{Q}]+1\}$ instead of 2. We note that Theorem 3.3 was proved in [10] in a more general form, for Thue-Mahler inequalities over algebaic number fields.

The proof of Theorem 3.3 is based on a sharp upper bound, derived by Evertse, Győry, Stewart and Tijdeman [13] for the number of solutions of unit equations in two unknowns.

References

[1] A. Baker, *Contributions to the theory of diophantine equations*, Phil. Trans. Roy. Soc. London A **263** (1968), 173-208.

[2] A. Baker, *Transcendental number theory*, Cambridge 1979, 2nd edn.

[3] E. Bombieri and W.M. Schmidt, *On Thue's equation*, Invent. Math. **88** (1987), 69-81.

[4] B. Brindza, J.H. Evertse and K. Győry, *Bounds for the solutions of some diophantine equations in terms of discriminants*, J. Austral. Math. Soc., to appear.

[5] B. Brindza', A.J. van der Poorten and M. Waldschmidt, *On the distribution and number of solutions of Thue's equation*, in preparation.

[6] J. Coates, *An effective p-adic analogue of a theorem of Thue II. The greatest prime factor of a binary form*, Acta Arith **16** (1970), 399-412.

[7] J.H. Evertse, *Upper bounds for the numbers of solutions of diophantine equations*. MC-tract **168**, Centre of Mathematics and Computer Science, Amsterdam 1983.

[8] J.H. Evertse, *On equations in S-units and the Thue-Mahler equation*, Invent. Math. **75** (1984), 561-584.

[9] J.H. Evertse and K. Győry, *Decomposable form equations*, in: "New advances in transcendence theory", A. Baker, ed., Cambridge 1988, 175-202.

[10] J.H. Evertse and K. Győry. *Thue-Mahler equations with a small number of solutions*, J. Reine Angew. Math. **399** (1989), 60-80.

[11] J.H. Evertse and K. Győry. *Thue inequalities with a small number of solutions*. "The Mathematical Heritage of C. F. Gauss". G. R. Rassias, ed., World Scientific Publ. Comp., to appear.

[12] J.H. Evertse and K. Győry, *Effective finiteness results for binary forms with given discriminant*, Compositio Math., to appear.

[13] J.H. Evertse, K. Győry, C L. Stewart and R. Tijdeman, *On S-unit equations in two unknowns*, Invent. Math. **92** (1988), 461-477.

[14] K. Győry, *Sur les polynômes à coefficients entiers et de discriminant donné II*. Publ. Math. Debrecen **21** (1974), 125-144.

[15] K. Győry, *Sur les polynômes à coefficients entiers et de discriminant donné III*. Publ. Math. Debrecen **23** (1976), 141-165.

[16] K. Győry, *On the solutions of linear diophantine equations in algebraic integers of bounded norm*, Ann. Univ. Sci. Budapest. Eötvös Sect. Math **22-23** (1979-1980), 225-233.

[17] K. Győry, *Explicit upper bounds for the solutions of some diophantine equations*, Ann. Acad. Sci. Fenn., Ser. A. Math. **5** (1980), 3-12.

[18] K. Győry, *On discriminants and indices of integers of an algebraic number field*, J. Reine Angew. Math **324** (1981), 114-126.

[19] K. Győry and Z.Z. Papp, *Effective estimates for the integer solutions of norm form and discriminant form equations*, Publ. Math Debrecen **25** (1978), 311-325.

[20] J. Mueller and W.M. Schmidt, *Thue's equation and a conjecture of Siegel*, Acta Math. **160** (1988), 207-247.

[21] P. Philippon and M. Waldschmidt, *Lower bounds for linear forms in logarithms*, in: New advances in transcendence theory, A. Baker, ed., Cambridge 1988, 280-312.

[22] T.N. Shorey and R. Tijdeman, *Exponential diophantine equations*, Cambridge 1986.

[23] V.G. Sprindžuk, *Classical diophantine equations in two unknowns*, (in Russian), Nauka, Moskva 1982.

On the solution of the diophantine equation $G_n = P(x)$ with sieve algorithm

I. Nemes *

Abstract. We consider the Diophantine equation $G_m = P(x)$, with G_m a second order linear recurrence sequence and $P(x)$ a polynomial over the integers and search the integral solutions x and m. With the presented algorithm it is possible to exclude the majority of non solutions m, which are not greater than a given limit.

1 Introduction

Let A, B, G_0, G_1 be integers. The relation

$$G_{n+1} = AG_n + BG_{n-1}, \quad n > 1$$

defines the terms of the binary recurrence sequence G_n for $n > 1$.

Let $P(x)$ be a polynomial with integer coefficients having degree at least two.

In this paper we show, how computers can be used to find the n, x integer solutions of the equation

$$G_n = P(x). \tag{1}$$

Further we give two applications of the method in algebraic number theory. Results concerning (1) were extensively referred to in [5]. We cite here only those papers, where the predecessors of our method were investigated. A special case of (1), when G_n is the Fibonacci sequence and $P(x) = x^3$, was solved in three quite different ways [2], [4] and [7]. In [7] Pethö, rediscovering the sieve method of Wunderlich [10], and combining it with an upper bound for n, $|x|$, used a computer to find all the solutions. With this method he determined all the fifth powers in the Fibonacci sequence as well [6].

Following his method we developed a computer program for more general recurrences and polynomials. The program takes A, B, G_0, G_1 and the coefficients of $P(x)$ as input and gives as output the possible solutions in n of (1) not greater than a parameter C depending on the computer and the precision of the arithmetic.

*Research supported by Hungarian National Foundation for Scientific Research Grant No. 273/86

Computational Number Theory

2 The sieve problem

The principle of sieving is quite ancient in mathematics, but the sieve problem in the following general form was stated by D.H. Lehmer [3].

Given a, b two integers, and $m_1, m_2, ..., m_k$ moduli, k other integers $n_1, ..., n_k$, $n_i < m_i$ and finally $\{a_{ij}\}$ $i = 1, ..., k$; $j = 1, ..., n_i$. The problem is to find all integers x, that satisfy

$$x \equiv a_{i1} \text{ or } x \equiv a_{i2} \text{ or } \dots \text{ or } x \equiv a_{in_i} \pmod{m_i} \quad i = 1, .., k$$

with $a < x < b$.

Many problems of the number theory can be considered as a special case of this problem: finding all primes in a given intervall (sieve of Eratosthenes), the Chinese remainder problem, quadratic Diophantine equations. We show in the next section how (1) can be reduced to a sieve problem.

3 Reducing the equation to a sieve problem

Let $m \geq 2$ be an integer and define $G_n^{(m)} \equiv G_n \pmod{m}$, with $0 \leq G_n^{(m)} < m$ for any $n \geq 0$. It is well known [9], for every $m \geq 2$ $\{G_n^{(m)}\}$ is periodic and has length of period not greater than m^2. This length will be denoted by $l(m)$.

Let us consider the following sets,

$$Q_m = \{u \mid u \equiv P(x) \bmod m \text{ for some } x, 0 \leq u < m\},$$

$$R_m = \{u \mid u = G_n^{(m)},\ n \geq 0,\ 0 \leq u < m\},$$

$$S_m = \{a \mid G_a \equiv u \bmod m,\ u \in R_m \cap Q_m\}$$

for each $m \geq 2$. For given $\{G_n\}$ and $P(x)$ there are some values of m with $R_m \subset Q_m$. This occurs for example if $P(x) = x^3$ and $m \equiv 2 \pmod{3}$. Such moduli are not convenient for our purpose, hence we exclude them. Let $M(G, P)$ denote the set of those $m \geq 2$ for which $R_m \not\subset Q_m$. For each $m \in M(G, P)$ we see that $R_m \cap Q_m$ is a proper subset of R_m. This means that if n is a solution of (1), then

$$n \equiv a_1 \text{ or } n \equiv a_2 \text{ or ... or } n \equiv a_r \pmod{l(m)}, \tag{2}$$

where $S_m = \{a_1, \dots, a_r\}$ and $r < l(m)$ holds for all $m \in M(G, P)$.

To find all n for which $0 \leq n < C$ and (2) hold for some $m \in M(G, P)$, is a sieve problem as defined above. The difficulties are, that the number of moduli in our examples could be arbitrary large and we are not able to select the moduli freely.

4 Main ideas of the algorithm

We chose the values of m from the set of primes sequentially, but we kept only those, for which $l(m)$ had prime factors only from among the first s prime numbers, where s was a parameter of the algorithm depending on the size of the computer. The selection of the primes and the computation of the corresponding period lengths $l(m)$ was continued, till the least common multiple of all the lengths was greater than C.

In the next phase the sets S_m, corresponding to the chosen moduli, were computed.

We called a set of natural numbers containing exactly h elements, none of them is greater than l, a *sieve* and denoted it by $T_{h,l}$. It is clear, that $S_m = T_{|S_m|,l(m)}$. The most frequently used procedure in our program computed the result of the following operation on the set of sieves:

$$T_{h_1,l_1} \circ T_{h_2,l_2} = \{x \mid 0 \leq x \leq \text{ LCM}(l_1, l_2); \exists i_1 \in T_{h_1,l_1}, \exists i_2 \in T_{h_2,l_2},$$
$$x \equiv i_1 \pmod{l_1} \quad \text{and} \quad x \equiv i_2 \pmod{l_2}\}.$$

$\text{LCM}(l_1, l_2)$ stands for the least commom multiple of l_1 and l_2 . We considered only those sieves, that were derived from the initial $S_m = T_{|S_m|,l(m)}$ sieves with $m \in M(G, P)$ using the $\circ$ operation. Obviously, if n is a solution of (1), then there exists $i \in T_{h,l}$ with $n \equiv i \pmod{l}$ for each considered $T_{h,l}$. That means that all the solutions for n in (2), not greater than l, are among the elements of $T_{h,l}$.

Our aim was to increase l to be as large as possible, but at the same time to keep h as small as possible, since all the costs of computation (time, space) depend on h. It is difficult to reach this double aim. By considering the relation $T_{h_1,l_1} \circ T_{h_2,l_2} = T_{h_3,l_3}$, where $l_3 = \text{ LCM}(l_1, l_2)$, and using the Chinese remainder theorem we see that when l_3 is large, i.e. $l_3 = l_1 \cdot l_2$, then $h_3 = h_1 \cdot h_2$, which means h_3 is large too. Similarly, when h_3 is small, then $l_3 < l_1 \cdot l_2$ must hold, which means that the resulting sieve is not the biggest possible .

In order to optimize the deriving process of the sieve sets, we used the following method. We ordered the moduli and the corresponding S_m sets, lexicographically, ordering the $s + 1$ dimensional vectors consisting of the exponents of the prime numbers in the factorization of $l(m)$. The first component of such a vector was the largest index of the prime factor having non-zero exponent. In the next step we iterated $\circ$ operation on the ordered sequence of sieves. The first iteration began with the first sieve and continued until the computational time of $\circ$ was reasonable. The next iteration began with the first non iterated sieve. The result of each iteration was stored and the iteration sequence finished with the last sieve. This process was repeated with the stored sieves until the constant C was reached by a sieve, i.e. the length of the sieve was greater than C and there were only a few elements in the sieve less than C. These numbers formed the output of the algorithm.

5 Applications

Following the ideas of Pethö, we used our method in the solutions of the following two problems.

5.1 Finding all integer points on elliptic curves

We considered a class of equations:

$$y^2 = x^3 + ax^2 + bx \qquad a, b \in \mathbb{Z}. \tag{3}$$

Pethö [8] derived from (3) finitely many equations of the form of (1). With the sieve algorithm it is possible to find all the solutions n under an upper bound. From this upper bound can be derived an upper bound on the absolute values of the solutions of (3) and from the solutions n of (1) the solutions x, y of (3) can be derived.

5.2 Resolution of index form equations in biquadratic number fields

In [1] it is shown, how index form equations in biquadratic number fields can be reduced to equations of the form $G_n = y^2 + \delta$ and all the solutions are given, when $|n| \le 37$. The solutions were derived by explicitly computing the values of G_n.

With the sieve program it is possible to find the solutions for $|n| < 10^7$, since in this case the algorithm use the values of n instead of the values of G_n. We applied the algorithm to one of the equations in [1]. We considered the field $Q(\sqrt{2+\sqrt{2}})$ and the index form equation $I(x_2, x_3, x_4) = \pm 1$ over it. As it is shown in [1] from this equation can be derived:

$$G_n = y^2 \tag{4}$$

and

$$G_n = y^2 \pm 2 \tag{5}$$

with integers n and y, furthermore for G_n hold $G_0 = 2$, $G_1 = 10$ and $G_{n+1} = 6G_n - G_{n-1}$ with $n \ge 1$.

The equation (4) can never hold, because $G_n \equiv 2 \pmod 4$. Applying the algorithm to (5) with $C = 10^7$ we get

$$|x_3|, |x_4| > \frac{\sqrt{2}}{4}(1+\sqrt{2})^C$$

for the solutions not determined by the algorithm. Using the arguments of [1] a lower bound for $|x_2|$ can be derived with the same order. The algorithm did not find any solutions under this greater limit different from the small solutions given in [1].

In order to illustrate how the algorithm works we give some partial results that were computed by the algorithm when it was excluding non-solutions of (5)

Clearly, the lengths of periods depend only on the recurrence sequence. The other part of the partial results depend on the polynomial in (5). We shall give partial results only for the case with $y^2 - 2$.

Table 1 contains the chosen moduli, the lengths of periods and the exponents of prime factors from their factorization. The table is ordered as it was described in section 4 and divided into subtables (rows 1 - 5, 6 - 12, ... 97 - 101). Each iteration was carried out on sieves corresponding to the elements of a subtable. For example, the sieves taken by the first iteration are:

$S_3 = \{ 0, 1, 2, 3 \}$,
$S_{17} = \{ 1, 3, 4, 5, 6, 8 \}$,
$S_{577} = \{ 1, 2, 7, 8, 9, 10, 15, 16 \}$,
$S_{257} = \{ 1, 3, 5, 6, 7, 9, 12, 13, 17, 18, 22, 23, 24, 26, 28, 29, 31, 32, 33, 34, 36, 37, 39, 41, 42, 43, 47, 48, 52, 53, 56, 58, 59, 60, 62 \}$,
$S_{1409} = \{ 1, 4, 7, 9, 10, 11, 13, 14, 16, 18, 19, 23, 25, 28, 30, 31, 32, 33, 34, 35, 37, 40, 42, 46, 47, 49, 51, 52, 54, 55, 56, 58, 61, 64 \}$.

The result of the first iteration is:

$$L_{64,6} = \{ 1, 9, 32, 33, 56, 64 \}.$$

The result of the second iteration is:

$$L_{324,28} = \{ 1, 19, 36, 37, 54, 73, 82, 90, 109, 126, 127, 135, 144, 162, 163, 181, 190, 198, 199, 216, 235, 243, 252, 271, 288, 289, 306, 324 \}.$$

The result of sieving $L_{64,6}$ and $L_{324,28}$ is:

$$L_{5184,36} = \{ 1, 73, 288, 289, 576, 649, 864, 1153, 1440, 1656, 1729, 1801, 2017, 2232, 2304, 2305, 2377, 2592, 2593, 2808, 2880, 2881, 2953, 3168, 3384, 3456, 3529, 3745, 4032, 4321, 4536, 4609, 4896, 4897, 5112, 5184 \} .$$

Table 1. Lengths of all periods

	m	$l(m)$	2	3	5
1	3	4	2		
2	17	8	3		
3	577	16	4		
4	257	64	6		
5	1409	64	6		
6	7	3	0	1	
7	199	9	0	2	
8	487	81	0	4	
9	5	6	1	1	
10	197	18	1	2	
11	53	54	1	3	
12	11	12	2	1	
13	73	36	2	2	
14	179	36	2	2	
15	107	108	2	3	
16	971	972	2	5	
17	1153	24	3	1	
18	1009	72	3	2	
19	433	216	3	3	
20	1297	648	3	4	
21	97	48	4	1	
22	193	96	5	1	
23	760	384	7	1	
24	41	5	0	0	1
25	31	15	0	1	1
26	151	75	0	1	2
27	751	75	0	1	2
28	1201	75	0	1	2
29	271	135	0	3	1
30	29	10	1	0	1
31	1549	50	1	0	2
32	269	30	1	1	1
33	149	150	1	1	2
34	19	20	2	0	1
35	59	20	2	0	1
36	499	500	2	0	3
37	601	60	2	1	1
38	1499	1500	2	1	3
39	1801	900	2	2	2

	m	$l(m)$	2	3	5	7	11
40	1619	1620	2	4	1		
41	241	40	3	0	1		
42	401	200	3	0	2		
43	1601	200	3	0	2		
44	641	320	6	0	1		
45	239	7	0	0	0	1	
46	1471	49	0	0	0	2	
47	71	35	0	0	1	1	
48	127	63	0	2	0	1	
49	631	315	0	2	1	1	
50	13	14	1	0	0	1	
51	293	98	1	0	0	2	
52	349	350	1	0	2	1	
53	113	28	2	0	0	1	
54	337	28	2	0	0	1	
55	139	140	2	0	1	1	
56	281	140	2	0	1	1	
57	419	140	2	0	1	1	
58	83	84	2	1	0	1	
59	251	84	2	1	0	1	
60	587	588	2	1	0	2	
61	1259	252	2	2	0	1	
62	673	336	4	1	0	1	
63	449	224	5	0	0	1	
64	23	11	0	0	0	0	1
65	353	11	0	0	0	0	1
66	727	363	0	1	0	0	2
67	463	231	0	1	0	1	1
68	991	495	0	2	1	0	1
69	783	891	0	4	0	0	1
70	1693	1694	1	0	0	1	2
71	109	110	1	0	1	0	1
72	461	462	1	1	0	1	1
73	1321	330	1	1	1	0	1
74	43	44	2	0	0	0	1
75	89	44	2	0	0	0	1
76	1451	484	2	0	0	0	2
77	307	308	2	0	0	1	1
78	617	308	2	0	0	1	1

Table 1 (continued).

	m	$l(m)$	2	3	5	7	11	13
79	881	220	2	0	1	0	1	
80	131	132	2	1	0	0	1	
81	1187	132	2	1	0	0	1	
82	659	660	2	1	1	0	1	
83	1979	1980	2	2	1	0	1	
84	79	13	0	0	0	0	0	1
85	599	13	0	0	0	0	0	1
86	911	455	0	0	1	1	0	1
86	313	39	0	1	0	0	0	1
87	1951	975	0	1	2	0	0	1
88	1013	338	1	0	0	0	0	2
89	181	182	1	0	0	1	0	1

	m	$l(m)$	2	3	5	7	11	13
90	521	130	1	0	1	0	0	1
91	1949	130	1	0	1	0	0	1
92	1429	1430	1	0	1	0	1	1
93	389	78	1	1	0	0	0	1
94	1637	1638	1	2	0	1	0	1
95	701	702	1	3	0	0	0	1
96	571	572	2	0	0	0	1	1
97	1091	156	2	1	0	0	0	1
98	467	468	2	2	0	0	0	1
99	937	468	2	2	0	0	0	1
100	1249	312	3	1	0	0	0	1
101	1873	936	3	2	0	0	0	1

It is not possible to give here all the derived sieves, but Table 2 gives their main characteristics and with the capital letters the order of the sievings is shown as well. The * denotes a sieve derived by iteration and the two numbers in front of the sieve in the first column indicate which sieves according to Table 1 were used in the iteration.

The program stopped after computing $L_{64929664800,3456}=\{1, 42162120, 51531481, 90810720, 136216081, \ldots\}$. With the polynomial y^2+2 the program returned the sieve $L_{190512000,48}=\{1, 12700801, 15876000, 16669801, \ldots\}$.

6 Remarks

The algorithm was implemented in FORTRAN-77 on a personal computer and the implementation does not yet contain any multiprecision arithmetic. Each of the results was computed in less than 10 minutes running time.

Although we used a heuristic arrangement of sievings to decrease the total cost of sieving, it would be interesting to find a method for determining the optimal arrangement.

A more detailed description, the correctness and the complexity of the algorithm will be given in a future paper.

Table 2. Lengths and cardinalities of the derived sieves

<table>
<tr><th></th><th>l</th><th>h</th><th>l</th><th>h</th><th>l</th><th>h</th><th>l</th><th>h</th><th>l</th></tr>
<tr><td>1</td><td></td><td>*</td><td rowspan="5">5184</td><td rowspan="5">36
A</td><td rowspan="10">31104</td><td rowspan="10">20
C</td><td rowspan="20">190512000</td><td rowspan="20">200
G</td><td rowspan="40">389577988800</td></tr>
<tr><td>5</td><td>64</td><td>6</td></tr>
<tr><td>—</td><td>——</td><td>——</td></tr>
<tr><td>6</td><td>324</td><td>28</td></tr>
<tr><td>12</td><td></td><td>*</td></tr>
<tr><td>13</td><td></td><td>*</td><td rowspan="5">31104</td><td rowspan="5">240
B</td></tr>
<tr><td>19</td><td>1944</td><td>92</td></tr>
<tr><td>—</td><td>——</td><td>——</td></tr>
<tr><td>20</td><td>10368</td><td>748</td></tr>
<tr><td>23</td><td></td><td>*</td></tr>
<tr><td>24</td><td></td><td>*</td><td rowspan="5">13500</td><td rowspan="5">40
D</td><td rowspan="10">31752000</td><td rowspan="10">520
F</td></tr>
<tr><td>30</td><td>1350</td><td>60</td></tr>
<tr><td>—</td><td>——</td><td>——</td></tr>
<tr><td>31</td><td>1500</td><td>52</td></tr>
<tr><td>39</td><td></td><td>*</td></tr>
<tr><td>40</td><td></td><td>*</td><td rowspan="5">31752000</td><td rowspan="5">9068
E</td></tr>
<tr><td>44</td><td>648000</td><td>6252</td></tr>
<tr><td>—</td><td>——</td><td>——</td></tr>
<tr><td>45</td><td>4410</td><td>104</td></tr>
<tr><td>51</td><td></td><td>*</td></tr>
<tr><td>52</td><td></td><td>*</td><td rowspan="5">352800</td><td rowspan="5">860
H</td><td rowspan="10">384199200</td><td rowspan="10">936
J</td><td rowspan="20">64929664800</td><td rowspan="20">3456
N</td></tr>
<tr><td>58</td><td>2100</td><td>106</td></tr>
<tr><td>—</td><td>——</td><td>——</td></tr>
<tr><td>59</td><td>14112</td><td>448</td></tr>
<tr><td>63</td><td></td><td>*</td></tr>
<tr><td>64</td><td></td><td>*</td><td rowspan="5">1372140</td><td rowspan="5">148
I</td></tr>
<tr><td>70</td><td>686070</td><td>3866</td></tr>
<tr><td>—</td><td>——</td><td>——</td></tr>
<tr><td>71</td><td>50820</td><td>752</td></tr>
<tr><td>78</td><td></td><td>*</td></tr>
<tr><td>79</td><td></td><td>*</td><td rowspan="5">11711700</td><td rowspan="5">3280
K</td><td rowspan="10">70270200</td><td rowspan="10">128
M</td></tr>
<tr><td>83</td><td>13860</td><td>410</td></tr>
<tr><td>—</td><td>——</td><td>——</td></tr>
<tr><td>84</td><td>177450</td><td>1556</td></tr>
<tr><td>89</td><td></td><td>*</td></tr>
<tr><td>90</td><td></td><td>*</td><td rowspan="5">1081080</td><td rowspan="5">610
L</td></tr>
<tr><td>96</td><td>540540</td><td>3942</td></tr>
<tr><td>—</td><td>——</td><td>——</td></tr>
<tr><td>97</td><td>936</td><td>44</td></tr>
<tr><td>101</td><td></td><td>*</td></tr>
</table>

References

[1] I. Gaál, A. Pethö and M. Pohst, *On the resolution of index form equation in biquadratic number fields*, to appear.

[2] J.C. Lagarias and D.P. Weisser, *Fibonacci and Lucas cubes*, Fibonacci Quart. **18** (1981), 39-43.

[3] D.H. Lehmer, *A history of sieve process* in A history of computing in the twentieth century, Academic Press, New York 1980, 445-456.

[4] H. London and R. Finkelstein, *On Fibonacci and Lucas numbers which are perfect powers*, Fibonacci Quart. **7** (1969), 476-481, 487. errata ibid **8** (1970), 248.

[5] I. Nemes and A. Pethö, *Polynomial values in linear recurrences II*, J. of Number Theory, **24** (1986), 47-53.

[6] A. Pethö, *Perfect powers in second order recurrences*, Colloquia Mathematica János Bolyai **34**, Topics in classical number theory, Budapest, 1981, 1217-1227.

[7] A. Pethö, *Full cubes in the Fibonacci sequence*, Publ. Math. Debrecen, **30** (1983), 117-127.

[8] A. Pethö, Manuscript, 1988.

[9] M. Ward, *The arithmetical theory of linear recurring series* , Trans. Am. Math. Soc. **35** (1933), 600-628.

[10] M.C. Wunderlich, *On the existence of Fibonacci square*, Math. Comp. **17** (1963), 455-457.

On Thue equations associated with certain quartic number fields

R.J. Stroeker

Abstract. Let **K** be a quartic number field, embedded in the reals but not totally real and let $\mathbf{L} = \mathbb{Q}(\sqrt{d})$ be its real quadratic subfield, with squarefree $d \equiv 3 \pmod 4$. Further, let $\xi > 1$ generate the subgroup of units relative to **L** in the group of positive units of **K**. Under certain conditions, which can be explicitly checked, and for suitable linear forms $X = X(u, v)$ and $Y = Y(u, v)$ with coefficients in $\mathbb{Z}[\sqrt{d}]$, the diophantine equation

$$Norm_{\mathbf{K}/\mathbb{Q}}(X + Y\xi^2) = 1,$$

which is a quartic Thue equation in the indeterminates u and v, can be explicitly solved. The method is illustrated by a few examples. Detailed proofs are given in [12].

1 Introduction

Diophantine equations have been around for a long time. The history of the subject stretches back a few thousand years and the methods for solving polynomial equations in integers are deeply rooted in the mathematics of antiquity. Bašmakova [1] contains much information of historical interest on Diophantus and diophantine equations. Most results on diophantine problems prior to 1920 can be found in Dickson's History [2], Volume II. For more recent results one may consult Skolem [9], Mordell [4] and Shorey & Tijdeman [8]. Apart from these there is a vast literature scattered over many different journals and conference proceedings. See [11] and [14] and the references mentioned there.
We shall restrict ourselves to polynomial diophantine equations, i.e. equations of type $f = 0$, where $f \in \mathbb{Z}[x_1, x_2, \ldots, x_n]$ and $n \geq 2$. Solutions are those $(x_1, x_2, \ldots, x_n)$ with integer components x_i that satisfy $f = 0$.

Roughly speaking, when dealing with diophantine equations, one has the following options:

- (Geometric approach) Interpret the equation as the algebraic representation of a curve, of a surface or of some other geometric object and extract information on its solutions by geometric means.

Computational Number Theory

- (Algebraic approach) Reduce the equation by factorization to a finite number of expressions in integers of a certain number field associated in some sense with the defining polynomial f, and try to solve these expressions algebraically.
- (Approximation approach) Rewrite the equation in such a form that results on the approximation of algebraic numbers by rationals can be applied.

In this paper we shall deal exclusively with a special type of polynomial equation, the so-called *Thue equation*, which is an equation of the form

$$f(x, y) = m, \tag{1}$$

where m is a fixed non-zero rational integer, and f is a binary form of degree at least 3 with rational integer coefficients. In addition one usually assumes f to be irreducible. Axel Thue proved in 1909 (see [16]) that equation (1) has at most a finite number of solutions in rational integers x and y, but his method of proof does not allow for the factual determination of all such solutions.
As a universal algorithm by which every diophantine equation can be solved cannot exist (see [3]), when divising methods of solution, one is doomed to consider individual equations or at best classes of similar equations. The recently applied approximation method of Tzanakis and de Weger [17] deals with individual Thue equations. It uses a combination of Baker's theory and recent computational diophantine approximation techniques; by Baker's method an unwieldy upperbound for $max(|x|, |y|)$ is found, which is subsequently reduced to manageable proportions by clever approximation techniques. In [17] the authors claim that their method can deal with any individual Thue equation. In [15] it is made plausible that every quartic Thue equation can be solved algebraically, at least in principle. Here we shall merely focus on the algebraic approach. Rewriting (1) as a norm form equation yields

$$Norm_{\mathbf{K}/\mathbb{Q}}(x + y\theta) = m, \tag{2}$$

where θ is a root of the equation $f(t, 1) = 0$ and $\mathbf{K}$ is the number field generated by θ over the rationals $\mathbb{Q}$. The field $\mathbf{K}$ is often referred to as a field *associated* with the Thue equation (1).

In order to solve (2) one needs detailed information on the field $\mathbf{K}$, in particular on its unit structure. This usually involves a substantial calculation effort. It is generally accepted that both the approximation and the algebraic approaches require a great many hard calculations for each individual equation. Why not try to combine the efforts? Consider a *class* of equations, instead of a single one, chosen in such a way that the associated number fields share relevant properties so that a considerable calculation gain can be made. Of course, these classes have to be chosen such that they contain more than merely one interesting equation. This is what we plan to do. In fact we consider norm form equations (2) with $m = 1$ so that

$$x + y\theta = \text{ unit of } \mathbf{K}. \tag{3}$$

In the following section we shall give a description of our procedure to solve (3) in case $\mathbf{K}$ is restricted to a special class of quartic number fields. Crucial to our method is that we use a set of fundamental units of the relevant number field **K** without having to explicitly calculate the coefficients of these units in terms of an integer basis of **K**; the number fields considered share a common unit structure. The proofs are omitted; the interested reader may find complete proofs in [12].

2 Outline of the method

Throughout the sequel **K** shall be a number field of absolute degree 4, with a real quadratic subfield $\mathbf{L} = \mathbb{Q}(\sqrt{d})$, where d is squarefree and $d \equiv 3 \pmod 4$. Moreover, it is assumed that **K** is not totally real, so that the absolute discriminant $\mathcal{D}_{\mathbf{K}}$ of **K** is negative.
The unit group of **L** is denoted by $\mathcal{U}_{\mathbf{L}} = \langle \pm 1 \rangle \times \mathcal{U}_{\mathbf{L}}^{+}$, where $\mathcal{U}_{\mathbf{L}}^{+} = \langle \varepsilon \rangle$ with fundamental unit $\varepsilon > 1$ of **L**. Likewise, the unit group of **K** is denoted by $\mathcal{U}_{\mathbf{K}} = \langle \pm 1 \rangle \times \mathcal{U}_{\mathbf{K}}^{+}$, where $\mathcal{U}_{\mathbf{K}}^{+}$ is a free abelian group of rank two.
Next we define $\mathcal{U}_{\mathbf{K}/\mathbf{L}}^{+} = \{\eta \in \mathcal{O}_{\mathbf{K}} \mid Norm_{\mathbf{K}/\mathbf{L}}(\eta) = 1, \, \eta > 0\}$, where $\mathcal{O}_{\mathbf{K}}$ denotes the ring of integers of **K**. Obviously, this relative unit group $\mathcal{U}_{\mathbf{K}/\mathbf{L}}^{+}$ has rank 1. Further, we assume that $\mathcal{U}_{\mathbf{K}/\mathbf{L}}^{+}$ is generated by $\xi > 1$, so that $\mathcal{U}_{\mathbf{K}/\mathbf{L}}^{+} = \langle \xi \rangle$.

It is well-known (see [5] or [7]) that $\mathcal{U}_{\mathbf{K}}^{+} = \langle \varepsilon, \xi \rangle$, provided both $\sqrt{\varepsilon}$ and $\sqrt{\varepsilon\xi}$ do not belong to **K**. It is rather easy to show that $\sqrt{\varepsilon} \notin \mathbf{K}$. Also $\sqrt{\varepsilon\xi} \notin \mathbf{K}$ if we assume $\xi + \xi^{-1} = a + b\sqrt{d}$ for odd a and b (see [12]). Note that the minimal polynomial for ξ is the monic reciprocal polynomial

$$x^4 - sx^3 + tx^2 - sx + 1, \text{ with } s = 2a \text{ and } t = a^2 - db^2 + 2. \tag{4}$$

Assuming a and b to be odd, equation (3) may be written as

$$x + y\theta = \pm \varepsilon^m \xi^n, \tag{5}$$

where θ is a root of the equation $f(t,1) = 0$ for irreducible f. We may drop the $\pm$ sign in (5) by adjusting the signs of x and y, if necessary.
Obviously, $(x,y) = (1,0)$, corresponding with $(m,n) = (0,0)$ is a solution of (5). Suppose (5) has another, less trivial solution $(x,y) = (u,v) \in \mathbb{Z}^2$ with $v \neq 0$. As f is irreducible, we must have

$$u + v\theta = \varepsilon^l \xi^k, \text{ with } k \neq 0, \tag{6}$$

and without loss of generality we may assume $k > 0$. Combining (5) and (6) leads to the following expressions:

$$X + Y\xi^k = v\varepsilon^m \xi^n, \tag{7}$$

with $X = xv - yu$ and $Y = y\varepsilon^l$.

As the conjugacy map, characterized by $\xi \mapsto \xi^{-1}$ leaves the elements of **L** unchanged, the conjugate equation to (7) is

$$X + Y\xi^{-k} = v\varepsilon^m \xi^{-n},$$

so that

$$-X\varepsilon^{-m} = \frac{\xi^{n-k} - \xi^{-n+k}}{\xi^k - \xi^{-k}} v, \quad Y\varepsilon^{-m} = \frac{\xi^n - \xi^{-n}}{\xi^k - \xi^{-k}} v. \tag{8}$$

The elements $s_n := (\xi^n - \xi^{-n})/(\xi - \xi^{-1})$ of the sequence $(s_n)_{n \in \mathbf{Z}}$ are **L**-integers and system (8) reads in terms of the elements of (s_n):

$$- X\varepsilon^{-m} = \frac{s_{n-k}}{s_k}v \, , \, Y\varepsilon^{-m} = \frac{s_n}{s_k}v. \tag{9}$$

Obvious solutions of (9) are given by $n = 0$ and $n = k$, because $s_0 = 0$, so that $X = 0$ and $Y = \varepsilon^m$ or $X = \varepsilon^m$ and $Y = 0$. These solutions shall be referred to as the *trivial solutions.* Clearly, it depends on the divisibility properties of the elements of the sequence (s_n) and the value of v whether any nontrivial solutions exist. In the special case that k is even but not divisible by 4, we can give information on the exponent to which XY is divisible by the unique prime ideal divisor of 2 in $\mathcal{O}_{\mathbf{L}}$, provided $XY \neq 0$.

A precise formulation is given in the following theorem.

Theorem 2.1. *Let $\xi > 1$ be real with minimal polynomial* (4), *where a, b and d are any three* odd *positive integers such that*

- *d is squarefree and $d \equiv 3 \pmod 4$, and*
- *$\mid a - b\sqrt{d} \mid < 2$.*

Let $\mathbf{K} = \mathbf{Q}(\xi)$ and let $\mathbf{L} = \mathbf{Q}(\sqrt{d})$ be the real quadratic subfield of $\mathbf{K}$. Further, suppose that ξ generates the relative unit group $\mathcal{U}^+_{\mathbf{K}/\mathbf{L}}$.

If $Norm_{\mathbf{K}/\mathbf{Q}}(X + Y\xi^k) = 1$, with $k \equiv 2 \pmod 4$ and $X, Y \in \mathcal{O}_{\mathbf{L}}$ then either $XY = 0$ or $0 \neq XY$ is exactly divisible by an odd power of the unique prime ideal divisor $\wp$ of 2 in $\mathcal{O}_{\mathbf{L}}$.

Remarks 2.2. A complete proof of the theorem is given in [12].

- The condition $\mid a - b\sqrt{d} \mid < 2$ guarantees that $\mathbf{K}$ is not totally real.
- It is clear that for any squarefree positive $d \equiv 3 \pmod 4$, there are infinitely many odd a and b with $\mid a - b\sqrt{d} \mid < 2$. Also for such d there is a unique prime ideal divisor $\wp$ of 2 in $\mathcal{O}_{\mathbf{L}}$ and $\wp^2 = (2)$.
- If $(t_n)_{n \in \mathbf{Z}}$ is defined by $t_n := \xi^n + \xi^{-n}$, and s_n is as above, then it is easily checked that
  - if n is odd then $\wp \not| \, s_n$ and $\wp \| t_n$,
  - if $n \equiv 2 \pmod 4$ then $\wp \| s_n$ and $\wp^3 \| t_n$,
  - if $n = 2^e n'$ with odd n' and $e \geq 2$ then $\wp^{2e} \| s_n$ and $\wp^2 \| t_n$.
- Now suppose that $v = \pm 1$ in (7) or, more generally that (7) can be reduced to a similar equation with $v = \pm 1$ by direct divisibility considerations. Further, let the corresponding X and Y be such that either the product XY vanishes or $ord_{\wp}(XY)$ is even — this can be checked if (u, v) is a known solution — then the only solutions of (7) are those corresponding to $XY = 0$.

3 Examples

Unfortunately, it is difficult to define the classes of equations mentioned in section 1 directly in terms of the coefficients of the equations, or in terms of a, b and d. This means that for any given quartic Thue equation it is not immediately clear whether the method just described will work for this equation or not. Below we give a few examples to illustrate how the theorem given in the previous section may be applied.

The procedure is this. First, choose values for a, b and d and check that an appropriate quartic field is obtained. As things are set up, there is a correspondence between Thue equations and pairs of linear forms X and Y with coefficients in $\mathbf{K}$. Next choose a pair of linear forms X and Y such that $ord_{\wp}(XY)$ is even for any choice of rational integral values of the variables x and y. As one solution is known, this should be rather easy and many 'good' choices can be made. It turns out that one can generate in this way many Thue equations associated with the single field $\mathbf{K}$. These equations, linked to the same quartic field, build a class as refered to in the introductory section.

Example 1. Let $d = 3$ and $a = b = 1$. Then $\xi > 1$ is a root of

$$x^4 - 2x^3 - 2x + 1 = 0,$$

in fact $2\xi = 1+\sqrt{3}+\sqrt{2\sqrt{3}}$. Moreover, $\mathbf{K} = \mathbb{Q}(\xi)$ and $\mathbf{L} = \mathbb{Q}(\sqrt{3})$. Clearly $\xi \epsilon \mathcal{U}^{+}_{\mathbf{K}/\mathbf{L}}$ and if ξ is not a perfect power then ξ generates $\mathcal{U}^{+}_{\mathbf{K}/\mathbf{L}}$. This can be checked by considering all perfect powers $\xi = \eta^c$ with $c < c_0(\mathbf{K})$, where the upperbound $c_0(\mathbf{K})$ can be effectively calculated by Nakamula's lemma (see [7] or [12]). We find $c_0(\mathbf{K}) = 1.6$, so that $\mathcal{U}^{+}_{\mathbf{K}/\mathbf{L}} = \langle\xi\rangle$.

Now any choice of $X = x\varepsilon^p$, $Y = y\varepsilon^q$ with rational integral x and y results in a Thue equation

$$Norm_{\mathbf{K}/\mathbb{Q}}(X + Y\xi^2) = 1, \tag{10}$$

in indeterminates x and y with solutions given by $xy = 0$ and no others. Indeed, for these choices of X and Y the product XY is never exactly divisible by an odd power of $\wp$.

For $p = q = 0$ one has the equation

$$x^4 + 4x^3y - 6x^2y^2 + 4xy^3 + y^4 = 1 \text{ or } (x + y)^4 - 12x^2y^2 = 1,$$

for $p = 1$, $q = 0$ this is

$$x^4 - 4x^3y + 6x^2y^2 + 20xy^3 + y^4 = 1 \text{ or } (x - y)^4 + 24xy^3 = 1,$$

and for $p = 0$, $q = 1$ we get a similar equation. See also [10].

Example 2. Choose $a = 3$, $b = 1$ and $d = 7$. Then ξ satisfies

$$x^4 - 6x^3 + 4x^2 - 6x + 1 = 0,$$

and hence $2\xi = 3 + \sqrt{7} + \sqrt{12 + 6\sqrt{7}}$, as ξ is real and larger than 1. Further, $\mathbf{K} = \mathbb{Q}(\xi)$. Applying Nakamula's lemma, we find this time that $c < 3.5$ if $\xi = \eta^c$.

Checking for $c = 2$ and $c = 3$ leads in both cases to a contradiction. So again ξ generates $\mathcal{U}^+_{\mathbf{K}/\mathbf{L}}$.
Experimenting with various choices for X and Y, one finds that

$$X = 3x + (7 + \sqrt{7})y,\ Y = (2 - \sqrt{7})y \tag{11}$$

leads to

$$Norm_{\mathbf{K}/\mathbb{Q}}(X + Y\xi^2) = 81(x^4 - 10x^2y^2 - 3y^4).$$

Now let $(x, y) \in \mathbb{Z}^2$ be a solution of the equation

$$x^4 - 10x^2y^2 - 3y^4 = 1. \tag{12}$$

Then it is easy to see that $y \equiv 0 \pmod 3$. For these x and y the expressions (11) satisfy $\frac{X}{3}, \frac{Y}{3} \in \mathcal{O}_{\mathbf{L}} = \mathbb{Z}[\sqrt{7}]$. Moreover, as y is even, the expression Y from (11) is exactly divisible by an even power of $\wp$, unless $Y = 0$. This shows that the only solutions of (12) are given by $y = 0$. See also [13].

In [12] and [13] a couple of tables of quartic Thue equations are included that can be solved by the method described.

References

[1] I.G. Bašmakova, *Diophant und diophantische Gleichungen*, Uni-Taschenbücher 360, Birkhäuser Verlag 1974.

[2] L.E. Dickson, *History of the theory of numbers, Vol. II*, Chelsea Publ. Co. 1971 (first published in 1920 by the Carnegie Institute of Washington).

[3] Yu.V. Matiyasevich, *Diophantine representation of enumerable predicates*, Izv. Akad. Nauk. SSSR Ser. Mat. **35** (1971), 3-30.

[4] L.J. Mordell, *Diophantine equations*, Pure Appl. Math. Vol. 30, Academic Press 1969.

[5] Trygve Nagell, *Sur quelques questions dans la théorie des corps biquadratiques*, Arkiv för Math. **4**(26) (1962), 347-376.

[6] Ken Nakamula, *Class number calculation and elliptic unit II*, Proc. Japan Acad. **57A** (1981), 117-120.

[7] Ken Nakamula, *Class number calculation of a quartic field from the elliptic unit*, Acta Arithm. **XLV** (1985), 215-227.

[8] T.N. Shorey and R. Tijdeman, *Exponential diophantine equations*, Cambridge Tracts in Maths. 87, Cambridge Un. Press 1986.

[9] T. Skolem, *Diophantische Gleichungen*, Erg. Math. Grenzgeb. Bd. 5, Heft 4, Springer Verlag 1938. (Reprinted by Chelsea, 1950).

[10] R.J. Stroeker, *On the diophantine equation* $(2y^2 - 3)^2 = x^2(3x^2 - 2)$ *in connection with the existence of non-trivial tight* 4*-designs*, Indag. Math. **43**(3) (1981), 353-358.

[11] R.J.Stroeker, *How to solve a diophantine equation, a number-theoretic excursion*, Am. Math. Monthly **91**(7) (1984), 385-392.

[12] R.J. Stroeker, *On quartic Thue equations with trivial solutions*, Math. of Comput. **52**(185) (1989), 175-187.

[13] R.J. Stroeker, *On diophantine equations of type* $X^4 - 2aX^2Y^2 - bY^4 = 1$, Proceedings of the NATO ASI on Number Theory and Applications (editor: R.A. Mollin) (1989), Banff Centre, 27 April - 5 May 1988, 547-553.

[14] R.J. Stroeker and R. Tijdeman, *Diophantine equations*, In: Computational methods in number theory, Part II, MC Tracts **155** (1982), Centre Math. Comp. Sci., Amsterdam, 321-369.

[15] R.J. Stroeker and N. Tzanakis, *On the application of Skolem's p-adic method to the solution of Thue equations*, J. Number Th. **29**(2) (1988), 166-195.

[16] A. Thue, *Über Annäherungswerte algebraischer Zahlen*, J. reine angew. Math. **135** (1909), 284-305.

[17] N. Tzanakis and B.M.M. de Weger, *On the practical solution of the Thue equation.*, J. Number Th. **31**(2) (1989), 99-132.

KANT - a tool for computations in algebraic number fields

*J. Graf v. Schmettow**

Abstract. KANT is an ANSI Fortran 77 subroutine library which has been in development in Düsseldorf since 1987. Its main task is performing arithmetic in algebraic number fields, based on algorithms from the geometry of numbers and partly supported by a multiprecision package. Until now it has been implemented on several systems. It can be used for treating number fields of degree up to 20.

1 Survey

The intention of this article is to sketch the motivation, facilities and structure of the subroutine library KANT.

The abbreviation KANT stands for

Computational **A**lgebraic **N**umber **T**heory

which includes a slight hint of its German origin.
There are several aspects to the motivation for the development of KANT:

- Proving or disproving conjectures concerning number fields and related topics
- Getting a better view on the behavior of number fields (computing large number field tables)
- Developing tools for solving (systems of) nonlinear diophantine equations

KANT is neither an interactive system nor a tool for symbolic computation. It is rather a subroutine library for number theoretical applications with a strong accent on arithmetic in arbitrary algebraic number fields. The implementation language of KANT is ANSI Fortran 77. This seems to be somewhat old-fashioned but on the other hand it makes KANT highly portable.

Although KANT should be regarded as a "subroutine pool" it contains several main programs which can be used in dialog and which are compatible with each other in a certain way which will be explained below.

*supported by the Friedrich-Naumann-Stiftung

Computational Number Theory

KANT is subdivided into six parts:

- *Basic tools*:
  - I/O routines
  - elementary vector and matrix operations
- *Extended arithmetic tools*:
  - interval arithmetic
  - polynomial arithmetic
  - multiprecision arithmetic for integers and real numbers
- *Higher level tools*:
  - Complex roots of polynomials with real coefficients
  - Determinants and inverses of real and integer matrices
  - Hermite and Smith normal form
  - LLL reduction [14]
  - Modified LLL reduction for computing a LLL reduced basis of a lattice when only a system of generators is given [16]
  - Enumeration of lattice points in ellipsoids
  - Computation of short and shortest vectors in lattices [11]
  - Finding integer relations among real numbers [8]
  - Factorization of polynomials modulo a prime (Berlekamp)
  - Irreducibility tests for polynomials
- *Number field tools*:
  - Arithmetic of algebraic numbers
  - Integral bases in algebraic number fields [19]
  - Torsion units [18]
  - Systems of independent units [3, 4, 18]
  - Various regulator estimates [18]
  - Systems of fundamental units [1, 2, 10, 18]
  - Solution of norm equations in orders [9]
  - Ideal arithmetic
  - Ideal class arithmetic
  - Class groups [5, 17]
  - Index form equations in cubic number fields [13]
  - Detection of number field isomorphisms [15]

- *Internal tools*:
  - Internal event handling and performance control tools
  - System time [1]
- Main programs and drivers for diverse applications

2 Realization

The implementation of the basic, extended arithmetic and higher level tools need not be explained in detail because it is straightforward. So the emphasis of this section will be on the realization of the *number field* arithmetic, i.e. we are going to explain

- How number field "objects" are represented in the computer
- How the operations between those objects are implemented

Let $F = \mathbb{Q}(\rho)$ be an algebraic number field of degree n and $f(x) \in \mathbb{Z}[x]$ (monic and irreducible) the minimal polynomial of the algebraic integer ρ with s real and $2t$ non-real zeros (the field degree and the coefficients of f are the "minimal input data" KANT requires for computations). Let $\omega_1, \ldots, \omega_n$ be an integral basis of the maximal order o_F of F. We have:

$$\omega_j = \frac{1}{d_j} \sum_{i=1}^{n} w_{ij} \rho^{i-1} \qquad (d_j, w_{ij} \in \mathbb{Z}).$$

The computation of d_j, w_{ij} is done by the method of Ford and Zassenhaus [12, 19]. — Let $\alpha \in o_F$ be an arbitrary algebraic integer with

$$\alpha = \sum_{j=1}^{n} a_j \omega_j \qquad (a_j \in \mathbb{Z}).$$

In KANT the integral basis is stored via the integer transition matrix

$$\begin{pmatrix} w_{11} & \cdots & w_{1n} \\ \vdots & \ddots & \vdots \\ w_{n1} & \cdots & w_{nn} \\ d_1 & \cdots & d_n \end{pmatrix} \in \mathbb{Z}^{(n+1) \times n} \tag{1}$$

and α can be represented by the integer vector

$$(a_1, \ldots, a_n)^t.$$

Once the coefficients of f and the integral basis $\omega_1, \ldots, \omega_n$ are known, every algebraic integer can be stored like this (algebraic *numbers* are viewed as algebraic

[1] Besides the INCLUDE-statement the system time is the only system dependent and non-standard Fortran feature of KANT

integers with an integer denominator). Furthermore, ideals and modules are stored as vectors of algebraic integers.

For doing arithmetic in o_F we need the so-called multiplication table $\Gamma \in \mathbb{Z}^{n \times n \times n}$ which is defined by

$$\omega_i \omega_j = \sum_{k=1}^{n} \Gamma(k, i, j) \omega_k.$$

The existence and uniqueness of Γ can be derived from the fact that o_F is an order. Γ can be computed in an elementary fashion from the definition of w_{ij}, d_j and the fact that $f(\rho) = 0$.

Once Γ is computed we have an easy method for multiplying algebraic integers: For $\alpha = \sum_{i=1}^{n} a_i \omega_i$ and $\beta = \sum_{i=1}^{n} b_i \omega_i$ ($a_i, b_i \in \mathbb{Z}$) we obtain

$$\alpha\beta = \sum_{i=1}^{n} \omega_i \sum_{j,k=1}^{n} a_j b_k \Gamma(i, j, k).$$

The norm of an algebraic integer α can be computed as the determinant of its representation matrix with respect to $\omega_1, \ldots, \omega_n$:

$$M_\alpha(k, j) := \sum_{i=1}^{n} \Gamma(k, i, j) a_i.$$

Moreover, the first column of M_α^{-1} (in $\mathbb{Q}^{n \times n}$) is the coefficient vector of α^{-1}, provided that $\omega_1 = 1$.

At this point we should mention also that multiplication, inversion and norm computation of algebraic integers can be done faster by using conjugates, provided that the accuracy of the floating point operations is sufficient.

As we have indicated in the first section, KANT includes a multiprecision package which is also written completely in Standard Fortran. For reasons of CPU time, memory occupation and readability of the sources it is recommended these routines be used only at some strictly defined "critical points". In many situations it is possible to avoid large numbers by just slight modifications. One example is transforming the input data: For a given number field F defined as above we consider the monomorphism [18]

$$\varphi : o_F \to \mathbb{R}^n : \alpha \mapsto (\alpha^{(1)}, \ldots, \alpha^{(s)}, \Re\alpha^{(s+1)}, \ldots, \Re\alpha^{(s+t)}, \Im\alpha^{(s+1)}, \ldots, \Im\alpha^{(s+t)})^t,$$

where $\alpha^{(i)}$ are the conjugates of α. Obviously, $\varphi(o_F)$ is a lattice with basis

$$\varphi(\omega_1), \ldots, \varphi(\omega_n).$$

Applying LLL-reduction to this basis gives us a transformed integral basis of o_F which usually is much better conditioned than the original one. Although the theory of [14] only guarantees

$$|\omega_i^{(j)}| < 2^{n/4} \sqrt{|d_F|}$$

(d_F = field discriminant) for a LLL-reduced integral basis, this estimation turns out to be much too pessimistic in practice. Moreover, the behavior of the multiplication table changes in a crucial way: the maximal entry in the multiplication table of a LLL-reduced integral basis seldom exceeds two digits, even for number fields of degree > 10.

Nevertheless, there are cases where it is not possible to keep all of the numbers which appear small, for example if the regulator of a number field with unit rank one exceeds 200 there is no chance of storing the coefficients of the fundamental unit in single-word integers.

The **ideal arithmetic** is one of the central features of KANT. It is built on the fact that every ideal $\mathbf{a}$ can be expressed in two different ways:

- via a $\mathbb{Z}$-basis: $\mathbf{a} = \bigoplus_{i=1}^{n} \mathbb{Z}\alpha_i \ (\alpha_i \in o_F)$
- via a two-element presentation: $\mathbf{a} = a o_F + \alpha o_F \ (a \in \mathbb{Z}^{>0}, \alpha \in o_F)$

where the latter can be sharpened to a so-called P-normal presentation [17, 18] which allows multiplication via the generators, i.e.

$$(a o_F + \alpha o_F)(b o_F + \beta o_F) = ab o_F + \alpha\beta o_F.$$

As already stated above, an ideal which is given in its $\mathbb{Z}$-basis presentation is stored as an array of algebraic integers, i.e. a two-dimensional integer array. If we apply Hermite reduction to this matrix the resulting presentation can be made unique and we have an easy method for comparing ideals. Moreover, an inclusion test (whether a given algebraic integer is contained in the ideal) turns out to be very easy.

The second method for storing ideals which is supported by KANT saves memory and gives us a fast way for multiplying ideals. The ideal is represented by an integer array of length $n+1$:

$$(a, a_1, \ldots, a_n)^t$$

where $\alpha = \sum_{i=1}^{n} a_i \omega_i$ is the second generator of the ideal.

KANT also provides tools for "switching" between the two presentations although we should mention that the conversion "two-element to $\mathbb{Z}$-basis" is much easier than the other way around.

Based on the ideal arithmetic is the arithmetic of **ideal classes**. We have implemented several ways for reducing ideals, i.e. substituting an ideal by an ideal of the same class which has got a "small" norm. Moreover, it is possible to check ideals for equivalence (i.e. check whether their classes are equal).

3 How to use KANT

As an example for a typical KANT main routine we present a program which solves norm equations in a number field[2] [9, 18]: For given $k \in \mathbb{Z}$ it searches for

[2] In KANT the routine `NEQ` does this job, for simplicity's sake we omitted some of the parameters and called it `NEQ1`.

an algebraic integer α with norm equal to k. The field data and k are chosen by the KANT user.

```
      PROGRAM SHOWIT
%include 'field.inc'
      INTEGER I, K, FLAG, COEF(20)
1     CALL COMINP(*9)
      CALL COMIN1(*1)
      WRITE(*,*) 'Give right side of norm equation: '
      READ(*,*) K
      CALL NEQ1(K,FLAG,COEF,*1)
      IF (FLAG.EQ.0) THEN
         WRITE (*,*) 'There is no solution.'
      ELSE
         WRITE (*,*) 'Solution: ',(COEF(I),I=1,N)
      ENDIF
      GOTO 1
9     STOP
      END
```

The use of the alternate return labels `1` and `9` will be discussed in the next section. – For the solution of norm equations by the method of Fincke/Pohst we need a lot of data which are initialized in the routines `COMINP` and `COMIN1`. The first routine has the task of reading data, in this case the polynomial coefficients, an integral basis and coefficients of a system of independent or fundamental units. The second routine, `COMIN1`, does a number of startup-computations: it determines the roots of the defining polynomial, the multiplication table etc. At this point the whole KANT runtime system is fully initialized and the application can start.

We should discuss more fully how KANT reads data (`COMINP`): Either (if all data are known) they can be read with the help of a simple command language:

```
(F=)
...          (coefficients of polynomial)
(ITB=)
...          (integral basis in terms of ρ)
(FUNIT=)
...          (coefficients of a system of fundamental units)
(END=)
```

The other way is to start only with the polynomial coefficients and to compute the integral bases and units with the main programs that KANT provides. The first of these programs is `MPITB1` which computes an integral basis if the polynomial coefficients are given. Since its output format can be understood by the input routine `COMINP` – which is also called by the unit program `MPFUN1` – it is easy to connect these routines. On a Unix system – by using the pipeline feature – the user only has to type

```
MPITB1|MPFUN1
(F=)
...
(END=)
```

and the output is exactly what is needed as input for the subsequent program.

4 KANT in depth: The internals

For staying completely independent KANT provides an event handling and tracing system which is fully based on Standard Fortran. This system makes use of the "alternate return" feature which allows a Fortran subroutine to "choose" between several addresses to jump back. For reasons of a clear programming style, this feature was handled very carefully by the KANT group. Nevertheless, it turned out to support an elegant way of error handling.

We will explain this using an example:

```
      SUBROUTINE NEQ1(K,FLAG,COEF,*)
          .
          .
      CALL START('NEQ1  ')
          .
          .
      IF (...(error condition)...) THEN
         IER = 5
         ERRMSG = '... (plaintext error message)...'
         GOTO 9999
      ENDIF
          .
          .
9999  CALL END (*9998)
      RETURN
9998  RETURN 1
      END
```

Most of the more sophisticated subroutines use the internal KANT routines START and END which have the task of "tracing" the actual program run. So the first executable statement (CALL START('...')) and the last four statements look very similar in all routines. If an error occurs, as indicated in the example above, a global error parameter called IER is set to a nonzero value (the initialization to zero is done in START) and a jump to the END-call is executed. This routine now realizes that something has gone wrong and prints the error message:

```
NEQ1: ... (plaintext error message) ...
Nested subroutine calls:  MAIN   SUB1   NEQ1
```

The routine `END` now initiates the *alternate* return (`RETURN 1`) which is indicated by the asterisk in the `SUBROUTINE`-statement.

A hypothetic calling program (in this case `SUB1`) can handle this situation adequately without the use of any switches defined explicitly: it might have a calling statement like

```
CALL NEQ1(K,FLAG,COEF,*9999)
```

where at label `9999` the same process happens again as it did one level lower. If an error of `NEQ1` is not crucial for the calling program (e.g. if there is an alternate routine which can be called instead) the "cascade" of alternate returns can be interrupted easily by giving a different label in the `CALL`-statement. In our example program from the last section an error in `NEQ1` leads back to the input routine which orders a new number field.

The task of the routines `START` and `END` is not only to provide event handling. They can also be used for logging facilities: calling the evaluation routine `SUMARY` (Fortran only allows names with maximal six letters!) one gets the statistics of the logging of `START` and `END`:

```
AINORM     121 calls,     2.5978 sec;    average time:     0.0017
COMIN1       2 calls,     0.6260 sec;    average time:     0.3130
ELLCNT     189 calls,     1.1572 sec;    average time:     0.0061
MATQF       48 calls,     0.1151 sec;    average time:     0.0024
NEQ1         2 calls,     2.7160 sec;    average time:     1.3580
SUB1         2 calls,     3.1129 sec;    average time:     1.5565
```

To accelerate a KANT program run it is possible to omit the logging.

5 Results and Conclusion

The development of KANT started in 1987 on a Siemens 7.580-S mainframe (operating system BS2000), but right from the beginning many of the routines were written and tested on Atari ST personal computers (CPU Motorola 68000). Since 1989 the development has been done only on Apollo workstations (68020/68030). Also KANT has been implemented on a Sun 3/60 and (in a smaller version) on an IBM PC AT.

The KANT group at Düsseldorf doesn't regard KANT as the "ultimate" number theory package. It should rather be viewed as an "experimental machine" which is to be developed further. The actual experimentation field of KANT are number fields with "small" discriminants: Tables of fundamental units and class groups of all 13073 totally real quartic fields with discriminant below 10^6 [6] and of all 81322 totally complex quartic fields with discriminant below the same bound have been computed, similar tables for quintic and octic fields are in progress.

References

[1] J. Buchmann, *A generalization of Voronoi's unit algorithm I,II*, J. Number Theory **20** (1985), 177–209.

[2] J. Buchmann, *The computation of the fundamental unit in totally complex quartic orders*, Math. Comp. **48** (1987), 39–54.

[3] J. Buchmann, *On the computation of units and class numbers by a generalization of Lagrange's algorithm*, J. Number Theory **26** (1987), 8–30.

[4] J. Buchmann and A. Pethö, *Computation of independent units in number fields by Dirichlet's method*, Math. Comp. **52** (1989), 149–159.

[5] J. Buchmann and M. Pohst, *On the complexity of computing class groups of algebraic number fields*, in "Applied Algebra, Algebraic Algorithms and Error-Correcting Codes", Proceedings AAECC-6, Rome 1988, Springer Lecture Notes in Comp. Sc. **357** (1989), 122–131.

[6] J. Buchmann, M. Pohst and J. v. Schmettow, *On the computation of unit groups and class groups of totally real quartic fields*, Math. Comp. **53** (1989), 387–397.

[7] J. Buchmann and H.C. Williams, *On principal ideal testing in algebraic number fields*, J. Symbolic Computation **4** (1987), 11–19.

[8] H.R.P. Ferguson, *A new integral relation finding algorithm involving partial sums of squares and no square roots*, Abstracts of the American Mathematical Society **9** (March 1988), p. 214.

[9] U. Fincke, *"Ein Ellipsoidverfahren zur Lösung von Normgleichungen"*, Dissertation, Düsseldorf 1984.

[10] U. Fincke and M. Pohst, *A new method for computing fundamental units in algebraic number fields*, Proceedings EUROCAL 85, Springer Lecture Notes in Computer Science **204** (1985), 470–479.

[11] U. Fincke and M. Pohst, *Improved methods for calculating vectors of short length in a lattice, including a complexity analysis*, Math. Comp. **44** (1985), 463–471.

[12] D.J. Ford, *The construction of maximal orders over a Dedekind domain*, J. Symbolic Computation **4** (1987), 69–75.

[13] I. Gaál and N. Schulte, *Computing all power integral bases of cubic fields*, Math. Comp. **53** (1989), 689–696.

[14] A.K. Lenstra, H.W. Lenstra Jr. and L. Lovasz, *Factoring polynomials with rational coefficients*, Math. Ann. **261** (1982), 515–534.

[15] M. Pohst, *On computing isomorphisms of equation orders*, Math. Comp. **48** (1987), 309–314.

[16] M. Pohst, *A modification of the LLL reduction algorithm*, J. Symbolic Computation **4** (1987), 123–127.

[17] M. Pohst and H. Zassenhaus, *Über die Berechnung von Klassenzahlen und Klassengruppen algebraischer Zahlkörper*, J. Reine Angew. Math. **361** (1985), 50–72.

[18] M. Pohst and H. Zassenhaus, "Algorithmic Algebraic Number Theory", Cambridge University Press 1989.

[19] H. Zassenhaus, *On the second round of the maximal order program*, in: Applications of Number Theory to Numerical Analysis, Academic Press, New York 1972, 398–431.

SIMATH - a computer algebra system *

C. Hollinger and P. Serf

Abstract. SIMATH is a computer algebra system focusing on algebraic number theory. It is based on the programming language C. A list system serves as a foundation for all SIMATH types such as integers, rationals, polynomials, algebraic numbers, algebraic functions, matrices and vectors. In addition to source libraries which contain all SIMATH functions, the system is equipped with a calculator called simcalc, by which most of the functions can be handled in interactive mode.

SIMATH has been implemented on the Siemens PC's MX2 and MX2+ as well as on SUN and Apollo workstations.

1 An overview

SIMATH, meaning SInix MATHematik, was developed in cooperation between the SIMATH group at the Universität des Saarlandes and Siemens AG.

It was first designed for a Siemens PC MX2 with operating system SINIX and then ported on the PC MX2+ as well as on the SUN workstation 3/100, and Apollo workstations DN3000 and DN4500 with operating system UNIX.

SIMATH is completely written in the programming language C. Thus, as opposed to many other computer algebra systems, the present one does not compel the user to learn a new system-specific language. He can simply work with C and incorporate the SIMATH functions into his own C-programs.

Furthermore, all functions may also be used in Fortran-programs by way of converting routines.

SIMATH is an open system; that is, on the one hand, the user can look at the source codes of the general algorithms and adjust them to his special purposes, and that on the other, he is given the possibility of adding his own modules to the system.

The connection between SIMATH and the operating system (SINIX/UNIX) is achieved by the SIMATH menu, which facilitates administration of libraries, the compiling, printing etc.

Besides, it contains two important on-line help features:

*A report on this work was delivered to the conference by the first author.

Computational Number Theory

- **the keyword index**:
  The user can obtain a list of all functions which are available in connection with a given keyword, e.g. Galois-field, by typing "?Galois-field".
- **the explanatory documentation**:
  If he wishes to get some more detailed information on a certain function, e.g. gfsprod (Galois-field with single characteristic, product of two elements), he types "~gfsprod" and is shown which input parameters the function requires and which output parameters it computes.

The system also provides a method of finding programming errors:
If the user types 't' or 'T' (instead of 'c' or 'C') to compile his main program, all SIMATH functions will test whether their input parameters are correct, and the user will eventually obtain an error message.

A detailed description of SIMATH, including some information on how to install the system and how to work with it, is given in the SIMATH manual.

2 The structure of SIMATH

The following diagram exhibits the structure of the system:

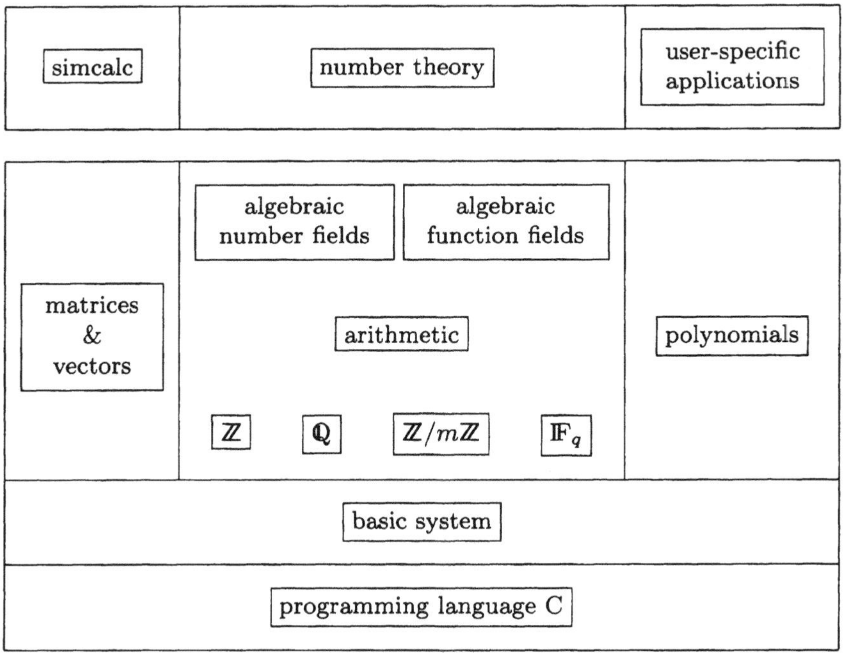

Figure 1.

Now, we shall describe the various subsystems.

2.1 The basic system

The basic system mainly consists of a list system. At the moment, there are about 90 list-functions relating to:

- list administration
- sorting
- searching
- input and output etc.

But it also contains programs for system initialization and administration.

However, since these routines are carried out below the user level, the user does not have to worry about internal details.

The space is organized in a dynamic manner, i.e. it will be automatically augmented if necessary, and administered by an automatic garbage collector.

2.2 The arithmetic system

Based on the list system, we built up an arithmetic for the domains

$$\mathbb{Z},\ \mathbb{Q},\ \mathbb{Z}/m\mathbb{Z} \text{ and } \mathbb{F}_q,$$

where m is a positive integer and $\mathbb{F}_q$ a finite field with q elements, as well as for

number fields
and function fields.

The different types are realized as follows:

- The zero element in all structures mentioned above is simply the atom 0.
- Integers are represented with reference to the constant BASIS $= 2^{30}$. Every nonzero integer $n \in \mathbb{Z}$ has a unique expansion

$$n = \sum_{i=0}^{k} a_i \cdot \text{BASIS}^i$$

with

$$k \geq 0,\ a_i \in \mathbb{Z},\ a_k \neq 0,\ |a_i| < \text{BASIS}$$
$$\text{and} \quad a_i = 0 \text{ or } \operatorname{sign}(a_i) = \operatorname{sign}(n).$$

n is represented by the atom a_0, if $k = 0$, and by the list of atoms $(a_0, \dots, a_k)$, if $k > 0$.

- A nonzero rational number $r \in \mathbb{Q}$,

$$r = \frac{m}{n}, \qquad \text{where } \gcd(m,n) = 1 \text{ and } n > 0,$$

corresponds to the list of the two integers m and n.

- The elements of $\mathbb{Z}/m\mathbb{Z}$ are simply the integers $0, 1, \ldots, m-1$.

- A finite field with $q = p^n$ elements is isomorphic to

$$\mathbb{F}_q \cong (\mathbb{Z}/p\mathbb{Z})[x]/\big(g(x)\big),$$

where $g(x)$ is an arbitrary monic irreducible polynomial of degree n over $\mathbb{Z}/p\mathbb{Z}$. Thus, the elements of $\mathbb{F}_q$ are represented by polynomials over $\mathbb{Z}/p\mathbb{Z}$ of degree less than n.

- Let $k = \mathbb{Q}(\alpha)$ be an algebraic number field, α an algebraic integer and $f(x) \in \mathbb{Z}[x]$ its minimal polynomial. Then k is isomorphic to $\mathbb{Q}[x]/\big(f(x)\big)$, and the elements of k correspond to polynomials over $\mathbb{Q}$ whose degrees are less than $\deg(f)$.
Let $g(x)$ be the polynomial corresponding to an arbitrary element $\beta \in k \setminus \{0\}$,

$$g(x) = \sum_{i=0}^{d} a_i x^i \qquad \text{with } d = \deg(g) < \deg(f) \text{ and } a_i \in \mathbb{Q},$$

and let n be the least common denominator of all a_i. Then

$$n \cdot g(x) = \sum_{i=0}^{d} \underbrace{(n \cdot a_i)}_{=:a_i' \in \mathbb{Z}} x^i$$

is a polynomial over $\mathbb{Z}$, and the list

$$G = (n, d, a_d', \ldots, a_0'),$$

which consists of the least common denominator n and the "dense form" of $n \cdot g(x)$ (see 2.3), represents the element β.
Therefore, we calculate with algebraic numbers as with dense polynomials over $\mathbb{Z}$.

- Algebraic function fields F/k in r variables are finitely generated extensions of transcendence degree r, i.e. there exist $x_1, \ldots, x_r \in F$, transcendental and algebraically independent over k, so that the extension F/K for $K := k(x_1, \ldots, x_r)$

is finite and separable. Then, $F = K(\alpha)$ for a primitive element α.

$$F = K(\alpha)$$

finite separable

$$K = k(x_1, \ldots, x_r)$$

purely transcendental

$$k$$

Of course, a nonzero rational function, i.e. an element of $K^\times = k(x_1, \ldots, x_r)^\times$, can be written as the quotient of two relatively prime polynomials f_1 and f_2 in r variables over k. In SIMATH, it is represented as the list (f_1, f_2).
An algebraic function, i.e. an element of $F = K(\alpha)$, is regarded as a polynomial in one variable over K with degree less than $[F : K]$.
In the current version of SIMATH, we are able to calculate with

rational and algebraic functions in one variable over $\mathbb{F}_p$
and rational functions in arbitrarily many variables over $\mathbb{Q}$.

The arithmetic system consists of about 320 functions.

- For $\mathbb{Z}$, $\mathbb{Q}$, $\mathbb{Z}/m\mathbb{Z}$ and $\mathbb{F}_q$, we provide programs relating to standard arithmetic, i.e.

  - addition, subtraction, negation
  - multiplication, division, inversion
  - exponentiation

  and higher algorithms, e.g.

  - greatest common divisor and least common multiple
  - chinese remainder algorithm
  - primality tests
  - factorization

- At present, the package for number fields is still relatively small. In addition to standard arithmetic and I/O-functions, it only contains a dozen programs which are needed in the ORDMAX-algorithm (see 2.3) and programs for quadratic number fields, e.g.

  - discriminant

  - norm and trace
  - decomposition of primes
  - fundamental unit in real quadratic fields

- Some higher algorithms for quadratic congruence function fields of transcendence degree 1 (i.e. k is finite, $r = 1$ and $[F : K] = 2$) will be discussed in Section 2.6 (see also ORDMAX in 2.3).

2.3 The polynomial system

We take as ground domains

$$\mathbb{Z},\ \mathbb{Q},\ \mathbb{Z}/m\mathbb{Z} \text{ and } \mathbb{F}_q,$$

where m is a positive integer and $\mathbb{F}_q$ a finite field, as well as

algebraic number fields,

and calculate with univariate and multivariate polynomials over these domains.

Polynomials are given either in sparse or in dense representation. Both representations are recursive. They can be described as follows:

Let R be an arbitrary commutative ring and R_n be the polynomial ring over R in n variables, i.e. $R_n = R[x_1, \ldots, x_n]$ (for $n \geq 0$). Then, for $n > 0$,

$$R_n = R[x_1, \ldots, x_n] \cong R[x_1, \ldots, x_{n-1}][x_n] = R_{n-1}[x_n].$$

$P = 0$ is again represented by the atom 0.

- Any nonzero polynomial $P = P(x_1, \ldots, x_n) \in R_n$ can be written in the form

$$P = \sum_{i=1}^{k} P_i(x_1, \ldots, x_{n-1}) \cdot x_n^{e_i},$$

  with $0 \neq P_i \in R_{n-1}$ and decreasing exponents

$$e_1 > e_2 > \ldots e_k \geq 0.$$

  Then the list

$$P^{(sp)} := \left(e_1, P_1^{(sp)}, \ldots, e_k, P_k^{(sp)}\right)$$

  is the sparse representation of P.

- Dense representation lists the degree and all coefficients of P, i.e. for

$$P = \sum_{j=0}^{d} Q_j(x_1, \ldots, x_{n-1}) \cdot x_n^{j},$$

  where $d = \deg(P)$ and $Q_j \in R_{n-1}$; thus

$$P^{(dns)} := \left(d, Q_d^{(dns)}, Q_{d-1}^{(dns)}, \ldots, Q_0^{(dns)}\right)$$

  is the dense representation of P.

SIMATH contains about 330 polynomial functions for

- standard arithmetic
- chinese remainder algorithm
- derivation and integration
- evaluation and substitution
- resultant and discriminant
- factorization
  - of univariate polynomials over $\mathbb{F}_q$ by the Berlekamp algorithm
  - of univariate and multivariate polynomials over $\mathbb{Q}$ by the Berlekamp-Hensel algorithm, recently made more efficient by employing the bounds of Landau and Mignotte.

Factorization over algebraic number fields will be realized in the near future.

A high-level algorithm included in the polynomial system is the Ford-Zassenhaus algorithm ORDMAX for determining maximal orders in semi-simple polynomial algebras over $\mathbb{Q}$:

Let $f(x) \in \mathbb{Z}[x]$ be a monic separable polynomial with $\deg(f(x)) = n$,

$$A_f \; := \; \mathbb{Q}[x]/(f(x)) \; = \; \mathbb{Q}[\omega] \qquad \text{for } \omega := x \bmod f(x)$$

and $O_f := \mathrm{Cl}\,(\mathbb{Z}, A_f)$ the integral closure of $\mathbb{Z}$ in A_f.

ORDMAX determines polynomials $g_1(x), \ldots, g_n(x) \in \mathbb{Q}\,[x]$ which form a $\mathbb{Z}$-module basis of the maximal order O_f, i.e.

$$O_f = \sum_{i=1}^{n} \mathbb{Z} \cdot g_i(\omega).$$

If the polynomial $f(x)$ is irreducible, the algebra A_f becomes an algebraic number field k, and the algorithm determines an integral basis of $k/\mathbb{Q}$.

ORDMAX can also be applied to semi-simple polynomial algebras over $\mathbb{F}_p(y)$, where p is a prime. If the defining polynomial $f(x) \in \mathbb{F}_p[y][x]$, which is always considered to be monic and separable, is also irreducible, the algebra $A_f := \mathbb{F}_p(y)[x]/(f(x))$ becomes a congruence function field of transcendence degree 1.

It should also be mentioned that ORDMAX makes it relatively easy to compute the decomposition of primes in number fields and congruence function fields.

The ORDMAX-version for algebras over $\mathbb{F}_p(y)$ as well as the programs for calculating the decomposition law have been completed only recently, and will soon be added to the system.

2.4 Matrices and vectors

This package contains programs dealing with matrices and vectors over

$$\mathbb{Z},\ \mathbb{Q},\ \mathbb{Z}/m\mathbb{Z} \text{ and } \mathbb{F}_q$$

as well as over

polynomial rings
and rings or fields of rational functions

over these domains.

A vector of dimension $n > 0$ over a structure S is represented by a list

$$(a_1, \ldots, a_n) \qquad \text{with } a_1, \ldots, a_n \in S.$$

The lines l_i of a (m, n)-matrix (a_{ij}) over S are realized as vectors

$$l_i = (a_{i1}, \ldots, a_{in}) \qquad \text{for } i = 1, \ldots, m.$$

Thus, the matrix corresponds to the list of line vectors

$$(l_1, \ldots, l_m) = \big((a_{11}, \ldots, a_{1n}), \ldots, (a_{m1}, \ldots, a_{mn})\big).$$

Contrary to the zero element of all other subsystems, the zero vector and zero matrix are not represented by the atom 0 but consist of n and $m \cdot n$ zeros, respectively.

The system of matrices and vectors contains about 300 functions and is now nearly complete. In addition to the standard arithmetic, there are programs relating to:

- evaluation and substitution

- for vectors:
  linear combination
  scalar product

- for matrices:
  determinant
  inverse
  Hermite-normal-form
  elementary divisor form
  characteristic polynomial
  eigenvalues of matrices over $\mathbb{Z}$, $\mathbb{Q}$, $\mathbb{Z}/m\mathbb{Z}$

- systems of linear equations

2.5 simcalc

simcalc, the SIMATH calculator, works in interactive mode. Most of the SIMATH functions for numbers (integers, rationals, elements of $\mathbb{Z}/m\mathbb{Z}$), polynomials, matrices and vectors are available. simcalc is particularly suited for programmers with little experience.

Expressions are written just as they would be on a sheet of paper. In addition, one can

- attach results to variables
- calculate with variables in the same way as with numbers
- substitute variables in polynomials, matrices or vectors, even in a recursive manner
- get a short description of every simcalc function including examples of how to use it
- list all commands and results on a record file
- get information about the running time as well as the number of garbage collections and used cells
- branch to the operating system

It is possible to list the last 10 commands, to correct them if necessary, and to use them again as input.

Many different help functions will answer most of the user's questions, e.g.:

- "?func": Which functions are available in simcalc in general?
- "?ifunc": Which functions are available for integers?
- "?sqrt": How to use the function sqrt?
- "?a, b, c": What are the values of the variables a, b and c?
- "?subst": How to substitute variables in polynomials, matrices and vectors?
- "?vardel": How to delete variables?

2.6 Applications in constructive number theory

In this section, we give two examples of how SIMATH can be used to develop higher algorithms in constructive number theory.

- **Quadratic congruence function fields**:
  Let k be a prime field $\mathbb{F}_p$, $K = k(x) = \mathbb{F}_p(x)$ the rational function field in one variable over $\mathbb{F}_p$, and F be a quadratic extension of K, i.e.

$$F = \mathbb{F}_p \left(x, \sqrt{f(x)} \right),$$

  where $f(x)$ is a squarefree polynomial over $\mathbb{F}_p$.
  Within this project, algorithms were written for determining, e.g.,

  - the class number
  - the unit group and
  - the regulator

  of F. These have been added to the arithmetic system of SIMATH.

- **Elliptic curves**:
  In addition to standard arithmetic routines for elliptic curves over $\mathbb{Q}$ and over algebraic number fields, such as

  - discriminant and j-invariant
  - birational transformations
  - addition of points,

  algorithms for calculating, e.g.,

  - curves of high rank over number fields
  - torsion groups over number fields
  - minimal models over local and global fields
  - heights over local and global fields
  - L-series over number fields
  - conductor over number fields
  - tables of congruent numbers

  are available.
  We have just begun to incorporate most of these algorithms into SIMATH.

3 How to use SIMATH

In this final section, we will illustrate how simple the programming with SIMATH is. We will write a function to compute the least common multiple of two arbitrarily large integers a and b by using the formula

$$\operatorname{lcm}(a, b) = \frac{a \cdot b}{\gcd(a, b)}.$$

```
# include <_simath.h>                   /* system header */

int lcm(a,b)
int a,b;
{
    int c;
                                        /* a, b and c are entered in a  */
    init(c);                            /* reference list in order that */
    bind(a,b);                          /* they are not destroyed by a  */
                                        /* garbage collector            */
    /* special case a=0 or b=0 */
    if ( a==0 || b==0 ) c=0;

    /* general case a!=0 and b!=0 */
    /* c = a / gcd(a,b) * b */
    else {
        c = igcd(a,b);
        c = iprod(iquot(a,c),b);
    }

    return(c);
}
```

As is obvious from this example, the names of the SIMATH functions are almost self-explanatory: igcd, iquot, iprod computes the greatest common divisor, the quotient and the product of two integers, respectively. The corresponding programs for polynomials over integers say would be pigcd, piquot and piprod.

For instance, the main program for calculating the lcm could be written as follows:

```
# include <_simath.h>

main()
{
    int a,b;

    init(a,b);

    printf("Give in two integers!\n");
    printf("a b  :  "); a = geti(); b = geti();
    printf("lcm(%i,%i) = %i\n",a,b,lcm(a,b));
}
```

From this main program, it can be seen that SIMATH contains an extension of the C-function printf so that it is possible to give out types such as arbitrarily large integers (%i), rational numbers (%r) and lists (%l).

References

[1] A.V. Aho, J.E. Hopcroft, J.D. Ullman, "The design and analysis of computer algorithms", Addison Wesley, Reading, Massachusetts, Menlo Park, California 1974.

[2] E. Artin, *Quadratische Körper im Gebiet der höheren Kongruenzen I,II*, Mathematische Zeitschrift **19** (1924), 153-246.

[3] E.R. Berlekamp, *Factoring polynomials over large finite fields*, Math. Comp. **24** (1970), 713-735.

[4] R. Böffgen, M.A. Reichert, *Computing the decomposition of primes p and p-adic absolute values in semi-simple algebras over* $\mathbb{Q}$, J. Symb. Comp. **4** (1987), 3-10.

[5] G.E. Collins, R. Loos, *ALDES SAC-2 Description manuals*, University of Wisconsin, Madison, Wisconsin, USA 1980.

[6] B.W. Kernighan, D.M. Ritchie, "Programmieren in C", Carl Hanser Verlag, München 1983.

[7] D.E. Knuth, "The art of computer programming I-III", Addison Wesley, Reading, Massachusetts, Menlo Park, California 1973.

[8] R.S. Lehmann, *Factoring large integers*, Math. Comp. **28** (1974), 637-646.

[9] M. Mignotte, "Mathématiques pour le calcul formel", Presses Universitaires de France, Paris 1989.

[10] M. Pohst, H. Zassenhaus, "Algorithmic algebraic number theory", Cambridge University Press, Cambridge 1989.

Bei Fragen zur Produktsicherheit wenden Sie sich bitte an:
If you have any questions regarding product safety,
please contact:

Walter de Gruyter GmbH
Genthiner Straße 13
10785 Berlin
productsafety@degruyterbrill.com